Apatites (Structu Characterization a Applications): Scienti and Technologic Perspectiv

Apatites (Structure, Characterization and Applications): Scientific and Technological Perspectives

Edited by
Martin Brooks

MURPHY & MOORE
www.murphy-moorepublishing.com

Published by Murphy & Moore Publishing,
1 Rockefeller Plaza,
New York City, NY 10020, USA

ISBN: 978-1-63987-052-3

Cataloging-in-Publication Data

Apatites (structure, characterization and applications) : scientific and technological perspectives / edited by Martin Brooks.
p. cm.
Includes bibliographical references and index.
ISBN 978-1-63987-052-3
1. Apatite. 2. Apatite--Analysis. 3. Phosphate minerals. I. Brooks, Martin.
QE391.A6 A53 2022
549.72--dc23

For information on all Murphy & Moore Publications
visit our website at www.murphy-moorepublishing.com

MURPHY & MOORE

Contents

Preface.......VII

Chapter 1 **Utilization of Compounds of Phosphorus**.......1
Petr Ptáček

Chapter 2 **Rare-earth Element Bearing Apatites and Oxyapatites**.......51
Petr Ptáček

Chapter 3 **Phosphate Rocks**.......95
Petr Ptáček

Chapter 4 **Identification, Characterization and Properties of Apatites**.......142
Petr Ptáček

Permissions

Index

Preface

This book has been a concerted effort by a group of academicians, researchers and scientists, who have contributed their research works for the realization of the book. This book has materialized in the wake of emerging advancements and innovations in this field. Therefore, the need of the hour was to compile all the required researches and disseminate the knowledge to a broad spectrum of people comprising of students, researchers and specialists of the field.

Apatites are ionic compounds which have high concentrations of hydroxyl ions, fluoride ions and chlorine ions. They are phosphate minerals which exist as igneous rocks, metamorphic rocks, sedimentary rocks and volcanic rocks. They could also be found as gem stones which are usually green or blue in color. Apatites are classified into hydroxyapatite, fluorapatite and chlorapatite. They form a major component of the bone material and tooth enamel. Apatites are the most abundant source of inorganic phosphorus, thus making them extremely useful in the fertilizer industry. This book aims to shed light on some of the unexplored aspects of apatites and the recent researches in this field. The various studies that are constantly contributing towards advancing technologies and evolution of this field are examined in detail. This book is a vital tool for all researching or studying apatites as it gives incredible insights into their structure, characterization and applications.

At the end of the preface, I would like to thank the authors for their brilliant chapters and the publisher for guiding us all-through the making of the book till its final stage. Also, I would like to thank my family for providing the support and encouragement throughout my academic career and research projects.

Editor

1

Utilization of Compounds of Phosphorus

Petr Ptáček

Abstract

The last chapter of this book provides brief description of utilization of apatites and phosphorus-bearing compounds in industry and material science research. Since the chemistry of phosphorus is quite complicated and a quickly developing field of science, the topics described below are only limited insights to chemically bonded ceramics and refractories, dental phosphate cements, oil-well cements, phosphate glasses and glass ceramics. Chapter continues with description of functional phosphate materials applied as solid oxide fuel cells electrolytes, sensors, phosphors, catalysts and coatings. The chapter ends with introduction of basic ideas for biological apatite in bone tissue engineering, collagen apatite composites, apatite layers and biocoatings.

Keywords: Apatite, Chemically Bonded Ceramics, Phosphate Cement, SFOC Electrolytes, Sensors, Biological Apatite, Collagen, Biocoatings

In the previous chapter, the utilization of phosphate ore was described. Elemental phosphorus (**Section**) and phosphoric acid are used in the manufacturing of other phosphorus compounds (as illustrated in **Fig. 1**) and materials, which are briefly reported in this chapter. Since the chemistry of phosphorus is quite complicated and a quickly develop-ing field of science, the topics described below are only limited insights to phosphorus-bearing compounds utilized in the industry and material science research.

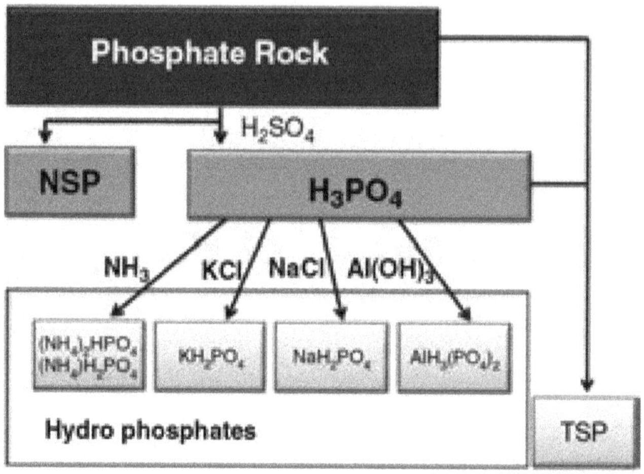

Fig. 1. Flow chart of production of acid phosphates from phosphate ores [1].

Chemically bonded phosphate ceramics

Ceramics are formed by the compaction of powders and their subsequent fusion at high to very high temperatures, ranging somewhere from ~700° to 2000°C. Once fused, the resulting ceramics are hard and dense and exhibit very good corrosion resistance [2]. Among the conventional ceramic bonds, the chemical (organic and inorganic) and hydraulic bonds were used during the preparation of ceramic materials.

Phosphate bond can be utilized for the preparation of hard and quick-setting ceramic materials, chemically bonded phosphate ceramics (PCBC) and ceramic composites [3],[4] via the reaction of metal cation with phosphate anions (**Fig. 2**). The reaction is attained by mixing a cation donor, generally an oxide (CaO [5],[6],[7],[8], MgO [2],[6],[9], ZnO [2],[8], Al_2O_3 [2], Fe_2O_3 [2],

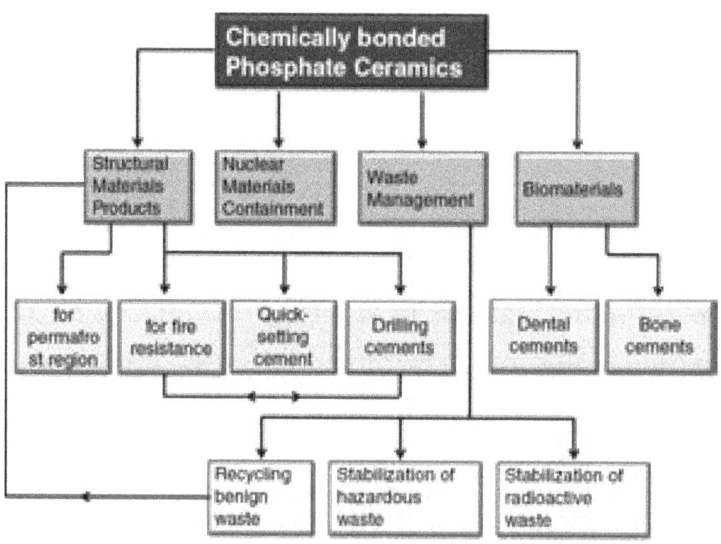

Fig. 2. Potential application of chemically bonded phosphate ceramics [1].

etc.), raw materials [10],or secondary raw materials (fly ash [11] and blast furnace slag [12]) with either phosphoric acid or acid phosphate such as ammonium phosphate solution or magnesium dihydrogen phosphate ($Mg(H_2PO_4)_2{\cdot}2H_2O$) and aluminum hydrogen phosphate ($AlH_3(PO_4)_2{\cdot}H_2O$). Typical acid-base reaction between a metal oxide and phosphoric acid can be written as [2],[13]:

$$MO_x + n\ H_3PO_4 + m\ H_2O \rightarrow MH_{3n-2x}\left(PO_4\right)_n + \left(m + x\right)\ H_2O \tag{1}$$

where x denotes half of valence of **M**, $n \geq 2x/3$, and m is an arbitrary integer that decides the amount of water to be added in the reaction.

In the case of magnesium phosphate chemically bonded ceramics (MPCBC), the acido-basic **reaction 1** of orthophosporic acid with MgO yields the following products [2]:

$$MgO + 2\ H_3PO_4 \rightarrow Mg\left(H_2PO_4\right)_2 \cdot H_2O \quad \Rightarrow x = 1,\ n = 2,\ \text{and} m = 0 \tag{2}$$

$$MgO + H_3PO_4 + 2\ H_2O \rightarrow Mg\left(H_2PO_4\right)_2 \cdot 3H_2O \quad \Rightarrow x = 1,\ n = 1,\ \text{and} m = 2 \tag{3}$$

$$3\ MgO + 2\ H_3PO_4 \rightarrow Mg3\left(PO_4\right)_2 + 3\ H_2O \quad \Rightarrow x = 1,\ n = \frac{2}{3},\ \text{and} m = 0 \tag{4}$$

If acidic phosphate is used, the reaction proceeds as follows [2]:

$$MgO + Mg\left(H_2PO_4\right)_2 \rightarrow 2\ MgHPO_4 + H_2O \tag{5}$$

and similar reaction can be also written for the reaction of $Al(H_2PO_4)_3$ with $AlO_{3/2}$:

$$2\ AlO_{3/2} + Al\left(H_2PO_4\right)_3 \rightarrow 3\ AlPO_4 + 3\ H_2O \tag{6}$$

Reactions **5** and **6** indicate that partially acidic phosphate salt is only an intermediate phase. The dissociation of orthophosporic acid and the pH stability range of formed ionic species are given by **Eqs. 11–13** in **Chapter**. General dissociation reaction for acidic dihydrogen phosphate and its dissociation constant in the form of pK_{dis} ($pK_{dis} = -log\ K_{dis}$) can be written as follows:

$$M(H_2PO_4)_n \leftrightarrow M^{n+} + n\ (H_2PO_4)^{n-} \quad (7)$$

and

$$pK_{dis} = -\log\left(\left[M^{n+}\right]\left[H_2PO_4\right]^{n-}\right) \quad (8)$$

The values of pK_{dis} for KH_2PO_4, $(NH_4)H_2PO_4$, $Mg(H_2PO)_2{\cdot}2H_2O$ and $Ca(H_2PO_4)_2{\cdot}H_2O$ are 0.15, -0.69, 2.97 and 1.15, respectively [2].

The preparation scheme for the synthesis of aluminum phosphate binder is shown in **Fig. 3**. The evolution of chemical composition of phosphate binder with Al:P ratio of 1.4:3 with temperature is introduced by **Table 1**.

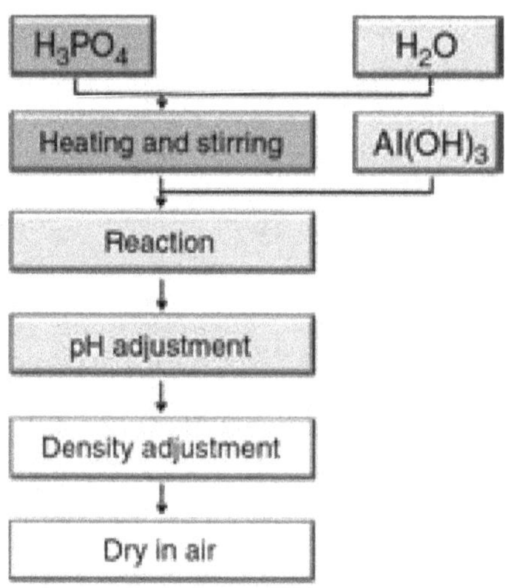

Fig. 3. Schematic procedure for the synthesis of aluminum phosphate binder [14].

Initially, the motivation for the development of these ceramic materials was the preparation of dental cements. Phosphate chemically bonded[1] ceramics (PCBC) find recently the application in diverse fields (**Fig. 2**), which include structural ceramics, refractory materials [15], toxic, radioactive [16],[17],[18],[19] or asbestos-containing[2] [20],[21],[22] waste management [23], oil drilling and bioceramics, pigments [1], etc. [2].

Ceramics and refractories

The broad term "ceramics" refers to any of a large family of materials that are usually inorganic and require high temperatures in their processing or manufacture. They are generally classified into glass, whitewares, including artware and structural ceramics, and refractories. In general,

[1] Chemical bonding as a means of solidification is very widely observed in nature. The formation of sedimentary rocks, such as carbonate rocks, lateritic soils and solidification of desert soils, are examples of this process [1].

Temperature [°C]	Dominant phase	Secondary phase
60	$AlH_3(PO_4)_2{\cdot}3H_2O$	—
105	$AlH_3(PO_4)_2{\cdot}3H_2O$ (monoclinic)	$Al(H_2PO_4)_3$ (hexagonal)
200	$Al(H_2PO_4)_3$ (hexagonal)	—
220	$AlPO_4$ (trigonal)	$AlH_2P_3O_{10}{\cdot}2H_2O$
250		
300		
400	$Al(PO_3)_3$	$AlH_2P_3O_{10}{\cdot}2.5H_2O$
500	$Al_2P_6O_{18}$	$Al(PO_3)_3$ (cubic)
600		
700		
800	$Al(PO_3)_3$ (cubic)	$AlPO_4$ (rhombic system)
900		
1000		

Table 1. The alteration of composition of phosphate binder with increasing temperature [14].

the term "glass" refers to an amorphous solid with non-directional properties, characterized by its transparency, hardness, rigidity at ordinary temperatures and capacity for plastic working at elevated temperatures. Major commercial uses of glass include plate or "float" glass, as used for windows and windshields; glass tubing, or formed shapes, used for electric lighting envelopes; and glass containers such as tumblers and bottles. Whiteware is characterized by a crystalline matrix held together by a glassy phase and usually covered by a glazed coating. Major classifications of whiteware are ceramic tiles, glazed and unglazed, for floors, walls and external use; sanitaryware in the form of toilets and lavatories; and tableware, from earthenware to fine china [24].

Kaolin is major raw material used for the fabrication of conventional ceramics. It is obtained from the alteration of granitoid rocks[3] [25]. It consists mostly of kaolinite and a small amount of impurities such as quartz, micas and other phyllosilicates. Firing of kaolinite induces numerous complex structural and microstructural transformations leading to the formation

---

[2] SEM picture of thin long fibrous structure of chrysotile ($H_4Mg_3Si_2O_9$ [20],[21],[22]) asbestos under the magnification of 98× (a), 5000× (b) and 10,000× (c).

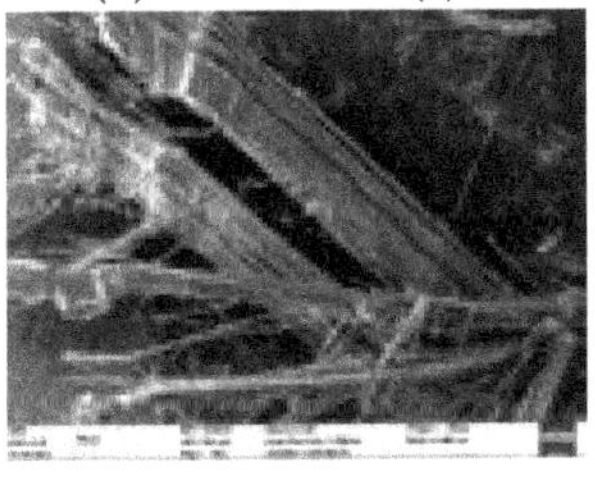
(a)

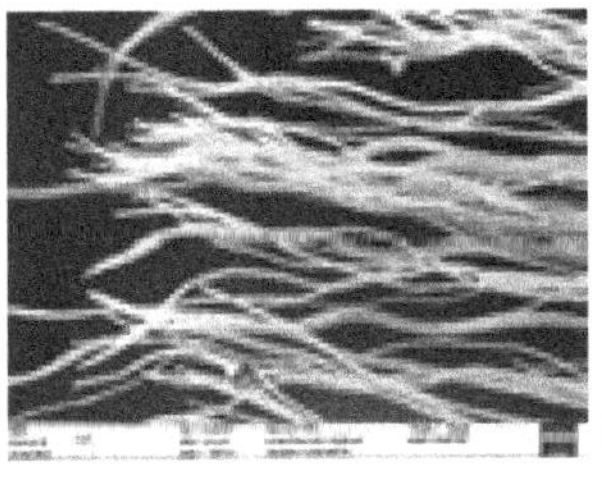
(b)

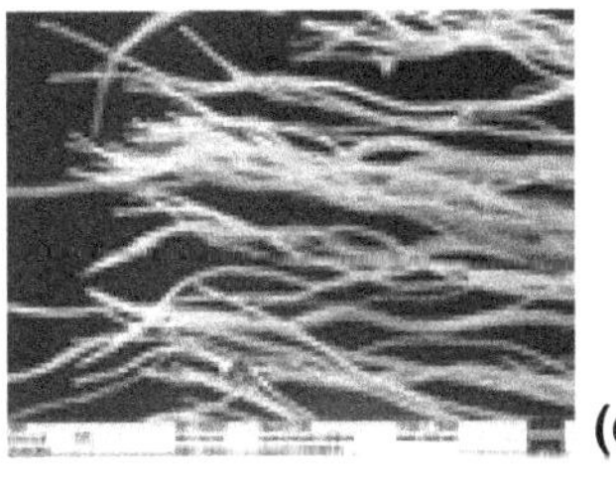
(c)

of mullite and silica (cristobalite) phase. Mullite phase is characterized by some advantageous properties such as good corrosion resistance, low dilatation coefficient, good creep and thermal shock resistances, thermal stability and high strength. These advantages make this phase favorable for different applications. Orthophosphoric acid reacts with aluminum from kaolin to provide new compounds, which are $Al(H_2PO_4)_3$ at room temperature and $AlPO_4$ when heated to temperatures above 800°C [10].

Approximately 70% of all refractories used in industry are in the form of bricks, which are cast in the shapes such as straights, soaps, splits, arches, wedges, keys, skews, jambs or other special and frequently patented shapes. Most industrial refractories are composed of metal oxides or of carbon, graphite or silicon carbide. Most refractory bricks are shaped by combining the size-graded refractory aggregate with a small amount of moisture and casting in a dry press. The important properties of any refractory, including its high-temperature strength, depend on its mineral makeup, the particle-size distribution of minerals and the way these materials react at high temperatures and in furnace environments. When a refractory is chosen for a particular service, the service conditions must be considered in the design. In proper selection of a refractory, these factors, together with an economic balance, must be considered so that the refractory ultimately produces the lowest cost per unit weight of product per unit weight of refractory consumed [26].

Fig. 4. Cutting plane throughout the block of chemically phosphate bonded refractory material containing large grains of calcined bauxite (a) and electron microscopy (SEM) picture of ceramic body (b).

The materials must be chosen with generous safety margins in the temperature capability and the refractory construction system, whether brick, monolithic or fiber, should be suitable not only for the operating conditions but also for the type of equipment concerned and the construction conditions. Refractory concretes and castables (monolithic refractories) are especially suitable for these small burner quarls. Fibers and refractory ceramic fibers (glass,

---

[3] Granular crystalline rock consisting essentially from quartz, orthoclase-feldspar and mica. Usually is light gray, white or light in color [25].

mineral, ceramic fibers and whiskers) become an important construction as well as insulation material, although the brickwork has the longest history of the development [27].

Aluminosilicate refractories are manufactured using refractory clays, sillimanite minerals, bauxite (**Fig. 4**) and mixtures of alumina and silica sand. They will refer, somewhat arbitrarily, to common crystalline compounds with melting temperatures of at least 1500°C. The major categories of traditional refractories are fire clays, high alumina and silica. The choice of material for traditional refractory applications, as well as for advanced material applications, was and is based on balancing the cost and the performance lifetime [28].

Industries involved in steel melting and casting of special alloys are always interested in using the maintenance-free, functional refractories to achieve energy-efficient metal processing. They seek for the development of new and advanced thermal ceramics for protective thermal insulation linings and molten metal handling crucibles. Conventionally, porous to dense Al_2O_3, SiO_2, MgO, ZrO_2, SiC[4] [29] and fireclay ceramic bodies were used as thermal insulation refractory liners [30]. Phosphate bond can also be utilized in manufacturing of refractory and wear-resistant and of protective coatings on metal [14] or ceramic surfaces [31]. Intensive development of ceramic materials has increased the availability of well-characterized engineering ceramics capable of the utilization over the range of temperatures and atmospheres [32].

Refractory castables can be classified according to different aspects including the content of calcium, binder source, overall chemical composition, bulk density, application method and others. Binders commonly used in monolithic products may be of various classes [33]:

1. Hydraulic (cement and hydratable alumina) [34]
2. Chemical (phosphoric acid, silicates, geopolymers, etc.) [33],[35]
3. Sol-gel (colloidal silica or alumina) [33]

Regarding the chemical ones, the bond strength can be provided by the addition of phosphates (dry or solution) or by in situ generation of phosphates (formed via the reaction with added H_3PO_4) in the refractory structure. The reaction of Al_2O_3 or $Al(OH)_3$ with orthophosphoric acid can be described by reactions [33]:

$$6\ H_3PO_4 + Al_2O_3 \rightarrow 2\ Al(H_2PO_4)_3 + 3\ H_2O \tag{9}$$

$$3\ H_3PO_4 + Al(OH)_3 \rightarrow Al(H_2PO_4)_3 + 3\ H_2O \tag{10}$$

A cold-setting refractory material was developed by Hipedinger et al [36] via the magnesia-phosphate reaction. A cement paste based on alumina, silica fume, magnesia and orthophosphoric acid or monoaluminum phosphate was designed to form cordierite-mullite during

[4] It was proved that H_3PO_4 is effective binder for SiC [29].

heating. This cement paste set at room temperature and $MgHPO_4 \cdot 3H_2O$ phase (newberyite) was observed, but amorphous phases were predominant. Two exothermic effects were detected during the setting process corresponding to the acido-basic reaction of magnesia with phosphates and to the formation of bonding hydrates. At 1100°C, *c*-$AlPO_4$ was formed by the reaction of alumina with orthophosphoric acid or monoaluminum phosphate. At 1350°C, the dominant crystalline phases were cordierite and mullite. A refractory concrete with obtained cement paste and a cordierite-mullite aggregate (scrap refractory material) was prepared.

The acid phosphate impregnation, with ozone pretreatment, improves the oxidation resistance of carbon materials (polycrystalline graphite and pitch-based carbon fiber), as shown by the weight measurement in air up to 1500°C. The impregnation involves using phosphoric acid and dissolved aluminum hydroxide in the molar ratio of 12:1 and results in rough, white and hard aluminum metaphosphate coating of the weight of about 20% of that of the carbon before the treatment. Without ozone pretreatment, the impregnation is not effective. Without aluminum hydroxide, the impregnation even degrades the oxidation resistance of carbon [37].

Dental phosphate cement

A variety of cements are used in modern clinical dentistry, such as glass ionomers, zinc phosphate and zinc polycarboxylate [38],[39]. Dental zinc phosphate cement is primarily used for the cementation of indirect restorations, such as crown and bridges. It has the longest record of any cement, approximate 100 years, and has remained popular throughout this time. Zinc phosphate cements are also considered the strongest among the dental cements. However, it is also applied for temporary fillings, cavity bases and buildings of teeth beneath crowns. Zinc phosphate cement is primarily in contact with the pulp-dentin system and in certain cases (e.g. temporary fillings) with the gingiva. A variety of cementing materials are currently used as the bases and luting[5] agents, but zinc phosphate cement has been used for many decades. Phosphoric acid-based cements originated from OSTERMANN´S formula from 1832, which was composed of calcium oxide and anhydrous phosphoric acid. In 1902, FLECK established a formula that is similar to that being in use today [40],[41],[42],[43],[44].

The powder is mainly a mixture of zinc oxide and up to 13% magnesium oxide. The liquid is an aqueous solution of phosphoric acid containing 38 – 59% H_3PO_4, 30 – 55% water and 0 – 10% zinc. Aluminum is essential to the cement-forming reaction, and zinc moderates the reaction between powder and liquid, allowing adequate working time and sufficient quantity of powder to be added for optimum properties of cement. When the powder is mixed with liquid, phosphoric acid attacks the surface of particles, dissolving zinc oxide, which releases zinc ions into the solution. Aluminum in the liquid reacts with phosphoric acid to form zinc aluminophosphate gel on remaining portion of particles. Thus, the cement reveals a cored structure consisting primarily of non-reacted zinc oxide particle core embedded in a cohesive amorphous matrix of zinc aluminophosphate (glasslike phosphate). Aluminum phos-

[5] The word 'luting' implies the use of a molded or moldable article to seal a space or to cement two components together [44].

phate, in addition to its role as a retarder, contributes to the increase in mechanical hardness of cement [40],[45].

Setting reactions and resultant structure of zinc phosphate cements are largely based on the formation of hopeite ($Zn_3(PO_4)_2{\cdot}4H_2O$) and/or zinc phosphate hydrate ($Zn_2P_2O_7{\cdot}3H_2O$) when using orthophosphoric acid (OPA) cement-forming liquids. OPA solutions buffered with aluminum and zinc ion produced better mechanical properties than non-buffered OPA solutions because of the formation of hopeite and amorphous phase. The development of crystalline forms of phosphate hydrates of zinc and magnesium was retarded and/or prevented by the incorporation of aluminum and zinc ion in the cement-forming liquid [43].

This mechanism is similar to the cement-forming reaction described by Wilson [43],[46] in a dental silicate cement (**Fig. 5**). When H^+ ions attack the glass powder, Al^{3+}, Ca^{2+}, Na^+ and F^- are liberated from the glass, leaving behind an ion-depleted layer of silicate gel at the surface of glass particles. Liberated ions migrate and react with $H_2PO_4^-$, and salts precipitate. The principal reaction is the formation of an insoluble aluminum phosphate, the gel matrix. Associated side reactions are the precipitation of calcium fluoride and the formation of soluble sodium dihydrogen phosphate.

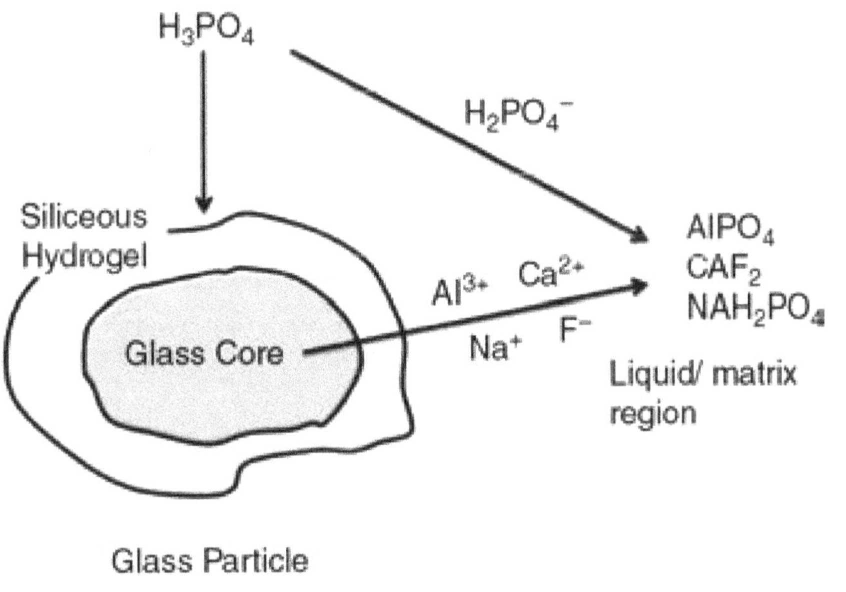

Fig. 5. Setting reaction of dental silicate cement [43].

Oil-well cement

The main application of the cement in an oil well is to stabilize the steel casing in the borehole and to protect it from corrosion. The cement is pumped through the borehole and is pushed upwards through the annulus between the casing and the formation. The cement is exposed to the temperature and pressure gradients of the borehole [47].

Various types of oil-well cements are distinguished. The most important is the phase composition of cement, primarily the C_3A phase content, which causes quick paste thickening. The main feature of oil-well cement is that it must remain sufficiently fluid for a long period, required for its pumping to deep well. Simultaneously, the temperature in the borehole increases with depth[6] [48],[49]. The types and properties of oil-well cements specified by the American Petroleum Institute (API) are introduced by Clinkers containing limestone loess, diatomite, pyritic ash and sand modified with gypsum and apatite were used for the manufacture of heat resistant oil-well cements. Apatite is also a good stabilizer for high belite cements [50].

API class	Special properties	Intended use
A	Same as ASTM Type I	Well depths up to 1800 m and temperatures of 27 – 77°C.
B	Similar to ASTM Type II, low C_3A, high sulfate resistance	
C	Similar to ASTM Type III, low C_3A, high sulfate resistance	Well depths up to 1800 m and temperatures of 80 – 170°C.
D	Low C_3A with set retarder	Depths up to 1800 – 3600 m and temperatures of 77 – 138°C.
E		Depths up to 1800 – 4200 m and temperatures of 77 – 138°C.
F		Depths up to 3000 – 4800 m and temperatures of 127 – 160°C.
G and H	Coarse-ground ASTM Types II and IV	Temperatures of 27 – 93°C.
J	Essentially β-C_2S and pulverized silica sand	Depths below 6000 m and temperatures > 177°C

Table 2 Types and properties of oil-well cements [49].

Phosphate glasses and glass-ceramics

In the past decades, optical waveguides have raised great interest, as they are the most fundamental and integral part of integrated optic circuits. Glass-based integrated optical devices have several obvious advantages over other technologies such as low intrinsic absorption in near-infrared region of the spectrum, minimized coupling losses to optical fibers and no intrinsic material birefringence compared to crystalline semiconductors. Phosphate glasses are regarded as excellent glass host for the waveguide laser fabrication mainly because of high solubility of rare-earth ions compared to other oxide glasses, which allows for high doping concentrations without significant lifetime reduction, resulting in high gain in short waveguides or cavities and a desirable feature in single-frequency lasers. High-performance

[6] The rate depends on the geothermal degree, which in Europe is about 33 $m \cdot °C^{-1}$ [48].

waveguide amplifiers or lasers are fabricated in various earth-ion-doped phosphate glasses and commercial phosphate glasses such as Kigre Q89 and Schott-IOG 1 [51].

Glass-ceramics are polycrystalline materials with an inorganic–inorganic microstructure, which are prepared from the base glass by controlled crystallization. This can be achieved by subjecting glasses to regulated heat treatment, which results in the nucleation and growth of one or more crystal phases within the glass [52]. Once a stable crystal nucleus has formed and begun to grow, there are a number of possible crystal growth mechanisms and these determine the final crystal morphology [53],[54]:

a. **Faceted crystal growth**: the favored growth sites on an atomic scale are the steps in rows of atoms. However, these sites can be easily eliminated by the very growth that they promote. Despite this, several types of lattice or crystal defects were shown to provide the growth sites that are impossible to be eliminated by the growth.

b. **Dendritic crystal growth**: formed dendritic crystals are characterized by their tree-like appearance. This type of solidification usually takes place in metals and sometimes occurs naturally in silicate minerals, especially in olivine during rapid cooling of lavas.

c. **Spherulitic crystal growth**: is most commonly associated with organic polymeric materials. The geological definition of a spherulite is: "A crystalline spherical body built of exceedingly thin fibers radiating outwards from a center and terminating on the surface of the sphere…".

Bioactive glass-ceramics are an alternative to synthetic HAP (**Section**) for the use in vivo both in restorative dental applications and in bone implantation [54],[55]. Artificial materials implanted into the bone defects are generally encapsulated by fibrous tissue isolating them from the surrounding bone. This is the normal response of the body towards inert artificial materials. However, some ceramics, such as bioglass, A-W glass-ceramics and sintered hydroxyapatite form a bone-like apatite on their surfaces in the living body and bond to living bone through this apatite layer. This bone-bonding ability is called the bioactivity [56].

These bioactive ceramics are already used clinically as important bone-repairing materials. Their bone-bonding ability is achieved by the formation of a biologically active apatite layer after the reaction of the ceramics with surrounding body fluid. Controlled surface reaction of the ceramics is an important factor governing its bioactivity as well as its biodegradability [56]. The preparation of glass-ceramics with high CaO/P_2O_5 ratio, containing large amounts of calcium phosphate crystals, is believed to be one of the best approaches to obtain the bioceramic implants suitable for bone replacement/regeneration [57].

Alkaline aluminum phosphate glasses

Alkaline aluminum phosphate glasses (NMAP) with excellent chemical durability for thermal ion-exchanged optical waveguide were investigated by Wang et al [51]. The transition temperature (T_g = 470°C) is higher than the ion-exchange temperature (390°C), which is favorable for sustaining the stability of the glass structure for planar waveguide fabrication. The glass-forming region in the Na_2O-Al_2O_3-P_2O_5 system is shown in **Fig. 6** [51],[58].

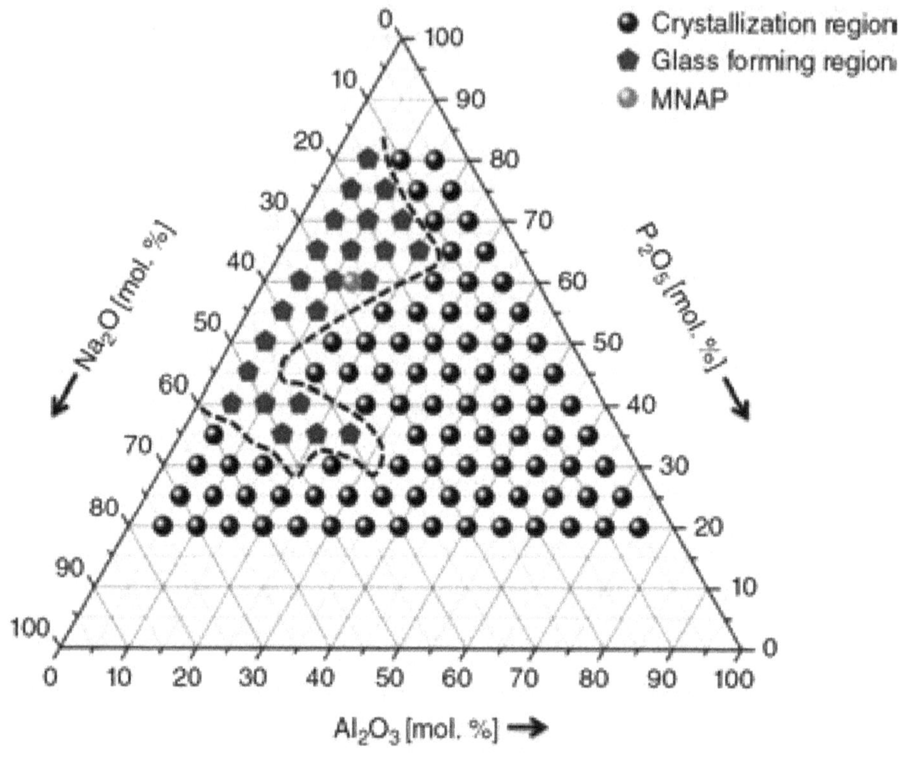

Fig. 6. Ternary phase diagram of Na_2O-Al_2O_3-P_2O_5 system [51],[58].

Iron phosphate glasses

Iron phosphate glass is a versatile matrix for the immobilization of various radioactive elements found in high-level nuclear waste (HLW). Among various compositions of iron phosphate glass, the one with 40 mol.% Fe_2O_3-60 mol.% P_2O_5 was found to be chemically durable. It also has the ability to accommodate large amounts of certain nuclear wastes, especially those that are not well suited for borosilicate glasses. Better chemical durability of iron phosphate glass is attributed to the presence of more hydration-resistant Fe-O-P bonds compared to P-O-P bonds available in other phosphate glasses [59],[60],[61].

Lithium vanado-phosphate glasses

Lithium vanado-phosphate (LiVP) glasses have been largely studied due to their potential application as cathode materials as a result of mixed electronic-ionic conductivity character.[7] Furthermore, lithium and vanadium structural rearrangements in the glass matrix could modify the transport properties of the systems. Interestingly, the modifier ions depolymerize the glass network, creating useful channels that enhance the ionic conductivity, but they can also break some $V^{4+}/O/V^{5+}$ linkages that are essential to the electronic conductivity because they are supposed to be the preferential path for small polaron hopping. The population of

[7] Fast ion conducting (FIC) phosphate glasses have become very important due to a wide range of applications in solid-state devices [63].

$V^{4+}/O/V^{5+}$ paths depends even on the V/P ratio and it can modify the main structure role of vanadium ions in glasses [62],[63],[64].

Apatite-wollastonite glass-ceramics

Wollastonite ($CaSiO_3$) is white glassy silicate mineral that occurs as masses or tabular crystals of metamorphosed limestone. A silica chain GC that contained crystalline apatite and wollastonite (AW) was introduced in MgO-CaO-SiO_2-P_2O_5 glassy matrix and it showed excellent bioactivity, biocompatibility, machinability and adequate mechanical properties such as Young's modulus (117 GPa), compressive strength (1080 MPa) and bending strength (215 MPa) [52],[65],[66],[67].

Wollastonite-2M and pseudowollastonite (low- and high-temperature forms of wollastonite, respectively) are the most common calcium silicate biomaterials proposed for bone tissue regeneration [68]. The major drawback of the $CaSiO_3$ bioceramics is their relatively fast dissolution rate that could reduce their mechanical strength. In addition, the pH of surrounding medium significantly increases, which could affect the osseointegration of the substitute material within the natural bone. The problem can potentially be solved by the development of multiphase materials containing highly dissolvable phases such as wollastonite, on one hand, and stable phases such as HAP, on the other hand [69].

From the bioactive ceramics, the A/W glass-ceramics show high bioactivity and high mechanical strength [56]. The A/W glass-ceramics composed of apatite and wollastonite crystalline phases in a glassy matrix was developed by Kokubo [70]. This bioceramics is highly bioactive and also mechanically strong in comparison with other glasses and glass-ceramics because of wollastonite and apatite crystals' presence.

The A/W glass-ceramics is used in some medical applications, either in powder form as a bone filler or as a bulk material. These materials are currently manufactured by the powder processing methods, providing uniform crystallization of apatite and wollastonite phases in the glassy matrix, as the crystallization of parent glass in a bulk form leads to the appearance of large cracks [69],[71]. The glass-ceramics exposed to the SBF releases predominantly Ca and Si ions, due to the dissolution of amorphous phases and wollastonite-2M, leading to the formation of an apatite-like layer on the surface of the material [69].

The addition of ZnO increased the chemical durability of A-W glass-ceramics, resulting in a decrease in the rate of apatite formation in simulated body fluid. On the other hand, the release of zinc from the glass-ceramics increased with increasing ZnO content. The addition of ZnO may provide bioactive CaO-SiO_2-P_2O_5-CaF_2 glass-ceramics with the capacity for appropriate biodegradation as well as the enhancement of bone formation [56]. The effect of MgO was investigated by Ma et al [72]. As the MgO content increased, the glass crystallization temperature increased and the crystallization of the glass-ceramics was changed from the bulk crystallization to the surface crystallization. The addition of MgO slowed down the rate of dissolution and retarded the formation of apatite layer.

Apatite-mullite glass-ceramics

Apatite-mullite glass-ceramics crystallize from the glass of generic composition SiO_2-Al_2O_3-P_2O_5-CaO-CaF_2 to form an osseoconductive apatite phase existing as spherulites within a mullite matrix. The formation of apatite spherulites is accompanied by a depletion zone, from which calcium, phosphate and fluorine is taken and added to the growing crystal. The depletion zone also inhibits the formation of further apatite crystals in the immediate vicinity due to the glass' compositional similarity with a mullite composition. As mullite begins to crystallize, there is interdependence between the growth of apatite and mullite. Spherulitic grain boundaries in partially devitrified apatite-mullite ceramics act as crack promoters, offering preferential paths to propagation due to the grain boundary interfacial surface energy [54],[73].

FAP-anorthite-diopside glass-ceramics

There is a considerable interest in oxyfluoride glasses and glass-ceramics for laser amplifiers and up-conversion processors. Fluoride-containing crystals have low phonon energies and apatite crystals, in particular, are good host phases to adsorb rare-earth ions. Fluoride-containing glass-ceramics also often crystallize on a nanoscale, which is an added advantage, since optically transparent materials are required for the applications such as fiber amplifiers. The evidence of the nanoscale crystallization in an FAP-anorthite-diopside-based glass-ceramics was found by Hill et al [74].

Apatite-wollastonite ceramics

Wollastonite-hydroxyapatite ceramics was successfully prepared by a novel method, corresponding to the thermal treatment of a silicone embedding micro- and nanosized fillers in air. $CaCO_3$ nanosized particles, providing CaO upon the decomposition, acted as "active" filler, whereas different commercially available or synthesized hydroxyapatite particles were used as "passive" filler. The homogeneous distribution of CaO, at a quasi-molecular level, favored the reaction with silica derived from the polymer, at only 900°C, preventing extensive decomposition of hydroxyapatite. Open-celled porous ceramics suitable for scaffolds for bone-tissue engineering applications were easily prepared from the filler-containing silicone resin mixed with sacrificial PMMA microbeads as the templates [75].

Oxyapatite glass-ceramics

The crystallization of oxyapatite of the composition $NaY_9(SiO_4)_6O_2$ ($P6_3/m$, $\boldsymbol{a}$ = 9.334 Å and $\boldsymbol{c}$ = 6.759 Å, *c:a* = 1:0.7241, *V* = 509.97 Å, **Fig.** 7) from three different glasses from the SiO_2-B_2O_3-Al_2O_3-Y_2O_3-CaO-Na_2O-K_2O-F glass-ceramics system with different F and B_2O_3 content after the

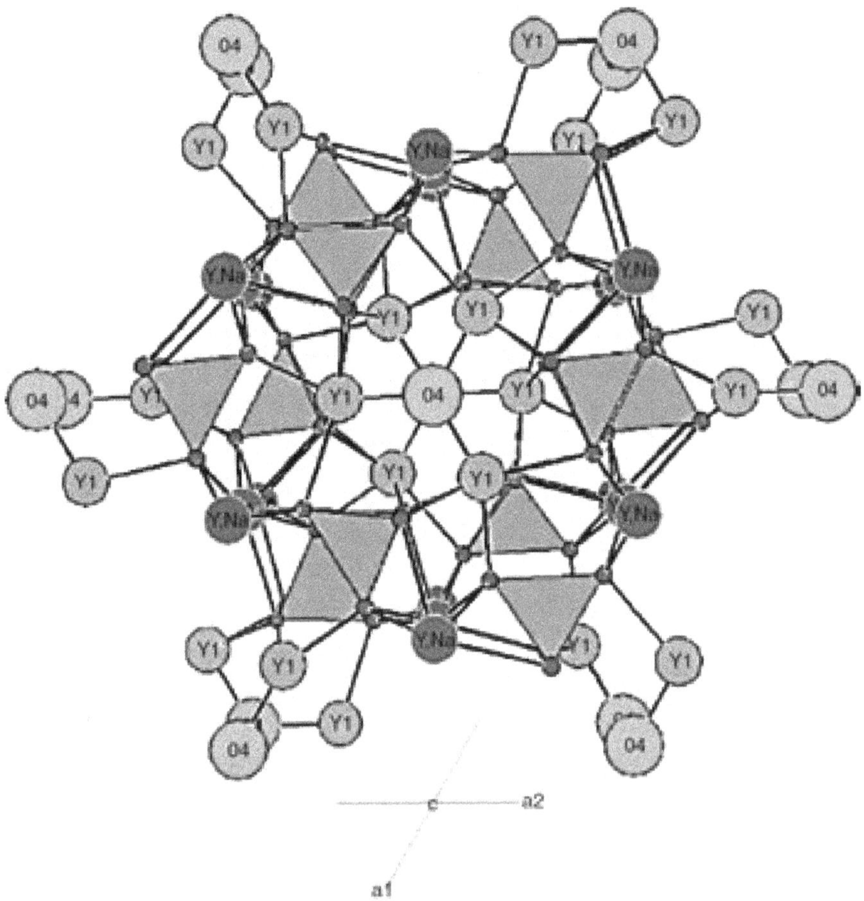

Fig. 7. Crystal structure of oxyapatite $NaY_9(SiO_4)_6O_2$ (perspective view along the *c*-axis) where the tetrahedra represents $[SiO_4]^{4-}$ structural units [76].

heat treatment was observed by VAN'T HOEN et al [76]. The formation of oxyapatite proceeds according to the mechanism of controlled surface nucleation and crystallization. Therefore, applying suitable chemical composition and heat treatment, it is possible to produce glass-ceramics containing oxyapatite crystals. This type of crystal exhibits similar crystal structure as that of fluorapatite. Therefore, there is an isotype relationship between those two phases: the differences in the crystal structures are found in the structural units, with fluorapatite containing $[PO_4]^{3-}$ and $[F]^-$, while oxyapatite containing $[SiO_4]^{4-}$ and $[O]^{2-}$. Because of its specific optical properties (high opacity), this new glass-ceramic material may be used as layering material for dental restoration [76].

Solid oxide fuel cells

Solid oxide fuel cell (SOFC) is an electrochemical energy-conversion device, which offers tremendous promise for delivering high electrical efficiency and significant environmental benefits in the terms of fuel flexibility (hydrocarbons and municipal waste) as well as clean and efficient (>70% with fuel regeneration) electric power generation. SOFC produces useful electricity by the reaction of fuel with an oxidant via the diffusion of oxide ions (or protons) through an ion-conducting solid-electrolyte layer [77].

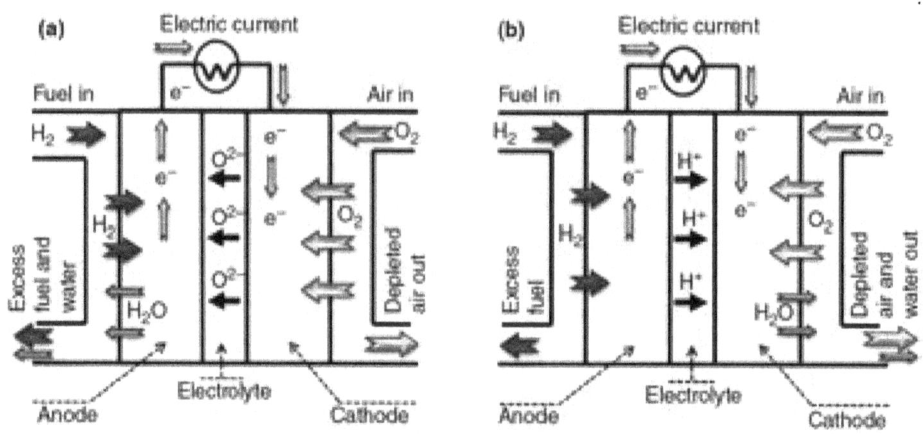

Fig. 8. Schematic diagram of solid oxide fuel cell (SOFC) showing non-ion-conducting electrolyte (a) and proton-conducting electrolyte during its operation (b) [77].

SOFC is composed of a dense electrolyte layer that is sandwiched between two porous electrodes (i.e. cathode and anode) as shown in **Fig. 8**. SOFC can use either oxide ion (a) and/or proton conduction through the electrolyte (b). Electrons generated through the oxidation of fuel on anode are accepted for the oxygen reduction on cathode, which completes the external circuit. The electricity is, thus, produced by the flow of electrons in the external circuit (from the anode to the cathode). Since the current is obtained via the diffusion of oxide ions (or protons) through a solid electrolyte, it becomes imperative to use high operating temperatures (~800 – 1000°C) for achieving high ionic conductivity (of ~0.1 $S \cdot cm^{-1}$) [77],[78].

The first fuel cell was invented in 1838 by an English scientist, William Grove. He named it "wet cell battery" or "Grove cell", which operated by reversing the electrolysis phenomena of water [77],[79]. The fuel cell, invented in 1839 by Grove, is an electrochemical device that converts the chemical energy of fuels directly into electricity and heat by electrochemically combining H_2, CO/H_2 or reformed hydrocarbons in fuel and an oxidant gas transported via an ion-conducting electrolyte. Direct combustion of fuels is eliminated here, which renders the fuel cells much higher conversion efficiencies compared to other conventional thermomechanical methods. Moreover, with fuel cells, the power generation is virtually noise-free and can produce 0.9 times lower emissions of NO_x and SO_x per unit of power output compared to that of conventional technologies. Additionally, it is possible to use the fuel cells for combined heat and power (CHP or cogeneration) generation [77],[80].

The development of high-performance SOFC involves the material selection and operation-related issues (of anode, cathode, electrolyte, sealant and interconnects). These challenges open up the myriad research opportunities for researchers in the field of SOFC. A list of various materials used in SOFC is presented in **Fig. 9**.

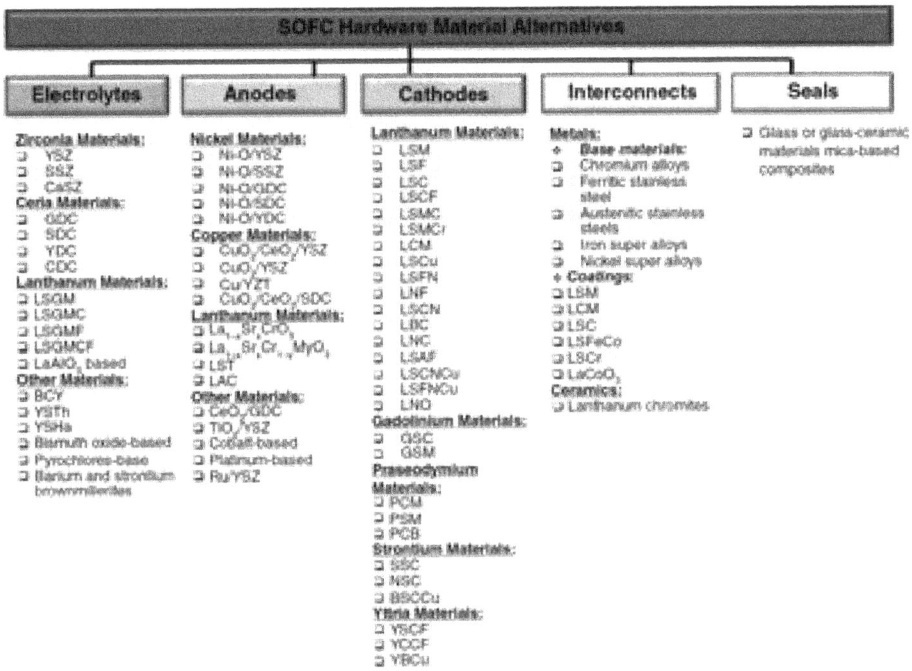

Fig. 9. Comprehensive list of various materials used in SOFC [77].

Apatite-type silicates are considered as promising electrolytes for solid oxide fuel cells. Lanthanum silicate with the composition of $La_{10}Si_{5.5}Al_{0.5}O_{26.75}$ was evaluated as an electrolyte with the electrode materials commonly used in SOFC, i.e. manganite, ferrite and cobaltite as cathode materials and NiO-CGO composite, chromium-manganite and Sr_2MgMoO_6 as anode materials by Marrero-López et al [81]. This electrolyte has conductivity values higher than those of YSZ and comparable to most important solid electrolytes proposed for the intermediate-temperature range, such as doped ceria and lanthanum gallate-based electrolytes. The chemical compatibility did not reveal appreciable bulk reactivity between silicate and many electrodes up to 1300°C [81]. Among several reported rare earth apatites, lanthanum silicates exhibit the ionic conductivity higher than their germinate counterparts [77].

On examining and modeling the probable conduction mechanism in apatite silicates, the atomistic simulation results suggest that the conduction in $La_{9.33}(SiO_4)_6O_2$ and $La_8Sr_2(SiO_4)_6O_2$ takes place via interstitial and vacancy mechanism, respectively [82],[83]. The predicted pathway appears to be a complex nonlinear "sinusoidal-like" process for the interstitial oxygen migration along the c-axis (**Fig. 10**), while a direct linear pathway is predicted for oxygen migration via the vacancy mechanism [77].

Zeng et al [84] developed a model capable of prediction of oxygen ion conduction from relative Coulomb electronic interactions in oxyapatites and rationalized observed experimental trends reported in the literature. Two types of fundamental chemical property, i.e. the electronegativities and ionic radii of the constituents, control oxygen ionic conduction in oxyapatites. Those two properties were used to represent the relative charge densities and the distances between charged units, respectively, and then to formulate the relative Coulomb energy. It was found that this relative Coulomb energy is linearly correlated to the oxygen ionic

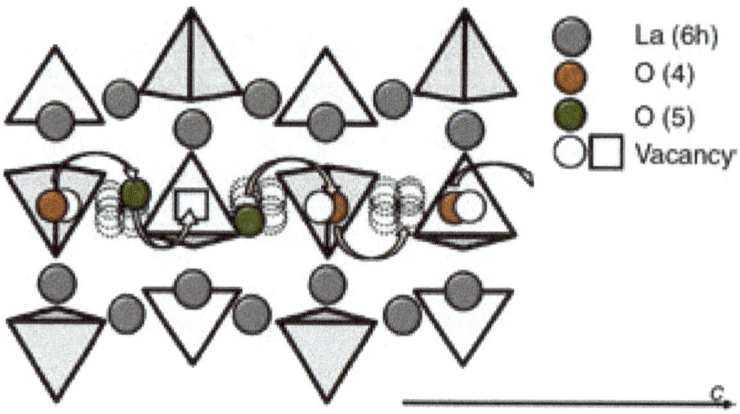

Fig. 10. Structural defect position and possible conduction mechanism along the c-axis representation of two adjacent unit-cells [77].

conductivity (in logarithmic form) in the oxyapatite systems. Doping a cation with large ionic radius and low electronegativity tends to increase the ionic conductivity of oxyapatite.

The c-axis-oriented apatite-type lanthanum silicate ($La_{10}Si_6O_{27}$) ceramics was prepared by NAKAYAMA et al [85] via the sintering process under high magnetic field. The degree of orientation in the $La_{10}Si_6O_{27}$ ceramics sintered at 1700°C was 48.1%. The conductivity of the c-axis-oriented ceramics is about 0.5 orders of magnitude higher than that of non-oriented ceramics. Higher conductivity is caused by the orientation of oxide ions in the grains composing the ceramics, which are located along the c-axis and are responsible for the ionic conduction.

Sensors

Compacted sinters of $Ln_{9.33+x/3}Si_{6-x}Al_xO_{26}$ ($0 \le x \le 2.0$, Ln = La, Nd and Sm) are composed of an apatite-like phase with a hexagonal structure. Compacted sinters were used as potentiometric oxygen sensors (**Fig. 11**). The concentration dependence of EMF was well expressed by the Nernst equation:

$$E_{obs} = \frac{RT}{nF} \ln \frac{p_{O_2}(I)}{p_{O_2}(II)} \tag{11}$$

Furthermore, the electron number n is comparable to the theoretical value of 4. The sensing characteristics of the sinters are comparable to those of the sensors with 3 and 8 mol.% YSZ [86].

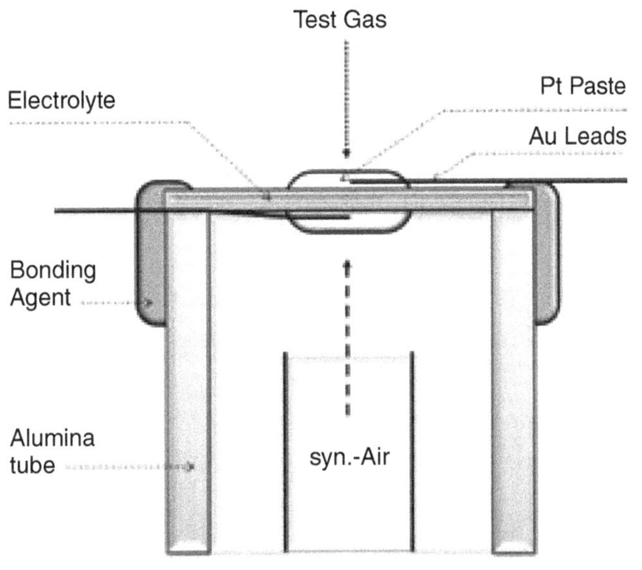

Fig. 11. Schematic picture of the O_2 concentration cell with the electrolyte of $Ln_{9.83}Si_{4.5}Al_{1.5}O_{26}$ where Ln = La, Nd and Sm [86].

Electrical properties and humidity sensor characteristics of lead hydroxyapatite material were reported by TUDORACHE et al [87]. The electrical characteristics of lead hydroxyapatite material treated at different temperatures made us focus on the analysis of the influence of water vapors upon the electrical characteristics. Thus, the electrical response to humidity adsorptive processes of lead hydroxyapatite material suggested that we analyze the material characteristics in terms of its use as a humidity sensor.

The humidity-sensitivity of yttrium-substituted calcium oxyhydroxyapatites was studied by OWADA et al [88]. The logarithm of electrical resistance of the present sensors decreased linearly with increasing relative humidity (RH) from 30 to 65%. The resistance of $[Ca_{9.0}Y_{1.0}](PO_4)_6[O_{1.5}\square_{0.5}]$ with the highest OH vacancy content was about one order of magnitude lower than that of calcium hydroxyapatite. It was found that the larger ratio of surface hydroxyl groups per unit surface area in the sample, the lower the resistance and the higher the amount of OH vacancies.

Using the hydroxylapatite ceramics as CO_2 sensor, which is based on electrical conductivity changes, was investigated by NAGAI et al [89]. Starting powders prepared by usual wet process were cast in film with an organic vehicle and fired on alumina substrates after the electrodes had been arranged. It was necessary to soak the samples in a $CaCl_2$ solution in order to make them reactive with CO_2. Both D.C. and A.C. measurements were carried out in various atmospheres including air, CO_2 and air containing different amounts of CO_2.

Phosphors

Fluorescent lamps typically have transparent glass envelope enclosing sealed discharge space containing an inert gas and mercury vapor. When subjected to a current provided by electrodes, mercury ionizes to produce the radiation having the primary wavelengths of 185 nm and 254 nm. This ultraviolet radiation, in turn, excites phosphors on the inside surface of the envelope to produce visible light that is emitted through the glass. Generally, the fluorescent lamp for illumination uses a phosphor that absorbs the 254 nm Hg-resonance wave; the phosphor is activated so as to convert the ultraviolet light into the visible light. In order to improve the color-rendering properties and the emission output of fluorescent lamps, efficient illumination of a white color has been recently provided using a three-band-type fluorescent lamp, which employs a mixture of red, green and blue-emitting phosphors. In such three-band-type phosphor lamp, the emitting colors of the respective phosphors are considerably different from one another. Therefore, if the emitting intensity of any of the three corresponding phosphors is decreased, the color deviation occurs, degrading the color-rendering properties of the lamp [90]. The literature dedicated to the preparation of apatite-type light-emitting phosphors is really abundant.

A series of orange-red-emitting $Ba_2Y_3(SiO_4)_3F{:}xSm^{3+}$ ($0.003 \le x \le 0.08$) fluorosilicate apatite phosphors were synthesized via the conventional solid-state reaction by Yu et al [91]. The emission spectra of the $Ba_2Y_3(SiO_4)_3F{:}Sm^{3+}$ phosphors contained some sharp emission peaks of Sm^{3+} ions centered at 564, 601, 648 and 710 nm. The strongest one is located at 601 nm. The optimum dopant concentration of Sm^{3+} ions in $Ba_2Y_3(SiO_4)_3F{:}xSm^{3+}$ is around 3 mol.% and the critical transfer distance of Sm^{3+} was calculated to be 26 Å. The quenching temperature is above 500 K.

Red-emitting phosphors $Ba_2Gd_8(SiO_4)_6O_2{:}Eu^{3+}$ (BGS:Eu^{3+}) with silicate apatite structure were prepared by Liu et al [92] via the conventional high-temperature solid-state reaction method. There are two different sites (4*f* and 6*h* [93]) for Eu^{3+} occupying the host. It was found that the phosphors BGS:Eu^{3+} exhibit red emission with high quenching concentration at ~70.75 at.%, and the critical transfer distance of Eu^{3+} in BGS:Eu^{3+} was calculated to be ~12.3 Å. More importantly, it has better CIE chromaticity coordinate for white light-emitting diode (w-LED) application in comparison with commercial phosphor $(Y,Gd)BO_3{:}Eu^{3+}$ (YGB:Eu^{3+}) under near-ultraviolet (n-UV) 393 nm excitation [92],[94]. White Tb^{3+}/Sm^{3+} ions co-doped $Ca_2La_8(GeO4)_6O_2$ (CLGO) phosphors prepared by Jeon et al [95] show observable emission spectra under 374 nm excitation.

A novel blue-emitting phosphor $Sr_8La_2(PO_4)_6O_2{:}Eu^{2+}$ was synthesized by Liu et al [96] via conventional high-temperature solid-state method and its photoluminescence (PL) properties were investigated for the applications in white light-emitting diodes. The phosphor exhibited strong broad absorption band in the near-ultraviolet (n-UV) range and generated bright-blue emission centered at 442 nm upon 365 nm excitation light. The critical Eu^{2+} quenching concentration (QC) mechanism was verified to be the dipole-dipole interaction.

A green-emitting phosphor of Eu^{2+}-doped $Ca_5(PO_4)_2SiO_4$ was prepared via a solid-state reaction by Roh et al [97]. The phosphor was excited at the wavelengths of 220 – 450 nm, which was suitable for the emission band of near-ultraviolet (n-UV) light-emitting diode (LED) (350 – 430 nm). In $Ca_5(PO_4)_2SiO_4{:}Eu^{2+}$ phosphor, there were three distinguishable Eu^{2+} sites, which resulted in a strong green emission peaking at 530 nm and broad bands up to 700 nm.

Catalysts

Catalysts are usually defined as the substances that increase the rate at which a chemical reaction approaches the equilibrium without becoming permanently involved in this reaction. Basically, the catalysis can be divided to [98]:

i. Homogeneous catalysis,

ii. Heterogeneous catalysis.

Heterogeneously catalyzed process is more complex because the catalyst is not uniformly distributed throughout the reaction medium. Considering a two-phase system, either vapor/solid or liquid/solid, with the catalyst in the solid phase, the several steps need to be realized to complete the catalytic cycle [98]:

1. Transport of the reactant to the catalyst;
2. Adsorption, i.e. the interaction of the reactant with the catalyst;
3. Reaction of adsorbed species to the products;
4. Desorption of the products from the catalyst;
5. Transport of the product away from the catalyst surface.

Due to their versatility in anionic and cationic composition and their ability to adsorb organic and organometallic molecules as well as metallic salts, the surface properties of apatites can be tuned and they can behave as powerful catalysts in a wide range of organic reactions. In many cases, the apatite-based catalysts can be used without a solvent and show good recycling capacity. The catalytic properties mainly arise from the acid-base character of the apatite's surface. In some cases, adsorbed moieties are responsible for the catalytic properties, apatites playing the role of a solid support. Finally, the catalytic activities can result from the combination of properties of the apatite's surface and of the adsorbed or anchored moieties [99].

The oxidative Glaser-Hay coupling reaction of terminal alkynes is a very important reaction in organic chemistry to achieve the synthesis of diyne compounds. In general, the reaction is performed under homogeneous conditions using Cu(I) or Cu(II) salts in the presence of a reagent such as tetramethylethylenediamine (TMEDA), which can bind to copper ions, an organic base and dioxygen. Although this reaction is known for a long time, the mechanism is still under the discussion. It is possible to catalyze the Glaser-Hay reaction under heterogeneous conditions using Cu-modified hydroxyapatite (Cu-HAp). With several para-substituted phenyl-acetylenes and alkynols, we can show that Cu-HAp acts as a catalyst for single-bond coupling reactions leading to diyne derivatives in high yields without using auxiliary chelating molecules and organic bases. These heterogeneous conditions allow easy recovery of the catalyst and simplify the purification work-up.

The oxidative Glaser-Hay coupling reaction of terminal alkynes (acetylenes) is a very important reaction in organic chemistry to achieve the synthesis of diyne compounds. In general, the reaction is performed under homogeneous conditions using Cu(I) or Cu(II) salts in the presence of base (ethanolic ammonia solution, tetramethylethylenediamine, pyridine, ...), which can bind to copper ions, an organic base and dioxygen [100],[101]:

$$R-C\equiv CH \xrightarrow[\text{base}]{\text{CuCl}} R-C\equiv C-C\equiv -R \tag{12}$$

Although this reaction is known for a long time, the mechanism is still under the discussion. It is possible to catalyze the Glaser-Hay reaction under heterogeneous conditions using Cu-modified hydroxyapatite (Cu-HAp). With several para-substituted phenyl-acetylenes and alkynols, where Cu-HAp acts as a catalyst for single-bond coupling reactions leading to diyne derivatives in high yields without using auxiliary chelating molecules and organic bases. These heterogeneous conditions allow easy recovery of the catalyst and simplify the purification work-up [100].

The apatite catalyst was utilized for the catalysis of the synthesis of n-butanol, 1,3-butadiene and high octane fuel from bioethanol. The process requires relatively low temperature. The synthesis shows significantly lower cost compared to n-butanol derived from petroleum-based processes. The technology offers a closed-loop system with no waste or emissions [102].

Phosphate conversion coatings

Conversion coatings provide the resistance to corrosive environments. Phosphate conversion coatings (PCC) bring about the transformation of metal substrates into new surfaces having non-metallic and non-conducting properties. The transformations occur in phosphating solution containing divalent metal phosphates and, in some cases, in solutions containing monovalent metal phosphates. Generally, the solutions are prepared from liquid concentrates containing one or more divalent metals (zinc, magnesium, calcium, etc., phosphates), free phosphoric acid and an accelerator. Three types of phosphate conversion coatings are currently being used [103],[104]:

1. **Zinc phosphate coatings** are often used as a pretreatment for painted parts. They are also used to impart the corrosion resistance and to aid in cold-forming operations.
2. **Iron phosphate coatings** are primarily used to form a passive substrate under paints.
3. **Manganese phosphate coatings** are primarily on machined parts such as gears and internal combustion engine components as an anti-scuff film for the break-in wear.

The addition of metal ions, such as cupric ions, to a conversion bath greatly reduces the formation time and the size and nonuniformity of coating crystals. Copper, which is cathodic to dissolving metals, deposits on the base metal to form many local cells and thus to increase the potential difference between local anode and cathode site. Nickel ions behave differently than cupric ions, and their benefit results from the catalytic action associated with the release of molecular hydrogen. Furthermore, the addition of Ni^{2+} and Mn^{2+} into the treating solution refines the grain size and reduces the porosity of phosphate conversion coatings on electrogalvanized steels. While Ni exists in both the zero-valance state and the two-valance state, Mn is mainly present in the two-valance state in the phosphate conversion coating [103],[105].

In fact, all of the chemical reactions of phosphating process are based on mutual interactions of metal immersed in phosphate bath and redox of the accelerators. Generally, phosphating proceeds in the acidic solution containing Zn^{2+}, Mn^{2+}, Ca^{2+}, Na^{+}, Fe^{2+} and Mg^{2+}. Phosphating with different metal substrates and types is not of the same reaction mechanism. For example, when a pure iron is immersed in the phosphating solution, iron dissolves on the micro-anodes through the following reaction [106]:

$$Fe + 2\ H^{+} \rightarrow Fe^{2+} + H_2 \tag{13}$$

The hydrogen evolution occurs at the micro-cathodic sites resulting in an increase of pH value at the metal-solution interface. This change in pH alters the dissociation equilibrium, which leads to the formation of PO_4^{3-}:

$$H_3PO_4 \rightarrow H_2PO_4^{-} + H^{+} \rightarrow HPO_4^{2-} + 2\ H^{+} \rightarrow PO_4^{3-} + 3\ H^{+} \tag{14}$$

When PO_4^{3-} and Me^{2+} (metal ion, e.g. Zn^{2+}, Mn^{2+}, Ca^{2+} and Fe^{2+}) in the solution reach the saturation, the deposition of insoluble phosphate will be achieved. Then, it can crystallize in PCC coating, as shown in **Fig. 12**.

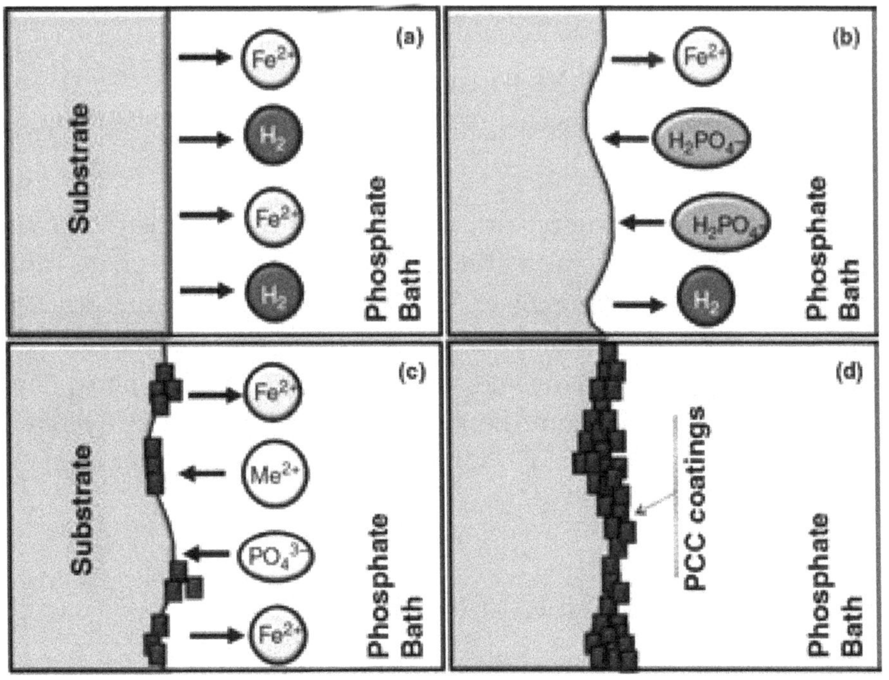

Fig. 12. Schematic representation of the deposition process of PCC coating on the surface of pure iron [106].

In the zinc phosphating solution, the addition of Zn^{2+} supports the formation of crystals of zinc phosphate. Zinc ions combine with phosphate ions to form an insoluble film. The formation of hopeite is described by the reaction [106]:

$$\begin{gathered} 3\ Zn^{2+} + 2\ H_2PO_4^- + 2\ H^+ + 4\ H_2O + 6\ e^- \rightarrow \\ Zn_3(PO_4)_2 \cdot 4H_2O + 3\ H_2 \end{gathered} \tag{15}$$

In some cases, Zn and ZnO were found in the phosphating process in reactions:

$$Zn + 2\ e^- \rightarrow Zn \tag{16}$$

$$Zn + 2\ H_2O \rightarrow Zn(OH)_2 + H_2 \tag{17}$$

$$Zn(OH)_2 \rightarrow ZnO + H_2O \tag{18}$$

In the phosphate solution of coexisting Zn^{2+} and Ca^{2+}, zinc calcium phosphate can be formed by the reactions:

$$Ca^{2+} + HPO_4^{2-} \rightarrow CaHPO_4 \tag{19}$$

$$Ca^{2+} + 2\ Zn^{2+} + 2\ H_2PO_4^- + 2\ H_2O \rightarrow CaZn_2(PO_4)_2 \cdot 2H_2O + 4\ H^+ \tag{20}$$

It is notable that the phosphating mechanism varies in different phosphating systems and materials [106].

The formation of conversion coating on zinc-coated samples under cathodic conditions was studied by Perrin et al [107] in a chromating bath containing phosphate (phosphate-chromate solution). Thick chromium phosphate (Cr-P) coating has two distinct layers: an outer porous layer and an inner and thinner pore-free adherent one. Both layers contain chromium phosphate as the main constituent and, to a lower extent, zinc phosphate species, the concentrations of which decrease from the metal-coating interface outwards. Formed zinc phosphates have a general formula of $x CrPO_4 \cdot y Zn_3(PO_4)_2 \cdot z\ H_2O$ where $x > (y, z)$.

Synthetic apatite analogues in tissue engineering

Biomaterials in general are based on the materials such as metals, polymers, and ceramics. Typical metallic biomaterials are based on stainless steel, cobalt-based alloys, titanium or titanium alloys and amalgam alloys. Polymeric biomaterial composites from monomers are based on amides, ethylene, propylene, styrene, methacrylates and/or methyl methacrylates. Biomaterials based on ceramics are found within all classical ceramic families (**Table 3**) including traditional ceramics, special ceramics, glasses, glass-ceramics, coatings and chemically bonded ceramics (CBCs) [13].

Ceramic material classification	Example
Traditional ceramics	Dental porcelain, leucite-based ceramics
Special ceramics	Al, Zr and Ti oxides
Glass	Bioglass (Na_2O-CaO-P_2O_5-SiO_2)
Glass-ceramics	Apatite-wollastonite, Li-silicate-based
CBCs	Phosphates, aluminates, silicates and sulfates

Table 3 Examples of biomaterials based on ceramics [13].

Whereas many chemists and materials scientists consider the biomaterial to be synthetically produced material, most biologists, geologists and mineralogists consider the materials such as bone and tooth, which are biologically produced, to be the biomaterials. Also a common reference to Ca:P (1.67, **Table 7** in **Chapter**) ratio is usually used in the biomaterials literature, which disregards the fact that different calcium phosphate phases have different crystalline structures.

There are many phosphate minerals and salts (**Table 4**) that do not have the crystalline structure of apatite [108],[109]. Apatite-based materials have attracted a considerable interest for orthopedic and dental applications because of their biocompatibility and tight bonding to bone, resulting in the growth of healthy tissue directly onto their surface. Several combinations of apatite and other phases were proposed in order to improve poor mechanical properties of apatite [110].

Typical acronym	Chemical name	Chemical formula	Mineral name	Structure	Ca/P ratio
HAP, HA	Tribasic calcium phosphate	$Ca_5(PO_4)_3OH$	Hydroxylapatite	Apatitic	1.67
ACP	Amorphous calcium phosphate	?	—	—	?
PCHA, PCA	Poorly crystalline hydroxylapatite	$Ca_5(PO_4)_3OH$	Hydroxylapatite	Apatitic	1.67
CAP	Carbonated apatite				
TCP	Tricalcium phosphate	$Ca_9(PO_4)_6$	Whitlockite	Non-apatitic	1.5
β-TCMP	Magnesium-substituted TCP	$Ca_9(PO_4)_6$	Whitlockite	Non-apatitic	≤1.5

Typical acronym	Chemical name	Chemical formula	Mineral name	Structure	Ca/P ratio
?	"Tricalcium phosphate"	$Ca_9(Mg,Fe^{2+})(PO_4)_6(HPO_4)$	Geologically occurring whitlockite	Non-apatitic	1.28
CPPD	Calcium pyrophosphate dihydrate	$Ca_2P_2O_7 \cdot 2H_2O$	–	Non-apatitic	1.0
γ-CCP	γ-Calcium pyrophosphate	$Ca_2P_2O_7$	–	Non-apatitic	1.0
OCP	Octacalcium phosphate	$Ca_8H_2(PO_4)_6 \cdot 5H_2O$	–	Non-apatitic	1.33
MON	Dibasic calcium phosphate	$Ca(HPO_4)$	Monetite	Non-apatitic	1.0
DCPD	Dicalcium phosphate dihydrate	$Ca(HPO_4) \cdot 2H_2O$	Brushite	Non-apatitic	1.0

Table 4 Different apatitic and non-apatitic calcium phosphates [108].

Small organic molecules incorporated in apatite crystals act as porogens that control the porous structure of apatite single crystal. The presence of amino acid under the apatite synthesis conditions leads to firm bindings and encapsulation of amino acid within apatite single crystals. The amino acid elimination by heating or electron beam irradiation enhances the pore formation in apatite single crystals. Moreover, the incorporation of acidic amino acid into apatite induces the peapod-like nanotubes in apatite single crystals. That suggests the potential of using small organics for nanostructural control of apatite single crystals, which would be valuable for enhancing the drug loadings or for modulating the material digestion in vivo [111].

Biological apatite in bone tissue engineering

Tissue engineering (TE) techniques were developed to recover or enhance lost tissue function and structure. In biological hard tissues, for example, lost portions can be effectively reconstructed by the control of environmental factors, physical stimulation, addition of growth factors and by the use of degradable materials. These factors strongly facilitate the regeneration of macroscopic shape of defected hard tissues. Nevertheless, the differences in microstructure, and also in mechanical and physical properties, between regenerated and original hard tissues must be examined prior to clinical application [112],[113],[114].

For in vitro engineering of living tissues, cultured cells are grown on bioactive degradable substrates (scaffolds[8]) that provide the physical and chemical cues to guide their differentiation and assembly into three-dimensional structures. One of the most critical issues in TE is the realization of scaffolds with specific physical, mechanical and biological properties. Scaffolds act as substrate for cellular growth, proliferation and the support for new tissue

[8] Scaffolds might be defined as artificial structure capable of supporting the three-dimensional tissue formation, which allows the cell attachment and migration, the delivery and retaining of cells and biochemical factors and enables the diffusion of vital cell nutrients and expressed products. In the case of bone, scaffolds should replicate its architecture and three-dimensional structure with predetermined density, hierarchical pore distribution and interconnected pathways [109],[115].

formation. Biomaterials and fabrication technologies play a key role in tissue engineering. Materials used for tissue engineering applications must be designed to stimulate specific cell response on the molecular level. They should elicit specific interactions with the cell and thereby direct cell attachment, proliferation, differentiation and extracellular matrix production and organization [109],[115]:

- **Biocompatibility** is defined as the ability to perform its function in the host tissue without eliciting any immune response.
- **Biodegradability** denotes tunable rate of degeneration to match the growth of new bone tissue as scaffold gets replaced by new bone.
- **Mechanical properties** include the properties such as sufficient mechanical strength to provide temporary support to the defect region and withstand in vivo loading forces.
- **Microarchitecture** is an interconnected scaffold structure that uniformly distributes stresses throughout the scaffold.
- **Osteoinductivity** includes osteoinductive properties that enable to recruit and differentiate the osteoprogenitors to the defect region.
- **Porosity** is large surface area, where the volume and the pore size allow the tissue in-growth, neovascularization, mass transport and the osteogenesis.
- **Surface properties** include appropriate chemical and topographical properties for influencing cellular adhesion, proliferation and differentiation.

Inorganic-organic composites aiming at mimicking the composite nature of real bone combine the toughness of the polymer phase with the compressive strength of an inorganic one to generate bioactive materials with improved mechanical properties and degradation profiles. Hydroxylapatite (HAP) is widely used as a biocompatible ceramic material in many areas of medicine, but mainly for the contact with bone tissue, due to its resemblance to mineral bone [115].

Under normal conditions, human body fluid is supersaturated with respect to apatite, so that, when apatite nuclei form, crystals can grow spontaneously. Chemical species that are capable of supporting the nucleation of apatite are calcium and silicate ions, while, in contrast, phosphate ion does not affect this process. Hence, ceramics that release the former mentioned ions are capable of developing the surface apatite layer when exposed to human body fluids. This explains why glasses that do not contain P_2O_5 are able to develop this surface layer and show greater bioactivity than glasses containing P_2O_5, even if the latter compositions approximate to that of hydroxyapatite [116].

Preparation of nanocrystalline apatites

Nanocrystalline calcium phosphate apatites play an important role in biomineralization[9] and in the field of biomaterials. Biological nanocrystalline apatites are the main inorganic compo-

[9] The enamel of vertebrate teeth, vertebrate bone and tooth-like microfossils of conodonts.

nents of hard tissues in mammals (bone[10] and tooth[11] (**Fig. 13**)) with the exception of enamel (which is closer to stoichiometric hydroxylapatite) and are involved in several pathological calcifications such as dental calculi, salivary stones and blood vessel calcification. In comparison with hydroxylapatite, which is a stoichiometric apatitic phase and is the most stable and the least soluble calcium phosphate at ambient conditions, nanocrystalline apatites are nonstoichiometric and calcium (and OH-) deficient and may incorporate substituted ions in their nanosized crystals. Their calcium and hydroxide deficiencies are responsible for higher solubility than HA. Besides, they have the ability to mature when submitted to humid environments [108],[112],[117],[118].

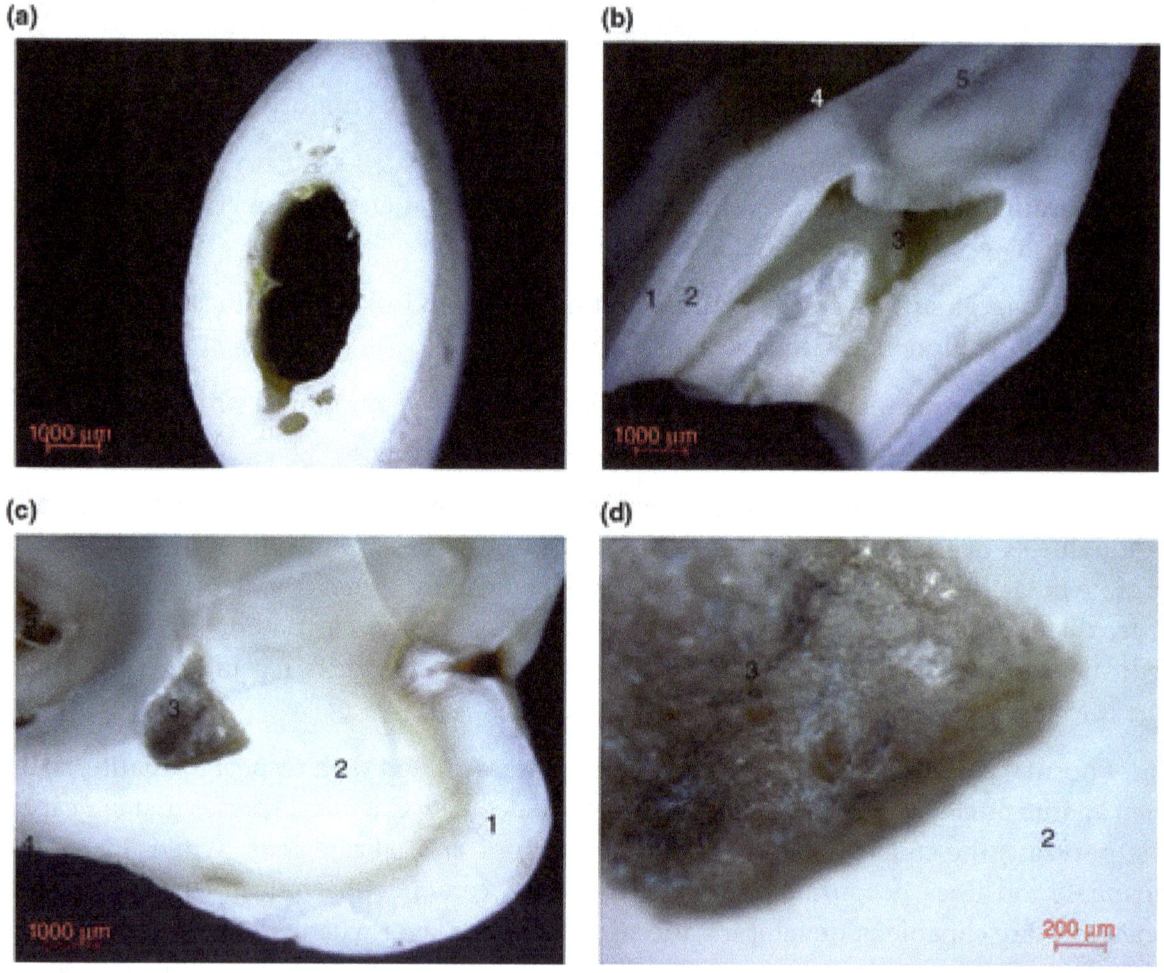

Fig. 13. The cross-section of lower jaw bone (a) and tooth (b) of European roe deer (*Capreolus capreolus*) and human tooth (c, d): enamel (1), dentin (2), pulp chamber (3), cementum (4) and root canal (5).

[10] Bones are rigid organs that form a part of the endoskeleton of vertebrates. Their function is to move, support and protect various organs of the body, produce red and white blood cells and store minerals. Bones appear in a variety of shapes and have a complex internal and external structure described by various hierarchical models [117]. Bone is a complex and hierarchical tissue consisting of nano-hydroxylapatite (70 wt.%) and collagen (30%) as major portions [113].

[11] Teeth consist of a bulk of dentin covered with (inorganic) enamel on the crown and cementum on the root surface. Thick collagen bundles, called periodontal ligaments (PDL), attach to cementum at one end and to the alveolar bone at the other end. The alveolar bone is supported by the jaw [117].

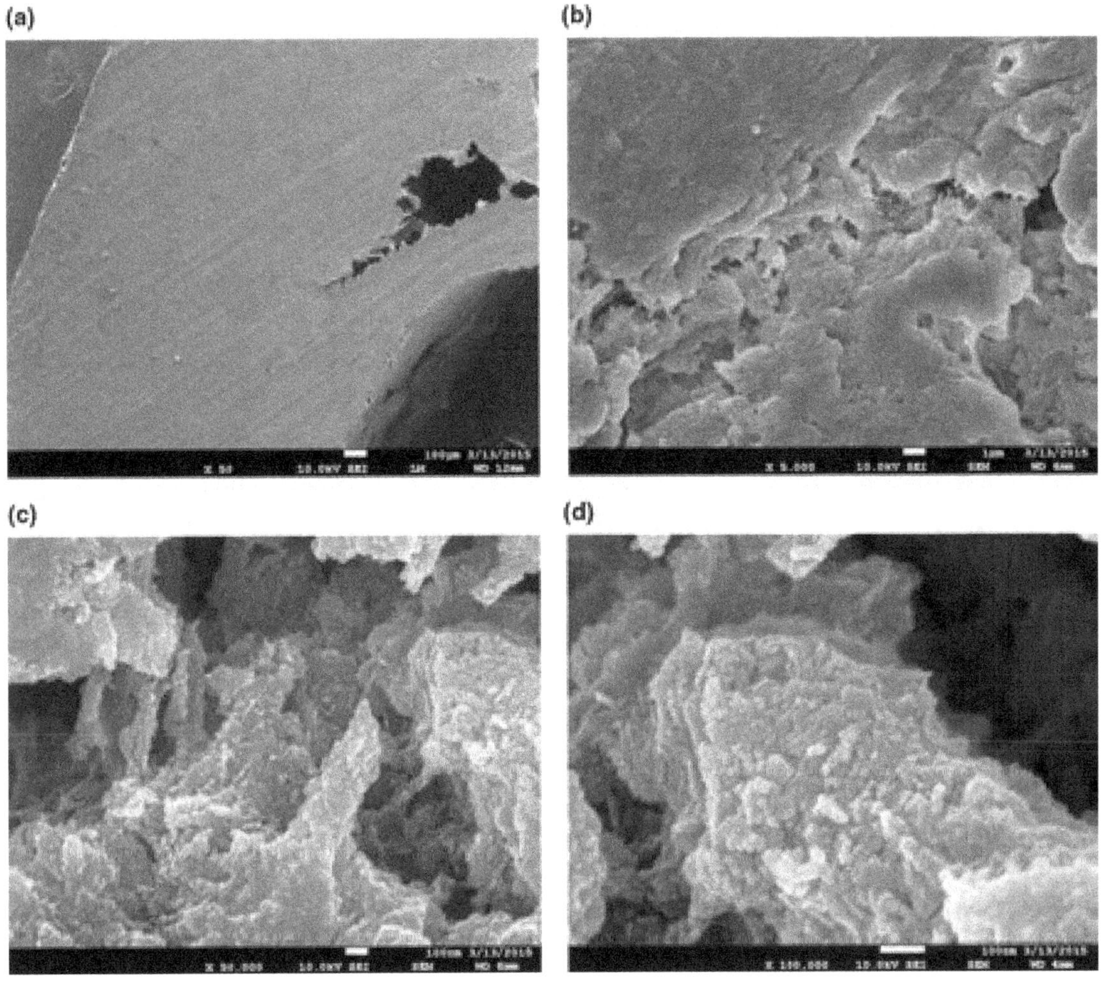

Fig. 14. Electron microscopy picture (SEM) of lower jawbone of European roe deer from **Fig. 13** under magnification of 50× (a), 5000× (b), 50,000× (c) and 100,000× (d).

Scanning electron microscopy of deer bone (**Fig. 14**) shows a spongiform texture (*a, b*) formed by crystals of carbonated hydroxylapatite with the size below 100 nm (*e, f*).

The inorganic portion of bone contains two major mineral phases

i. Non-crystalline calcium phosphate;

ii. Crystalline bone apatite.

In amorphous calcium phosphate, the Ca:P ratio is about 1.33. The non-crystalline or amorphous bone mineral is metastable with respect to bone apatite. Bone apatite crystals are calcium deficient due to the defects in the crystalline lattice and isomorphous substitutions, that is, the replacement of some ions by others in the crystal without disrupting the general symmetry. Young bone tissue was found to be richer in amorphous mineral than crystalline apatite [119].

In the body, only the bone collagen has the property of inducing the mineralization through in vitro; collagens of other tissues do not possess this property. One of the differences between nucleating and non-nucleating collagens might be the presence of a sort of inhibitor bound to the latter. This inhibiting substance was shown to be pyrophosphate. Calcium is accreted in bone tissue in the process of new bone formation or remodeling and resorbed from the bone tissue in the process of bone destruction. The loss of endogenous calcium in urine and feces is compensated for by an equivalent intake of this element (**Fig. 15**) [119].

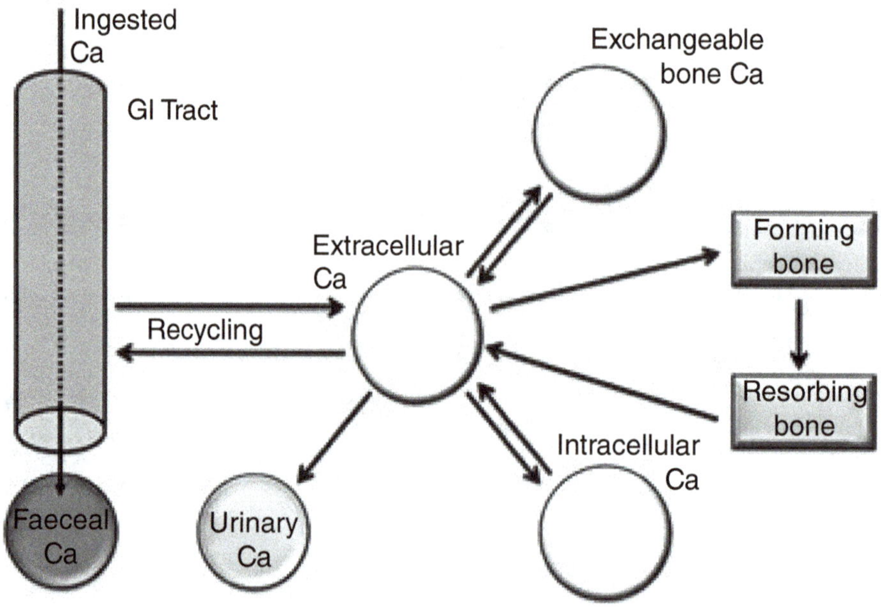

Fig. 15. General scheme of metabolism of calcium [119].

Enamel (**Fig. 17**) is normally the best preserved from hard tissues. It is almost completely a mineral, so the decomposition of organic matter has little effect on it. Archeological enamel nearly always yields good microscope sections, often indistinguishable from fresh enamel [120]. The structure of carbonate dental enamel refined by WILSON et al [121] is shown in **Fig. 16**.

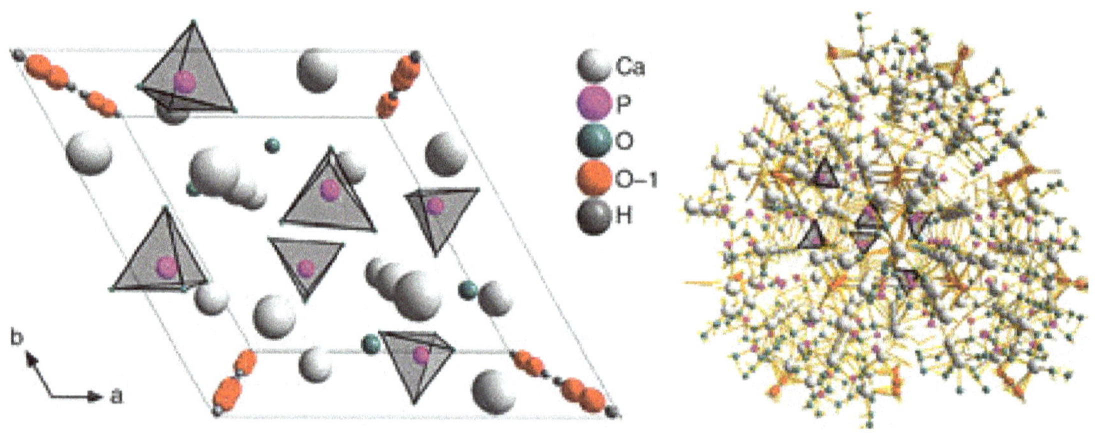

Fig. 16. Crystallographic structure of human dental enamel apatite (perspective view along the c-axis): $P6_3/m$, $\boldsymbol{a}$ = 9.4081 Å, $\boldsymbol{b}$ = 6.8887 Å, $\boldsymbol{c{:}a}$ = 0.7322 and $\boldsymbol{V}$ = 528.05 Å^3 [121].

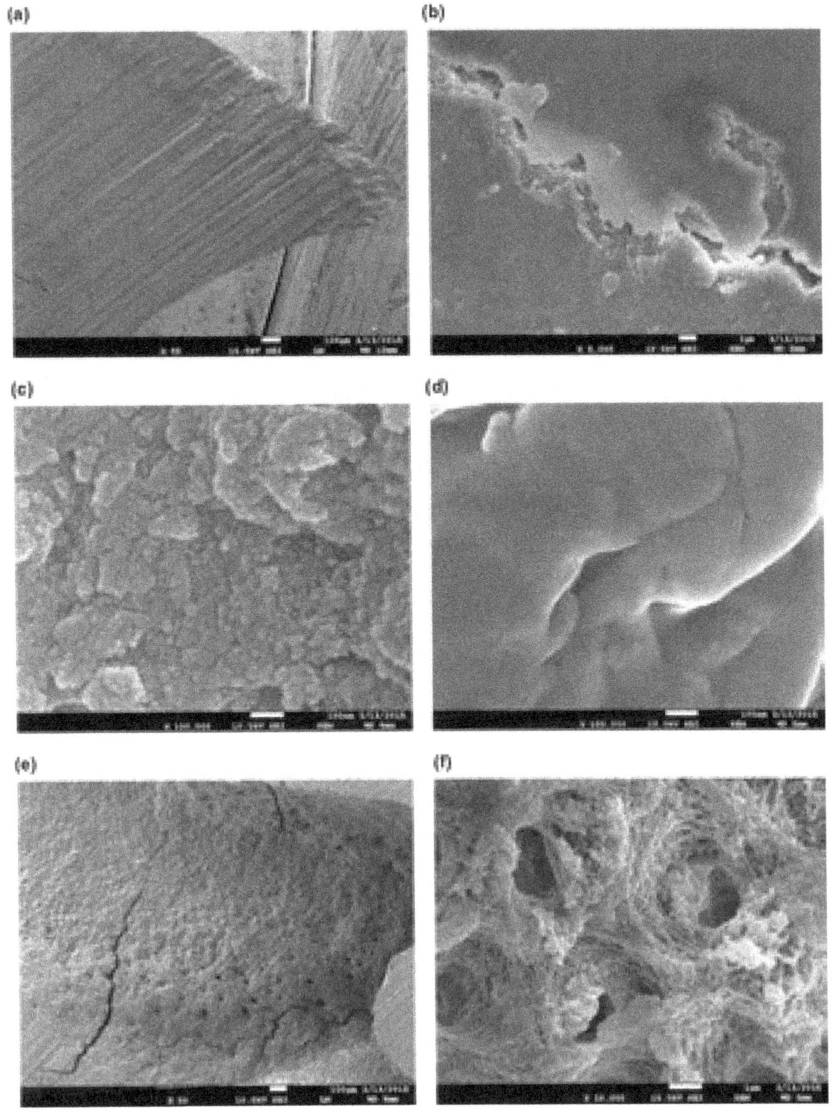

Fig. 17. Electron microscopy picture (SEM) of tooth from **Fig. 13**: the crown with enamel-dentin interface (*a, b*), nano-crystals of carbonated hydroxylapatite in the texture of dentin (*c*), glass-like texture of enamel (*d*) and surface of pulp chamber (*e, f*).

Among carbonate apatites, which are formed in mouth cavity, only dental enamel mine belongs to the B-type of carbonate-apatite (**Section**) and the others (of dentin, salivary and dental stones) belong to the AB-type (B > A). According to the variations in unit-cell parameters, isomorphic replacements in crystal structures of apatites of pathogenic origin (renal, salivary and dental stones) are more intensive in comparison with physiogenic dental enamel apatites. Among pathogenic apatites, the most considerable compositional variations are observed in renal stone apatites. That indicates strong variability of conditions of their formation. The changes in unit-cell parameters of bone apatites are not completely interpretable, because these apatites consist of nanosized crystals that are smaller than those of other biological apatites [122].

The age variations of the crystal lattice parameters of human enamel apatites are related to complicated processes of de- and remineralization, which result in the increase or reduction of vacancies in Ca positions and in the respective changes of CO_3^{2-}, H_2O and HPO_4^{2-} contents in the unit cell:

$$\begin{aligned}&(n+m/2)\ Ca^{2+}n\ PO_4^{3-}+m\ OH^- \rightarrow\\&(n+m/2)\ [\ ]_{Ca}+n\ A^{2-}+m\left([\ ]_{OH^-},H_2O\right)\end{aligned} \tag{21}$$

where $A^{2-} = CO_3^{2-}$ or HPO_4^{2-}, and $[]_{Ca}$, $[]_{OH-}$ are the vacancies. Until the age of 50 years, the values of $\boldsymbol{a}$ and c-parameter of enamel apatites change considerably without any dependence of particular age that may be explained by essential fluctuations of the content of Ca in human organism. After 50 years of age, significant direct correlation between the age and the $\boldsymbol{a}$-parameter appears [122].

The surface of apatite nanocrystals is possibly doped with foreign elements or functionalized with organic molecules [117],[123],[124]. The course of facile synthesis of B-type carbonated nanoapatite with tailored microstructure is described by GUALTIERI et al [125].

Collagen apatite composites

Tissue engineering techniques have been developed to recover or enhance lost tissue function and structure.[12] Collagen-apatite[13] (or collagen/apatite, **Col-AP**) composite resembling the composition of natural bone[10] has been studied extensively and considered as a promising bone tissue engineering material, which can be used to replace or regenerate damaged tissue, resulting from an accident, trauma or cancer. Such synthetic or hybrid biomaterials must have high porosity with interconnected pores to allow the vascularization as well as the nutrients and gases diffusion. Moreover, they should be biodegradable to act as temporary cellular support [114],[126],[127],[128].

The defining feature of collagen is a structural motif in which three parallel polypeptide strands in a left-handed, polyproline II-type (**PPII**) helical conformation coil around each other with a one-residue stagger to form a right-handed triple helix. Tight packing of **PPII** helices within the triple helix mandates that every third residue be **GLY**, resulting in a repeating **XAAYAAGLY** sequence, where **XAA** and **YAA** can be any amino acid. This repeating occurs in all types of collagen, although it is disrupted at certain locations within the triple-helix domain of nonfibrillar collagens. The amino acids in the **XAA** and **YAA** positions of collagen are often (2S)-

---

[12] One of the first polymers used for bone tissue engineering was based on a hydrolytically copolymer of polylactic-co-glycolic acid (PLGA) but its use for large bone defect regeneration was controversial as inflammatory events were observe. The utilization of chitosan and alginate was also investigated [128].

[13] Names such as collagen-hydroxylapatite or collagen-hydroxyapatite composite were also often applied in published literature.

proline (**Pro**, 28%) and (2S,4R)-4-hydroxyproline (**Hyp**, 38%), respectively. The **ProHypGly** is the most common triplet (10.5%) in collagen (**Fig. 18**) [129].

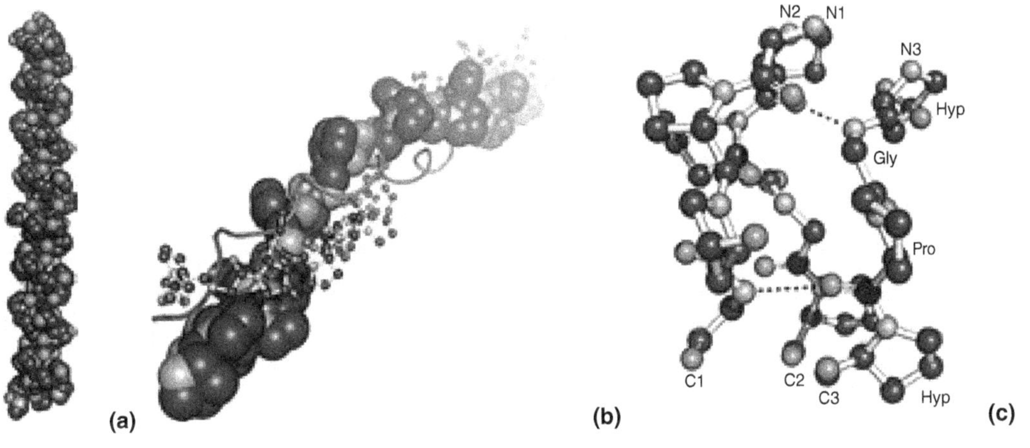

Fig. 18. The triple helix of collagen formed from $(ProHypGly)_4$-(ProHypAla)-$(ProHypGly)_5$ (a), the view down the axis of a $(ProProGly)_{10}$ triple helix (b) and the segment of triple helix (c) with hydrogen bonds (—) [129].

The categories of collagen[14] include the classical fibrillar and network-forming collagens, the FACITs (fibril-associated collagens with interrupted triple helices), MACITs (membrane-associated collagens with interrupted triple helices) and MULTIPLEXINs (multiple triple-helix domains and interruptions). The collagen of type I is the most abundant type used in tissue engineering. Natural polymer collagen that represents the matrix material of bone, teeth and connective tissue can be extracted from animal or human sources (skin, bones, tendons, ligaments and cornea). The treatment includes the separation and isolation (in soluble or insoluble form), decalcification, purification (purification is required to eliminate the antigenic component of protein), sterilization and chemical modification process to achieve polyanionic or purified protein. Type I polyanionic collagen was found to improve the cell adhesion [129],[130].

Collagen, the most abundant protein in extracellular matrix, is chemotactic to fibroblasts. It shows high affinity to cells and good resorbability in vivo. Nevertheless, its poor mechanical properties have restricted its usage in load-bearing applications. Carbonated apatite and collagen interact to form a composite material, the mechanical, physicochemical and biological properties of which differ considerably from those of either constituent considered separately. Collagen and non-collagenous proteins (**NCPs**) are thought to control the crystal deposition, size, crystallization and multiplication/maturation. On the other hand, the crystal deposition in intrafibrillary spaces is likely to modify the three-dimensional conformation of collagen [114],[131],[132],[133]. The solubility of apatite-collagen composites is significantly reduced by UV radiation [134],[135].

[14] There are 27 types of collagen described in literature composed of at least 46 distinct polypeptide (CRP, collagen-related peptide) chains [129], but the types I – V are the most common. More than 90% of collagen in human body is the fibrillar type I.

Calcium phosphates are available commercially, as hydroxylapatites are extracted from bones or they can be produced wet by direct precipitation from pH-adjusted solutions of calcium and phosphate salts [130]. The crystallographic c-axes of the plate-shaped apatite crystals are well aligned with long axes of collagen fibrils (**Fig. 19**), and this preferred orientation between the mineral and the organic framework is assumed to be the general feature of the calcium phosphate biomineralization process. Several attempts were made to mimic this lowest level of hierarchical organization of bone by using proteins as a site for the heterogeneous nucleation and subsequent growth of stoichiometric hydroxylapatite crystals. In special approaches, the biomimetic apatite coatings on surfaces were prepared by soaking the materials in simulated body fluid (SBF) solutions, which contained ions in the concentrations similar to those in inorganic part of human blood plasma. It is generally accepted that the in vitro apatite growth during the exposure to SBF is an indicator for the in vivo bioactivity of materials surface [136],[137].

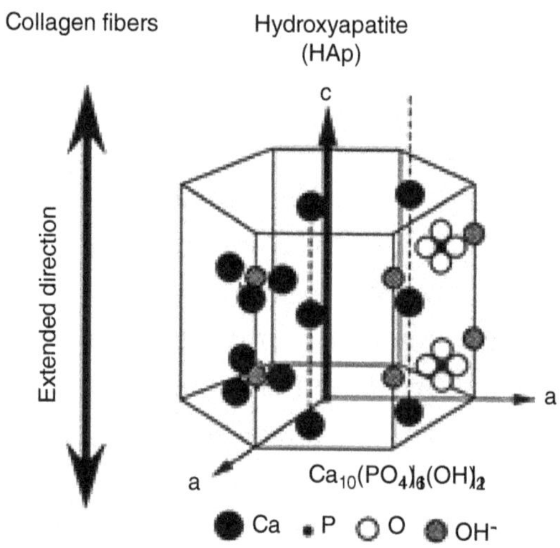

Fig. 19. The crystal structure of hydroxylapatite is oriented according to the c-axis with the extended direction of collagen fibrils [112].

Biomimetic materials are able to mimic the morphological and physicochemical features of biological apatite compounds, i.e. they are synthetic analogues of inorganic part of hard tissues [112],[117]. The biomimetic deposition is considered as an ideal method to produce calcium phosphate ceramics such as apatite coatings on titanium and its alloys for medical applications. It has also been proved that the chemical pretreatment in alkali solution can improve the bonding between the titanium substrate[15] and calcium phosphate coatings fabricated by subsequent biomimetic deposition in simulated body fluid (SBF) [138],[139],[140],[141]. The biomimetic growth of apatite was also described on hydrogen-implanted silicon [142], polyvinyl alcohol (PVA) [143], TiO_2 nanotubes [144], alumina [145], zirconia ceramics (Y-TZP) [146], forsterite [147], akermanite [148], magnetite [149], glasses [150] and

[15] Titanium and its alloys are widely used as orthopedic and dental implant materials because of their high mechanical strength, low modulus and good corrosion resistance [139].

bioglasses [151],[152], cements [153], geopolymers [154], carbon nanotubes [155] and microspheres [156].

The ability to form apatite [150],[153] from supersaturated solution has been widely used to imply the bioactivity of an implant in vivo. However, the method itself may provide at best incomplete information, primarily because it is determined only by the solution supersaturation, irrespective of biological processes. The bone regeneration is triggered mainly by the vitality of osteoblasts and regulated by the expression of growth factors such as estrogen, parathyroid hormone and bone morphogenetic proteins, while ions or other species released from an implant may affect the expression of such growth factors and so the bone resorption or formation. The misinterpretation of the outcome of such tests must result in the misunderstanding of the true effects and behavior of materials intended for use in embedded biological applications. Moreover, SBF may not be able to mimic properly the physiological conditions because it is based on analytical concentrations and not on the activity of key components [157].

The methods of the preparation of collagen-hydroxylapatite composites include the production of composite gels, films, collagen-coated ceramics, ceramic-coated collagen matrices and composite scaffolds for bone substitutes and hard tissue repair via the following techniques [130],[158],[159]:

- **In vitro collagen mineralization**: the method is based on direct mineralization of a collagen substrate (film) through which calcium and phosphate ions diffuse into the fibrils or as phosphate-containing collagen solution[16] (in situ precipitation [159]). **Fig. 20** introduces the experimental set-up for direct mineralization of a collagen sheet. The growth of HAP crystals with c-axis oriented along the collagen fibrils requires the pH in the range from 8 to 9 and the temperature of 40°C. These conditions promote the accumulation of calcium ions on the carboxyl groups of collagen molecules, leading to the nucleation of hydroxylapatite.

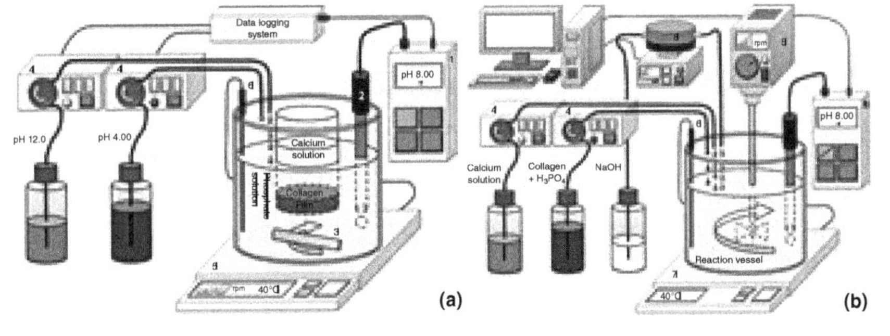

Fig. 20. Simplified scheme of apparatus for the preparation of collagen-hydroxylapatite composite by in vitro collagen mineralization method: modified scheme according to Wahl and Czernuszka [130] for in situ collagen/HAP precipitation (*a*); modified scheme according to Wang and Liu [159]: pH meter (1), pH electrode (2), stirrer (3), peristaltic pump (4), hot-plate magnetic stirrer (5a) or stirrer (5b), thermocouple (6), hot plate (7) and peristaltic pump driven by pH controller (8) (b).

[16] These techniques usually continue with freeze-drying.

- **Thermally triggered assembly of HA/collagen gels**: the solution of calcium and phosphate ions is encapsulated within the liposomes and next inserted into the acidic suspension of collagen. After the injection into a skeletal defect, the increasing temperature due to body heat initiates the gelation process, which leads to fibrous network. The mineralization occurs after reaching the liposome's transition temperature of 37°C.
- **Vacuum infiltration of collagen into a ceramic matrix**: ceramic scaffold is prepared by heating aqueous hydroxylapatite slurry containing poly(butyl methacrylate) (PBMA) spheres to high temperatures. The pyrolysis of PBMA particles leads to porous HAP green body. Pores in this matrix are then filled with collagen suspension under vacuum. The final composite was then freeze-dried to produce the microsponges within.
- **Enzymatic mineralization of collagen sheets**: is a method based on the cycle (**Fig. 21**) where collagen-containing alkaline (basic) phosphatase (ALP) is treated by aqueous solution of calcium ions and phosphate ester. The enzyme provides a reservoir for PO_4^{3-} ions for calcium phosphate to crystallize and the mineralization occurs on coated area. The sample is then coated again with collagen suspension, air-dried and cross-linked with UV irradiation. Repeating this cycle results in multilayered composite sheets of calcium/phosphate and collagen.

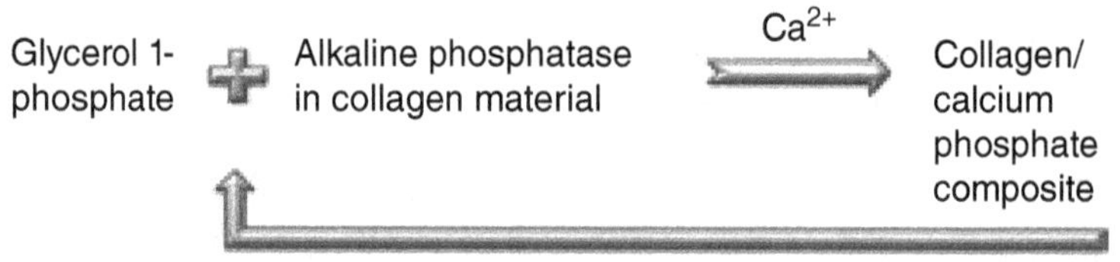

Fig. 21. The cycle of enzymatic mineralization of collagen sheets [130].

- **Water-in-oil emulsion system**: purified collagen suspension mixed with HAP powders at the temperature of 4°C is next dispersed in olive oil and stirred at 37°C. The collagen aggregates and reconstitutes into the aqueous droplets. The addition of phosphate-buffered saline[17] (PBS) leads to the gelation of Col-HAP microspheres (gel beads) of bone filler. The main disadvantages of this method are the problem with complete removing of the oil content from the composite and too low viscosity of the mixture.
- **Freeze-drying and critical point drying (CPD) scaffolds**: the ice crystals with collagen fibers at the interstices can be formed by freezing the suspension of collagen and HAP in water under controlled conditions. In the case when the freeze-drying is applied (temperature and pressure corresponding to CPD), ice crystals sublimate to water vapor.

  Under the condition of critical point,[18] the density of liquid and gas phase converge and become identical (supercritical fluid) as well as the surface tension is negligible.

[17] Aqueous solution of sodium phosphate and chloride, where the ion concentrations and osmolarity (osmotic concentration) correspond to human body. In some cases, potassium phosphate and chloride were used.

[18] Critical point is defined by the value of three parameters including critical temperature, critical pressure and critical volume. Critical temperature is the highest temperature at which pure matter can exist as liquid.

- Col-HAP was cast into the mould and frozen, ice crystals replaced with ethanol, ethanol-liquid CO_2 exchanged and critical point dried to finally arrive with an exact porous replica of the original bone. Solid freeform fabrication techniques have recently been developed with artificial polymers and ceramic materials. These have the ability to change the pore interconnectivity, pore size and pore shape but have the disadvantage of not having the affinity of collagen to cell attachment. Another major advantage of Col-HAP scaffolds produced through the SFF method is the ability to control variables such as the control of external and internal structure, porosity and cross-linking [130].

The collagen-hydroxyapatite/pectin (Col-HA/pectin) composite was prepared in situ by the introduction of pectin, a kind of plant polysaccharide, into the collagen-hydroxyapatite composite. The structure of composite consisted of hydroxylapatite of low crystallinity with particles uniformly dispersed in organic materials. There is strong bonding interaction between HAP, collagen and pectin. The mechanical properties, water absorption, enzyme degradation and cytotoxicity indicate a potential use in bone replacement for the new composite [160].

Most mammalian biofluids are supersaturated with respect to the bone and tooth mineral hydroxylapatite. Nevertheless, the biofluids, which are in contact with soft tissues, especially those like milk that must be stored for any length of time, should be highly stable with respect to calcium phosphate precipitation. In contrast, saliva and the extracellular matrix of hard tissues, especially near to the sites of mineralization or remineralization, are required not only to maintain the mineral phase with which they are in contact but also to deposit calcium phosphate in a highly controlled manner [161].

Collagen composites were also used as a scaffold for the repair of soft tissues. These materials provide analogous environment to extracellular matrix (ECM) and induced rate of synthesis or growth of new tissues. Several natural polymers as collagen, chitosan,[19] gelatin[20] and keratin[21] possess the ability to induce the proliferation of cells and hence find their use as biomaterials for a wide range of biomedical applications. Among all biopolymers, collagen is a widely accepted material for the tissue engineering applications in the view of its low antigenicity, excellent biocompatibility and biodegradability [162].

The preparation of collagen-based biocomposite constructed from micro-crimped long collagen fiber bundles extracted from a soft coral embedded in alginate hydrogel matrix was described by Sharabi et al [163]. This biocomposite demonstrated the hyperelastic behavior similar to human native tissues.

---

[19] Chitosan is (1–4)-linked 2-amino-2-deoxy-b-glucan, a byproduct of N-deacetylation of chitin. It is a major constituent of crab and shrimp shells and of cuticles of insects.

[20] Gelatin is a denatured form of collagen. Gelatin has low antigenicity and promotes the cell adhesion, differentiation and proliferation. Gelatin also possesses high cytocompatibility, which makes it a potential candidate as a biomaterial for various tissue engineering applications [162].

[21] Keratin is a family of fibrous proteins, which is found abundantly in nature. It is the main constituent of hair, wool, nails, horns and hooves of mammals, birds and reptiles [162].

Layers and biocoatings

The nanostructured coatings are expected to enhance the mechanical properties and improve the strength of bonding between coating and implant. These materials can also promote the deposition of calcium-containing minerals on their surface (bioactivity). Some typical examples of the research carried out in recent years, in the field of fabrication of nanostructured glass-ceramic coatings, utilize various techniques such as conventional enameling, sputtering, sol-gel processing, ion beam deposition, plasma spraying, electrophoretic deposition and pulsed laser deposition [164].

As was mentioned earlier, among the bioactive ceramics, the apatite/wollastonite (A/W) glass-ceramics, containing apatite and wollastonite crystals in the glassy matrix, has been largely studied because of good bioactivity and was used in some fields of medicine, especially in orthopedics and dentistry. However, medical applications of bioceramics are limited to non-load-bearing applications because of their poor mechanical properties. The solution of this problem can be the preparation of layers on the titanium alloy base material. Thermally sprayed layers by APS (atmospheric plasma spraying) on Ti-6Al-4V substrates combine good bioactivity of the bioceramics and good mechanical strength. The microstructure and the resulting properties were evaluated depending on the processing parameters and postprocessing thermal treatments. Thermal treatments decreased the bioactivity of the coatings, and after specific treatments, some bioactive materials were transformed into inert materials [67].

The biological performance of a porous apatite-mullite glass-ceramics, manufactured via the selective laser sintering (SLS) method, was investigated by Goodridge et al [165]. Laser-sintered A-M has shown similar cytotoxicity, in vitro bioactivity and in vivo results to cast A-M, indicating that the laser-sintering processing route does not alter the behavior of the material. The porous structure produced by SLS was seen by in vivo testing to be excellent for bone in-growth. However, the inability of the material to form apatite in vitro raised the concerns over the material's ability to prove bioactive in vivo, and further assessment is required to confirm whether this material is not bioactive or whether it is just not particularly suited for the characterization through SBF testing [165].

Author details

Petr Ptáček

Brno University of Technology, Czech Republic

References

[1] Meseguer S, Tena MA, Gargori C, Badenes JA, Llusar M, Monrós G. Structure and colour of cobalt ceramic pigments from phosphates. Ceramics International 2007;33(5) 843–849.

[2] Wagh AS. Introduction to Chemically Bonded Ceramics. Chemically Bonded Phosphate Ceramics, 2004. ISBN: 978-0-08-044505-2.

[3] Colorado HA, Hiel C, Hahn HT. Chemically bonded phosphate ceramics composites reinforced with graphite nanoplatelets. Composites Part A: Applied Science and Manufacturing 2011;42(4) 376–384.

[4] Dimitry SA. Characterization of reinforced chemically bonded ceramics. Cement and Concrete Composites 1991;13(4) 257–263.

[5] Yang L, Harink B, Habibovic P. Calcium phosphate ceramics with inorganic additives. Comprehensive Biomaterials 2011;1 299–312.

[6] Klammert U, Ignatius A, Wolfram U, Reuther T, Gbureck U. In vivo degradation of low temperature calcium and magnesium phosphate ceramics in a heterotopic model. Acta Biomaterialia 2011;7(9) 3469–3475.

[7] Chanda A, Dasgupta S, Bose S, Bandyopadhyay A. Microwave sintering of calcium phosphate ceramics. Materials Science and Engineering: C 2009;29(4) 1144–1149.

[8] Bandyopadhyay A, Withey EA, Moore J, Bose S. Influence of ZnO doping in calcium phosphate ceramics. Materials Science and Engineering: C 2007;27(1) 14–17.

[9] Formosa J, Chimenos JM, Lacasta AM, Niubó M. Interaction between low-grade magnesium oxide and boric acid in chemically bonded phosphate ceramics formulation. Ceramics International 2012;38(3) 2483–2493.

[10] Sahnoun RD, Bouaziz J. Sintering characteristics of kaolin in the presence of phosphoric acid binder. Ceramics International 2012;38(1) 1–7.

[11] Gardner LJ, Bernal SA, Walling SA, Corkhill CL, Provis JL, Hyatt NC. Characterisation of magnesium potassium phosphate cements blended with fly ash and ground granulated blast furnace slag. Cement and Concrete Research 2015;74 78–87.

[12] Liu Z, Qian G, Zhou J, Li C, Xu Y, Qin Z. Improvement of ground granulated blast furnace slag on stabilization/solidification of simulated mercury-doped wastes in chemically bonded phosphate ceramics. Journal of Hazardous Materials 2008;157(1) 146–153.

[13] Hermansson L. Nanostructural Bioceramics: Advances in Chemically Bonded Ceramics. CRC Press, 2014. ISBN: 978-9814463430

[14] Chen D, He L, Shang S. Study on aluminum phosphate binder and related Al_2O_3-SiC ceramic coating. Materials Science and Engineering: A 2003;348(1-2) 29–35.

[15] Hipedinger NE, Scian AN, Aglietti EF. Magnesia-ammonium phosphate-bonded cordierite refractory castables: Phase evolution on heating and mechanical properties. Cement and Concrete Research 2004;34(1) 157–164.

[16] Choi J, Um W, Choung S. Development of iron phosphate ceramic waste form to immobilize radioactive waste solution. Journal of Nuclear Materials 2014;452(1-3) 16–23.

[17] Metcalfe BL, Donald IW, Fong SK, Gerrard LA, Strachan DM, Scheele RD. Ageing of a phosphate ceramic used to immobilize chloride contaminated actinide waste. Journal of Nuclear Materials 2009;385(2) 485–488.

[18] Singh D, Mandalika VR, Parulekar SJ, Wagh AS. Magnesium potassium phosphate ceramic for ^{99}Tc immobilization. Journal of Nuclear Materials 2006;348(3) 272–282.

[19] Scheetz BE, Agrawal DK, Breval E, Roy R. Sodium zirconium phosphate (NZP) as a host structure for nuclear waste immobilization: A review. Waste Management 1994;14(6) 489–505.

[20] von Kobell F. Über den schillernden asbest von Reichenstein in Schlesien. Journal für Praktische Chemie 1834;2 297–298.

[21] Warren BE, Bragg WL. The structure of the chrysotile $H_4Mg_3Si_2O_9$. Zeitschrift für Kristallographie 1931;76 201–210.

[22] Falini G, Foresti E, Gazzano M, Gualtieri AF, Leoni M, Lesci IG, Roveri N. Tubular-shaped stoichiometric chrysotile nanocrystals. Chemistry — A European Journal 2004;10 3043–3049.

[23] Viani A, Gualtieri AF. Preparation of magnesium phosphate cement by recycling the product of thermal transformation of asbestos containing wastes. Cement and Concrete Research 2014;58 56–66.

[24] Ciullo PA. Industrial Minerals and Their Uses: A Handbook and Formulary. William Andrew, 1996. ISBN: 978-0815518082

[25] Clarke DB. Granitoid Rocks. International Studies in Economic Modelling — Volume 7. Topics in the Earth Sciences. Springer Science & Business Media, 1992. ISBN: 978-0412291708

[26] Ruh E. Concise Encyclopedia of Advanced Ceramic Materials. Advances in Materials Sciences and Engineering. Elsevier, 2012. ISBN: 978-0080983707

[27] Huggett LG. Refractories in the chemical industries. International Journal of Materials in Engineering Applications 1979;1(5) 280–294.

[28] Sadik C, El Amrani Iz-E, Albizane A. Recent advances in silica-alumina refractory: A review. Journal of Asian Ceramic Societies 2014;2(2) 83–96.

[29] Marghussian VK, Naghizadeh R. Chemical bonding of silicon carbide. Journal of the European Ceramic Society 1999;19(16) 2815–2821.

[30] Sujith SS, Kumar ASL, Mangalaraja RV, Mohamed AP, Ananthakumar S. Porous to dense $LaPO_4$ sintered ceramics for advanced refractories. Ceramics International 2014;40(9) 15121–15129.

[31] Souza TM, Luz AP, Santos T Jr, Gimenes DC, Miglioli MM, Correa AM, Pandolfelli VC. Phosphate chemical binder as an anti-hydration additive for Al_2O_3-MgO refractory castables. Ceramics International 2014;40(1) 1503–1512.

[32] Gale WF, Totemeier TC. Smithells Metals Reference Book. 8th ed., Butterworth-Heinemann, 2003. ISBN: 978-0080480961

[33] Luz AP, Gomes DT, Pandolfelli VC. High-alumina phosphate-bonded refractory castables: $Al(OH)_3$ sources and their effects. Ceramics International 2015;41(7) 9041–9050.

[34] García-Prieto A, Ramos-Lotito MD, Gutiérrez-Campos D, Pena P, Baudín C. Influence of microstructural characteristics on fracture toughness of refractory materials. Journal of the European Ceramic Society 2015;35(6) 1955–1970.

[35] Toy C, Whittemore OJ. Phosphate bonding with several calcined aluminas. Ceramics International 1989;15(3) 167–171.

[36] Hipedinger NE, Scian AN, Aglietti EF. Magnesia-phosphate bond for cold-setting cordierite-based refractories. Cement and Concrete Research 2002;32(5) 675–682.

[37] Lu W, Chung DDL. Oxidation protection of carbon materials by acid phosphate impregnation. Carbon 2002;40(8) 1249–1254.

[38] Smith BGN, Wright PS, Brown D. The Clinical Handling of Materials. 2nd ed., Oxford: Butterworth-Heinemann, 1994. ISBN: 978-0723610038

[39] Nicholson JW, Czarnecka B, Limanowska-Shaw H. A preliminary study of the effect of glass-ionomer and related dental cements on the pH of lactic acid storage solutions. Biomaterials 1999;20(2) 155–158.

[40] Liu Y, Yu Hai-Y. Does dental zinc phosphate cement really shrink in clinical applications? Medical Hypotheses 2009;73(2) 257–258.

[41] Schmalz G, Bindslev DA. Titul Biocompatibility of Dental Materials. Springer Science & Business Media, 2008. ISBN: 978-3540777823

[42] Paffenbarger GC, Sweeney WT, Isaacs A. Zinc phosphate cements: Physical properties and a specification. The Journal of the American Dental Association 1922;21(11) 1907–1924.

[43] Park C.-K, Silsbee MR, Roy DM. Setting reaction and resultant structure of zinc phosphate cement in various orthophosphoric acid cement-forming liquids. Cement and Concrete Research 1998;28(1) 141–150.

[44] Mitra SB. Dental cements: formulations and handling techniques. In: Curtis RV, Watson TF (Eds.), Dental Biomaterials. Elsevier, 2014. ISBN: 978-1845694241

[45] Jabri M, Mejdoubi E, El Gadi M, Hammouti B. Optimisation of hardness and setting time of dental zinc phosphate cement using a design of experiments. Arabian Journal of Chemistry 2012;5(3) 347–351.

[46] Wilson AD. The chemistry of dental cements. Chemical Society Reviews 1978;7(2) 265–296.

[47] Ramkumar N. Phosphate based oil well cements. ProQuest Dissertations and Theses; Thesis (Ph.D.)-University of Illinois at Chicago, 2005; Publication Number: AAI3199873; ISBN: 9780542467271; Source: Dissertation Abstracts International, Volume: 66-12, Section: B, page: 6863; 131 p.

[48] Kurdowski W. Cement and Concrete Chemistry. Springer Science & Business, 2014. ISBN: 978-9400779457

[49] Warner J. Practical Handbook of Grouting: Soil, Rock, and Structures. John Wiley & Sons, 2004. ISBN: 978-0471463030

[50] Ghosh SN. Cement and Concrete Science and Technology. Cement and Concrete Science & Technology — Part 1. Progress in Cement and Concrete. Thomas Telford, 1991. ISBN: 978-8185522005

[51] Wang F, Chen B, Pun EYB, Lin H. Alkaline aluminum phosphate glasses for thermal ion-exchanged optical waveguide. Optical Materials 2015;42 484–490.

[52] Saadaldin SA, Rizkalla AS. Synthesis and characterization of wollastonite glass-ceramics for dental implant applications. Dental Materials 2014;30(3) 364–371.

[53] Lewis M, Metcalf-Johansen J, Bell P. Crystallisation mechanisms in glassceramics. Journal of Non-Crystalline Solids 1979;62 5–6.

[54] Stanton KT, O'Flynn KP, Kiernan S, Menuge J, Hill R. Spherulitic crystallization of apatite-mullite glass-ceramics: Mechanisms of formation and implications for fracture properties. Journal of Non-Crystalline Solids 2010;356(35-36) 1802–1813.

[55] Hill RG. Bioactive Glass-Ceramics, 181-186. Comprehensive Biomaterials, Volume 1. In: Ducheyne P (Ed.). Elsevier, 2011. ISBN: 978-0-08-055294-1

[56] Kamitakahara M, Ohtsuki C, Inada H, Tanihara M, Miyazaki T. Effect of ZnO addition on bioactive CaO-SiO_2-P_2O_5-CaF_2 glass-ceramics containing apatite and wollastonite. Acta Biomaterialia 2006;2(4) 467–471.

[57] Zhang Y, Santos JD. Microstructural characterization and in vitro apatite formation in CaO-P_2O_5-TiO_2-MgO-Na_2O glass-ceramics. Journal of the European Ceramic Society 2001;21(2) 169–175.

[58] Kishioka A, Hayashi M, Kinoshita M. Glass formation and crystallization in ternary phosphate systems containing Al_2O_3. Bulletin of the Chemical Society of Japan 1976;49(11) 3032–3036.

[59] Joseph K, Jolley K, Smith R. Iron phosphate glasses: Structure determination and displacement energy thresholds, using a fixed charge potential model. Journal of Non-Crystalline Solids 2015;411 137–144.

[60] Stefanovsky SV, Stefanovsky OI, Kadyko MI, Presniakov IA, Myasoedov BF. The effect of Fe_2O_3 substitution for Al_2O_3 on the phase composition and structure of sodium-aluminum-iron phosphate glasses. Journal of Non-Crystalline Solids 2015;425 138–145.

[61] Premila M, Rajaraman R, Abhaya S, Ravindran TR, Kamali K, Amarendra G, Sundar CS. Spectroscopic evidence of irreversible changes in cesium loaded iron phosphate glasses under pressure. Journal of Non-Crystalline Solids 2014;406 111–118.

[62] Broglia G, Mugoni C, Du J, Siligardi C, Montorsi M. Lithium vanado-phosphate glasses: Structure and dynamics properties studied by molecular dynamics simulations. Journal of Non-Crystalline Solids 2014;403 53–61.

[63] Das SS, Singh NP, Srivastava PK. Ion conducting phosphate glassy materials. Progress in Crystal Growth and Characterization of Materials 2009;55(3-4) 47–62.

[64] Kishore MS, Pralong V, Caignaert V, Varadaraju UV, Raveau B. Synthesis and electrochemical properties of a new vanadyl phosphate: $Li_4VO(PO_4)_2$. Electrochemistry Communications 2006;8(10) 1558–1562.

[65] Kokubo T. Bioactive glass ceramics: properties and applications. Biomaterials 1991;12(2) 155–163.

[66] Cannillo V, Pierli F, Sampath S, Siligardi C. Thermal and physical characterisation of apatite/wollastonite bioactive glass-ceramics. Journal of the European Ceramic Society 2009;29(4) 611–619.

[67] Cannillo V, Colmenares-Angulo J, Lusvarghi L, Pierli F, Sampath S. In vitro characterisation of plasma-sprayed apatite/wollastonite glass-ceramic biocoatings on titanium alloys. Journal of the European Ceramic Society 2009;29(9) 1665–1677.

[68] Liu X, Ding C, Chu PK. Mechanism of apatite formation on wollastonite coatings in simulated body fluids. Biomaterials 2004;25(10) 1755–1761.

[69] Magallanes-Perdomo M, Luklinska ZB, De Aza AH, Carrodeguas RG, De Aza S, Pena P. Bone-like forming ability of apatite-wollastonite glass ceramic. Journal of the European Ceramic Society 2011;31(9) 1549–1561.

[70] Kokubo T. A/W glass-ceramic: Processing and properties. Chapter . In: Hench LL (Ed.). An Introduction to Bioceramics. Advanced Series in Ceramics — Volume 1. World Scientific, 1993. ISBN: 978-9810214005.

[71] Kokubo T, Shigematsu M, Nagashima Y, Tashiro M, Nakamura T, Yamamuro T, Higash S. Apatite- and wollastonite-containing glass ceramics for prosthetic applications. Bulletin of the Institute of Chemistry Research of Kyoto University 1982;60 260–268.

[72] Ma J, Chen CZ, Wang DG, Shao X, Wang CZ, Zhang HM. Effect of MgO addition on the crystallization and in vitro bioactivity of glass ceramics in the CaO-MgO-SiO_2-P_2O_5 system. Ceramics International 2012;38(8) 6677–6684.

[73] Clifford A., Hill R. Apatite-mullite glass-ceramics. Journal of Non-Crystalline Solids 1996;196 346–351.

[74] Hill RG, O'Donnell MD, Law RV, Karpukhina N, Cochrane B, Tulyaganov DU. The early stages of nucleation and crystallisation of an apatite glass-ceramic: Evidence for nano-scale crystallization. Journal of Non-Crystalline Solids 2010;356(52-54) 2935–2941.

[75] Bernardo E, Colombo P, Cacciotti I, Bianco A, Bedini R, Pecci R, Pardun K, Treccani L, Rezwan K. Porous wollastonite-hydroxyapatite bioceramics from a preceramic polymer and micro- or nano-sized fillers. Journal of the European Ceramic Society 2012;32(2) 399–408.

[76] van't Hoen C, Rheinberger V, Höland W, Apel E. Crystallization of oxyapatite in glass-ceramics. Journal of the European Ceramic Society 2007;27(2-3) 1579–1584.

[77] Mahato N, Banerjee A, Gupta A, Omar S, Balani K. Progress in material selection for solid oxide fuel cell technology: A review. Progress in Materials Science 2015;72 141–337.

[78] Fergus J, Hui R, Li X, Wilkinson DP, Zhang J. Solid Oxide Fuel Cells: Materials Properties and Performance. Chemistry and Chemical Engineering. CRC Press, 2008. ISBN: 978-1420088847.

[79] Grove WR. On voltaic series and combination of gases by platinum. Philosophical Magazine Series 3 1839;14(86) 127–130.

[80] Camaratta M. Microstructural engineering of composite cathode systems for intermediate and low temperature solid oxide fuel cells, Ph.D. Thesis. Gainesville, Florida: University of Florida, 2007.

[81] Marrero-López D, Martín-Sedeño MC, Peña-Martínez J, Ruiz-Morales JC, Núñez P, Aranda MAG, Ramos-Barrado JR. Evaluation of apatite silicates as solid oxide fuel cell electrolytes. Journal of Power Sources 2010;195(9) 2496–2506.

[82] Islam MS, Tolchard JR, Slater PR. An apatite for fast oxide-ion conduction. Chemical Communications 2003;(13) 1486–1487.

[83] Tolchard JR, Islam MS, Slater PR. Defect chemistry and oxygen ion migration in the apatite-type materials $La_{9.33}Si_6O_{26}$ and $La_8Sr_2Si_6O_{26}$. Journal of Materials Chemistry 2003;13(8) 1956–1961.

[84] Zeng Y, Mao Pei-L, Jiang SP, Wu P, Zhang L, Wu P. Prediction of oxygen ion conduction from relative Coulomb electronic interactions in oxyapatites. Journal of Power Sources 2011;196(10) 4524–4532.

[85] Nakayama S, Higuchi Y, Sugawara M, Makiya A, Uematsu K, Sakamoto M. Fabrication of c-axis-oriented apatite-type polycrystalline $La_{10}Si_6O_{27}$ ceramic and its anisotropic oxide ionic conductivity. Ceramics International 2014;40(1) 1221–1224.

[86] Takeda N, Itagaki Y, Aono H, Sadaoka Y. Preparation and characterization of $Ln_{9.33+x/3}Si_{6-x}Al_xO_{26}$ (Ln = La, Nd and Sm) with apatite-type structure and its application to a potentiometric O_2 gas sensor. Sensors and Actuators B: Chemical 2006;115(1) 455–459.

[87] Tudorache F, Petrila I, Popa K, Catargiu AM. Electrical properties and humidity sensor characteristics of lead hydroxyapatite material. Applied Surface Science 2014;303 175–179.

[88] Owada H, Yamashita K, Umegaki T, Kanazawa T, Nagai M. Humidity-sensitivity of yttrium substituted apatite ceramics. Solid State Ionics 1989;35(3-4) 401–404.

[89] Nagai M, Nishino T, Saeki T. A new type of CO_2 gas sensor comprising porous hydroxyapatite ceramics. Sensors and Actuators 1988;15(2) 145–151.

[90] Phosphor system for improved efficacy lighting sources. Patent US 8866372 B2, 2011.

[91] Yu R, Xue N, Wang T, Zhao Z, Wang J, Hei Z, Li M, Noh HM, Jeong JH. Photoluminescence characteristics of high thermal stable fluorosilicate apatite $Ba_2Y_3(SiO_4)_3F$: Sm^{3+} orange-red emitting phosphor. Ceramics International 2015;41(4) 6030–6036.

[92] Liu Y, Wang Z, Zhong J, Pan F, Liang H, Xiao Z. Synthesis and photoluminescence properties of red-emitting phosphors $Ba_2Gd_8(SiO_4)_6O_2$:Eu^{3+}. Materials Letters 2014;129 130–133.

[93] Sokolnicki J, Zych E. Synthesis and spectroscopic investigations of $Sr_2Y_8(SiO_4)_6O_2$:Eu^{2+},Eu^{3+} phosphor for white LEDs. Journal of Luminescence 2015;158 65–69.

[94] Zhang F, Liu B. An intense red-emitting phosphor $Ca_3Gd_7(SiO_4)_5(PO_4)O_2$: Eu^{3+} for NUV-LEDs application. Journal of Alloys and Compounds 2012;542 276–279.

[95] Jeon YI, Bharat LK, Yu JS. White-light emission of $Ca_2La_8(GeO_4)_6O_2$: Tb^{3+}/Sm^{3+} nanocrystalline phosphors for solid-state lighting applications. Journal of Luminescence 2015;166 93–100.

[96] Liu H, Zhang Y, Liao L, Guo Q, Mei L. Synthesis, broad-band absorption and luminescence properties of blue-emitting phosphor $Sr_8La_2(PO_4)_6O_2$:Eu^{2+} for n-UV white-light-emitting diodes. Ceramics International 2014;40(8) 13709–13713.

[97] Roh Hee-S, Hur S, Song HJ, Park IJ, Yim DK, Kim Dong-W, Hong KS. Luminescence properties of $Ca_5(PO_4)_2SiO_4$:Eu^{2+} green phosphor for near UV-based white LED. Materials Letters 2012;70 37–39.

[98] Augustine RL. Heterogeneous Catalysis for the Synthetic Chemist. CRC Press, 1995. ISBN: 978-0824790219

[99] Gruselle M. Apatites: A new family of catalysts in organic synthesis. Journal of Organometallic Chemistry 2015. In press. DOI: 10.1016/j.jorganchem.2015.01.018

[100] Maaten B, Moussa J, Desmarets C, Gredin P, Beaunier P, Kanger T, Tõnsuaadu K, Villemin D, Gruselle M. Cu-modified hydroxy-apatite as catalyst for Glaser-Hay Csingle bondC homo-coupling reaction of terminal alkynes. Journal of Molecular Catalysis A: Chemical 2014;393 112–116.

[101] Li JJ. Name Reactions for Homologation, Part I. John Wiley & Sons, 2009. ISBN: 978-0470487013.

[102] Edition D. Apatite catalyst used to synthesise n-butanol from ethanol. Focus on Catalysts 2005;4 5–6.

[103] Cubberly WH, Bakerjian R. Tool and Manufacturing Engineers Handbook Desk Edition. Society of Manufacturing Engineers, 1989. ISBN: 978-0872633513

[104] Wang LK, Hung Yung-T, Shammas NK. Handbook of Advanced Industrial and Hazardous Wastes Treatment. Advances in Industrial and Hazardous Wastes Treatment. CRC Press, 2009. ISBN: 978-1420072228

[105] Su Hsiang-Y, Lin Chao-S. Effect of additives on the properties of phosphate conversion coating on electrogalvanized steel sheet. Corrosion Science 2014;83 137–146.

[106] Liu B, Zhang X, Gui-y Xiao, Lu Yu-p. Phosphate chemical conversion coatings on metallic substrates for biomedical application: A review. Materials Science and Engineering: C 2015;47 97–104.

[107] Perrin FX, Gigandet MP, Wery M, Pagetti J. Chromium phosphate conversion coatings on zinc electroplates: cathodic formation and characterization. Surface and Coatings Technology 1998;105(1-2) 135–140.

[108] Wopenka B, Pasteris JD. A mineralogical perspective on the apatite in bone. Materials Science and Engineering: C 2005;25(2) 131–143.

[109] López-Álvarez M, Rodríguez-Valencia C, Serra J, González P. Bio-inspired ceramics: promising scaffolds for bone tissue engineering. Procedia Engineering 2013;59 51–58.

[110] Abdel-Hameed SAM, Hessien MM, Azooz MA. Preparation and characterization of some ferromagnetic glass-ceramics contains high quantity of magnetite. Ceramics International 2009;35(4) 1539–1544.

[111] Matsumotoa T, Uddina MH, Ana SH, Arakawab K, Taguchib E, Nakahirac A, Okazakid M. Modulation of nanotube formation in apatite single crystal via organic molecule incorporation. Materials Chemistry and Physics 2011;128(3) 495–499.

[112] Nakano T, Kaibara K, Ishimoto T, Tabata Y, Umakoshi Y. Biological apatite (BAp) crystallographic orientation and texture as a new index for assessing the microstructure and function of bone regenerated by tissue engineering. Bone 2012;51(4) 741–747.

[113] Venkatesan J, Bhatnagar I, Manivasagan P, Kang Kyong-H, Kim Se-K. Alginate composites for bone tissue engineering: A review. International Journal of Biological Macromolecules 2015;72 269–281.

[114] Liu Y, Lim J, Teoh Swee-H. Review: Development of clinically relevant scaffolds for vascularised bone tissue engineering. Biotechnology Advances 2013;31(5) 688–705.

[115] Armentano I, Dottori M, Fortunati E, Mattioli S, Kenny JM. Biodegradable polymer matrix nanocomposites for tissue engineering: A review. Polymer Degradation and Stability 2010;95(11) 2126–2146.

[116] Nicholson JW. The Chemistry of Medical and Dental Materials. RSC Materials Monographs — Volume 3. Royal Society of Chemistry, 2002. ISBN: 978-0854045723

[117] Gómez-Morales J, Iafisco Mi, Delgado-López JM, Sarda S, Drouet C. Progress on the preparation of nanocrystalline apatites and surface characterization: Overview of fundamental and applied aspects. Progress in Crystal Growth and Characterization of Materials 2013;59(1) 1–46.

[118] Rey C, Combes C, Drouet C, Sfihi H, Barroug A. Physico-chemical properties of nanocrystalline apatites: Implications for biominerals and biomaterials. Materials Science and Engineering: C 2007;27(2) 198–205.

[119] Szymendera J. Bone Mineral Metabolism in Cancer. Recent Results in Cancer Research — Volume 27. Springer Science & Business Media, 2012. ISBN: 978-3642999789

[120] Hillson S. Teeth. Cambridge Manuals in Archaeology. 2nd ed., Cambridge University Press, 2005. ISBN: 978-1139444057

[121] Wilson RM, Elliot JC, Dowker SEP. Rietveld refinement of the crystallographic structure of human dental enamel apatites Sample: RHB-enamel. American Mineralogist 1999;84 1406–1414.

[122] Krivovichev SV. Titul Minerals as Advanced Materials I. Springer Science & Business Media, 2008. ISBN: 978-3540771234

[123] Jamuna-Thevi K, Daud NM, Kadir MRA, Hermawan H. The influence of new wet synthesis route on the morphology, crystallinity and thermal stability of multiple ions doped nanoapatite. Ceramics International 2014;40(1) 1001–1012.

[124] Delgado-López JM, Iafisco M, Rodríguez I, Tampieri A, Prat M, Gómez-Morales J. Crystallization of bioinspired citrate-functionalized nanoapatite with tailored carbonate content. Acta Biomaterialia 2012;8(9) 3491–3499.

[125] Gualtieri ML, Ma Romagnoli, Hanuskova M, Fabbri E, Gualtieri AF. Facile synthesis of B-type carbonated nanoapatite with tailored microstructure. Journal of Solid State Chemistry 2014;220 60–69.

[126] Xia Z, Yu X, Jiang X, Brody HD, Rowe DW, Wei M. Fabrication and characterization of biomimetic collagen-apatite scaffolds with tunable structures for bone tissue engineering. Acta Biomaterialia 2013;9(7) 7308–7319.

[127] Sader MS, Martins VCA, Gomez S, LeGeros RZ, Soares GA. Production and in vitro characterization of 3D porous scaffolds made of magnesium carbonate apatite (MCA)/anionic collagen using a biomimetic approach. Materials Science and Engineering: C 2013;33(7) 4188–4196.

[128] Fricain JC, Schlaubitz S, Le Visage C, Arnault I, Derkaoui SM, Siadous R, Catros S, Lalande C, Bareille R, Renard M, Fabre T, Cornet S, Durand M, Léonard A, Sahraoui N, Letourneur D, Amédée J. A nano-hydroxyapatite — Pullulan/dextran polysaccharide composite macroporous material for bone tissue engineering. Biomaterials 2013;34(12) 2947–2959.

[129] Shoulders MD, Raines RT: Collagen structure and stability. Annual Review of Biochemistry 2009 ;78 929–958.

[130] Wahl DA, Czernuszka JT. Collagen-hydroxyapatite composites for hard tissue repair. European Cells and Materials 2006;11 43–56.

[131] Magne D, Pilet P, Weiss P, Daculsi G. Fourier transform infrared microspectroscopic investigation of the maturation of nonstoichiometric apatites in mineralized tissues: a horse dentin study. Bone 2001;29(6) 547–552.

[132] Li H, Guo Z, Xue B, Zhang Y, Huang W. Collagen modulating crystallization of apatite in a biomimetic gel system. Ceramics International 2011;37(7) 2305–2310.

[133] Su X, Sun K, Cui FZ, Landis WJ. Organization of apatite crystals in human woven bone. Bone 2003;32(2) 150–162.

[134] Okazaki M, Ohmae H, Hino T. Insolubilization of apatite-collagen composites by UV irradiation. Biomaterials 1989;10(8) 564–568.

[135] Okazakl M, Ohmae H, Takahashi J, Kimura H, Sakuda M. Insolubilized properties of UV-irradiated C03 apatite-collagen composites. Biomaterials 1990;11(8) 568–572.

[136] Müller FA, Müller L, Caillard D, Conforto E. Preferred growth orientation of biomimetic apatite crystals. Journal of Crystal Growth 2007;304(2) 464–471.

[137] Fernandez-Yague MA, Abbah SA, McNamara L, Zeugolis DI, Pandit A, Biggs MJ. Biomimetic approaches in bone tissue engineering: Integrating biological and physicomechanical strategies. Advanced Drug Delivery Reviews 2015;84 1–29.

[138] Chu CL, Pu YP, Yin LH, Chung CY, Yeung KWK, Chu PK. Biomimetic deposition process of an apatite coating on NiTi shape memory alloy. Materials Letters 2006;60(24) 3002–3006.

[139] Shi J, Ding C, Wu Y. Biomimetic apatite layers on plasma-sprayed titanium coatings after surface modification. Surface and Coatings Technology 2001;137(1) 97–103.

[140] Song Won-H, Jun Youn-K, Han Y, Hong Seong-H. Biomimetic apatite coatings on micro-arc oxidized titania. Biomaterials 2004;25(17) 3341–3349.

[141] Faure J, Balamurugan A, Benhayoune H, Torres P, Balossier G, Ferreira JMF. Morphological and chemical characterisation of biomimetic bone like apatite formation

on alkali treated Ti_6Al_4V titanium alloy. Materials Science and Engineering: C 2009;29(4) 1252–1257.

[142] Liu X, Fu RKY, Poon RWY, Chen P, Chu PK, Ding C. Biomimetic growth of apatite on hydrogen-implanted silicon. Biomaterials 2004;25(25) 5575–5581.

[143] Ye M, Mohanty P, Ghosh G. Biomimetic apatite-coated porous PVA scaffolds promote the growth of breast cancer cells. Materials Science and Engineering: C 2014;44 310–316.

[144] Kunze J, Müller L, Macak JM, Greil P, Schmuki P, Müller FA. Time-dependent growth of biomimetic apatite on anodic TiO_2 nanotubes. Electrochimica Acta 2008;53(23) 6995–7003.

[145] Rambo CR, Müller FA, Müller L, Sieber H, Hofmann I, Greil P. Biomimetic apatite coating on biomorphous alumina scaffolds. Materials Science and Engineering: C 2006;26(1) 92–99.

[146] Stefanic M, Krnel K, Pribosic I, Kosmac T. Rapid biomimetic deposition of octacalcium phosphate coatings on zirconia ceramics (Y-TZP) for dental implant applications. Applied Surface Science 2012;258(10) 4649–4656.

[147] Tavangarian F, Emadi R. Improving degradation rate and apatite formation ability of nanostructure forsterite. Ceramics International 2011;37(7) 2275–2280.

[148] Wu C, Chang J. Synthesis and apatite-formation ability of akermanite. Materials Letters 2004;58(19) 2415–2417.

[149] Múzquiz-Ramos EM, Cortés-Hernández DA, Escobedo-Bocardo J. Biomimetic apatite coating on magnetite particles. Materials Letters 2010;64(9) 1117–1119.

[150] Fujibayashi S, Neo M, Kim Hyun-M, Kokubo T, Nakamura T. A comparative study between in vivo bone ingrowth and in vitro apatite formation on Na_2O-CaO-SiO_2 glasses. Biomaterials 2003;24(8) 1349–1356.

[151] Berbecaru C, Alexandru HV, Ianculescu Ad, Popescu A, Socol G, Sima F, Mihailescu I. Bioglass thin films for biomimetic implants. Applied Surface Science 2009;255(10) 5476–5479.

[152] Bingel L, Groh D, Karpukhina N, Brauer DS. Influence of dissolution medium pH on ion release and apatite formation of Bioglass® 45S5. Materials Letters 2015;143 279–282.

[153] Gandolfi MG, Taddei P, Tinti A, De Stefano Dorigo E, Prati C. Alpha-TCP improves the apatite-formation ability of calcium-silicate hydraulic cement soaked in phosphate solutions. Materials Science and Engineering: C 2011;31(7) 1412–1422.

[154] Pangdaeng S, Sata V, Aguiar JB, Pacheco-Torgal F, Chindaprasirt P. Apatite formation on calcined kaolin-white Portland cement geopolymer. Materials Science and Engineering: C 2015;51 1–6.

[155] Akasaka T, Watari F, Sato Y, Tohji K. Apatite formation on carbon nanotubes. Materials Science and Engineering: C 2006;26(4) 675–678.

[156] Wu C, Chan J. Bonelike apatite formation on carbon microspheres. Materials Letters 2007;61(11-12) 2502–2505.

[157] Pan H, Zhao X, Darvell BW, Lu WW. Apatite-formation ability – Predictor of "bioactivity"? Acta Biomaterialia 2010;6(11) 4181–4188.

[158] Tien Wen-B, Chen Ming-T, Yao Pin-C. Effects of pH and temperature on microstructure and morphology of hydroxyapatite/collagen composites synthesized in vitro. Materials Science and Engineering: C 2012;32(7) 2096–2102.

[159] Wang J, Liu C. Biomimetic collagen/hydroxyapatite composite scaffolds: Fabrication and characterizations. Journal of Bionic Engineering 2014;11(4) 600–609.

[160] Wenpo F, Gaofeng L, Shuying F, Yuanming Q, Keyong T. Preparation and characterization of collagen-hydroxyapatite/pectin composite. International Journal of Biological Macromolecules 2015;74 218–223.

[161] Holt C, Lenton S, Nylander T, Sørensen ES, Teixeira SCM. Mineralisation of soft and hard tissues and the stability of biofluids. Journal of Structural Biology 2014;185(3) 383–396.

[162] Kakkar P, Verma S, Manjubala I, Madhan B. Development of keratin-chitosan-gelatin composite scaffold for soft tissue engineering. Materials Science and Engineering: C 2014;45 343–347.

[163] Sharabi M, Mandelberg Y, Benayahu D, Benayahu Y, Azem A, Haj-Ali R. A new class of bio-composite materials of unique collagen fibers. Journal of the Mechanical Behavior of Biomedical Materials 2014;36 71–81.

[164] Marghussian V. Nano-Glass Ceramics: Processing, Properties and Applications. Elsevier Science, 2015. ISBN: 9780323354325

[165] Goodridge RD, Wood DJ, Ohtsuki C, Dalgarno KW. Biological evaluation of an apatite-mullite glass-ceramic produced via selective laser sintering. Acta Biomaterialia 2007;3(2) 221–231.

2

Rare-earth Element Bearing Apatites and Oxyapatites

Petr Ptáček

Abstract

A number of prepared alkaline-earth-rare-earth silicates and germanates also have the structure of apatite type. The fifth chapter of this book then continues with description of synthetic compounds of apatite structure. Attention will be directed to description of rare-earth element bearing apatites and oxyapatites. The structure, properties and preparation of apatite-type silicates, germanates and borates were described. This chapter gives also description of oxygen-rich apatites, which are promising material for electrolytes in solid oxide fuel cells and sensors and explain the basic concepts between structure and conductivity of these compounds. The additional information about application of apatites is given in the last chapter of this book. Furthermore, N-apatite, REE vanadocalcic apatite and apatite type yttrium phosphates were described.

Keywords: Apatite, Oxyapatite, Conductivity, Silicates, Germanates, Borates, N-apatite, Apatite-type yttrium phosphates

In this chapter, the preparation and the properties of synthetic phases of apatite are given; the geological role is described in **Section**. The ideal general formula of an apatite-type oxide may be written as $M_{10}(XO_4)_6O_2$ (M = alkaline-earth and/or rare-earth element, X = Si, Ge, P, V, …). The structure (**Fig. 1**) can be described in terms of a "microporous"[1] [1] framework $(A(1)_4(XO_4)_6)$ composed of face sharing $M(1)O_6$ trigonal meta-prismatic columns, which are corner connected to MO_4 tetrahedra. This framework allows some flexibility to accommodate remaining $M(2)_6O_2$ units [2].

[1] Microporous material is defined as containing pores with the diameters >2 nm. The materials with the pore diameter in the range from 2 to 50 nm and higher than 50 nm are termed as mesoporous and macroporous, respectively. In combination with nanotechnology, the term nanoporous material is often used. Despite the fact that there is not clear definition, usually the pores with the size from 0.1 to 100 nm are considered. In other words, nanoporous covers the range from microporous to macroporous [1].

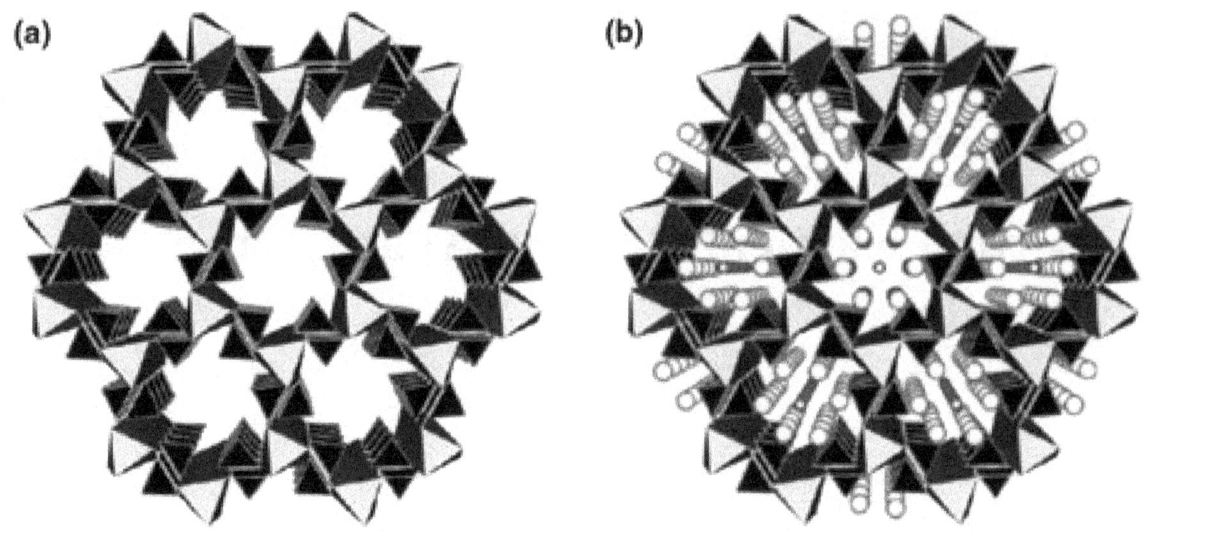

Fig. 1. Illustration of "microporous" $M(1)(XO_4)_6$ framework of the apatite ($M_{10}(XO_4)_6O_2$) structure (a): tetrahedra MO_4, M(1) cation at the center of trigonal meta-prism. Remaining $M(2)_6O_2$ units occupy the cavities within this framework (b): large spheres are M(2) cations and small spheres are O anions [2].

Apatite-type lanthanium silicates

During recent decades, oxyapatite[2]-type structure with the general formula: $REE_{9.33+x}Si_6O_{26+3x/2}$ (where REE is rare-earth element) [3],[4], $REE_{9.33}\square_{0.67}(SiO_4)_6O_2$ [5] or $REE_{10-x}(SiO_4)_6O_{2+y}$ [6] has attracted considerable attention as oxide ion conductors. Apatite-type oxides have attracted much attention as the material for electrolytes in solid oxide fuel cell and sensors.

In low atomic number rare-earth silicate systems, an apatite phase occurs with a range of stability extending from $Ln_{4.67}(SiO_4)_3O$ to $Ln_4(SiO_4)_3$. The stability decreases as the rare-earth atomic number increases, with a mixture of Ln_2SiO_5 and $Ln_2Si_2O_7$ replacing apatite as the preferred phase assemblage [7],[8],[9],[10],[11].

Apatite-type rare-earth element (REE) silicates of the composition of $REE_{10-x}(SiO_4)_6O_{2+y}$, where REE = La, Nd, Gd and Dy, were prepared by MARTÍNEZ-GONZÁLEZ et al [6] via the mechanochemical synthesis (stabilized zirconia planetary ball mill: ball-to-powder ratio ~10:1, 350 rpm for maximum time of 9 h) starting from the stoichiometric mixtures of constituent oxides, REE_2O_3 and SiO_2 (molar ratio = 4:5), followed by post-milling thermal treatment (1500°C for 3 h). The ionic conductivity increases with the increasing size of REE cations.[3] The mechano-

[2] Since the prefix "oxy-" can be explained as containing oxygen or additional oxygen, and the prefix "oxo-" is used for the functional group or substituent oxygen atom connected to another atom by a double bond (=O), the names oxyapatite, oxy-apatite, oxoapatite and oxo-apatite can be considered as synonyms. In the published literature, the name oxyapatite is the most frequently used (~90%), and the term oxy-apatite is the second (~8%). The names oxoapatite and oxo-apatite are used much rarely (only about ~2%).

Apatite-type silicates described in this chapter can be also named as oxy-britholites (oxybritholites) [84].

[3] This conclusion is in discrepancy with the findings of HIGUCHI et al [12] described below.

chemical synthesis of apatite-type lanthanum silicates from the mixture of La_2O_3 and amorphous silica without post-milling thermal treatment was described by FUENTES et al [4].

Rare-earth element-doped apatite-type lanthanum silicates of the composition of $La_9MSi_6O_{27}$, where M = Nd, Sm, Gd and Yb, were synthesized by the high-temperature solid-state reaction process by XIANG et al [3]. All rare-earth oxide powders (La_2O_3, Nd_2O_3, Sm_2O_3, Gd_2O_3 and Yb_2O_3) were firstly pre-calcined at 900°C for 2 h in order to achieve complete decarbonation and dehydroxylation before weighing. The stoichiometric mixtures were mechanically mixed in absolute ethanol for 24 h using zirconia milling media at the speed of 400 rpm and dried at 100°C in air. The powder mixture was calcined at 1350°C for 10 h and then ground by hand with an agate mortar and pestle to reduce the particle size. After that, the powders were uniaxially pressed at 20 MPa and then statically cold pressed at 200 MPa for 5 min. The compacts were pressureless sintered at 1650 K for 10 h in air.

Apatite-type phase	Lattice parameters [Å]		*M*	*V*	Density	*E*	σ_0
	a	*c*	[g·mol⁻¹]	[Å³]	[g·cm⁻³]	[eV]	[S·K·cm⁻¹]
$La_4SiO_3O_{12}$	9.376	6.761	831.87	514.73	–	–	–
$La_{10}Si_6O_{27}$	9.709	7.176	1989.552	585.82	5.168	0.76	$2.51\cdot10^4$
$La_9NdSi_6O_{27}$	9.729	7.191	1994.886	589.46	5.344	0.79	$4.78\cdot10^4$
$La_9SmSi_6O_{27}$	9.692	7.161	2001.046	582.55	5.405	0.82	$1.85\cdot10^4$
$La_9GdSi_6O_{27}$	9.723	7.187	2007.896	588.41	5.485	0.87	$3.24\cdot10^4$
$La_9YbSi_6O_{27}$	9.701	7.143	2023.686	582.16	5.459	0.98	$4.34\cdot10^5$

Table 1. The properties of apatite-type lanthanium silicates [3],[8].

The lattice parameters and the properties of prepared apatite-type lanthanium silicates are listed in **Table 1**. All prepared compounds possess hexagonal apatite structure with the space group $P6_3/M$. The temperature dependence of total electrical conductivity for different compositions is determined using the Arrhenius equation[4] [13],[14],[15] in the following form [3],[5],[16],[17]:

$$\sigma\, T = \sigma_0 \exp\left(-\frac{E}{k_B T}\right) = \sigma_0 \exp\left(-\frac{\Delta H_m + \Delta H_a}{k_B T}\right) \quad (1)$$

where σ is the total electrical conductivity, σ_0 is the pre-exponential factor related to the effective number of mobile oxide ions, E is the activation energy for the electrical conduction process, k_B is the Boltzmann constant and T is absolute temperature. ΔH_m and ΔH_a denote the

[4] The equation of SVANTE AUGUST ARRHENIUS [13],[14], which predicts that the rate constant k depends on the temperature: $k = A\, exp\,(-E_a/RT)$, where A is the frequency (pre-exponential factor), E_a is the activation energy, R is universal gas constant (8.314 J·K⁻¹·mol⁻¹) and T is the thermodynamic temperature [13].

migration enthalpy of oxygen ion and the association enthalpy of defects, respectively. The determined activation energy and pre-exponential factor are listed in **Table 1**. It can be seen that the activation energy gradually increases from $La_{10}Si_6O_{27}$ to $La_9GdSi_6O_{27}$. Total electrical conductivity can be calculated from the following equation:

$$\sigma = \frac{h}{RS} \tag{2}$$

where *h* is the thickness of the specimen, *S* is the electrode area of the specimen surface and *R* is the total resistance including grain and grain boundary resistance. Lanthanum silicates doped with Nd or Yb cations exhibit higher total electrical conductivity than undoped lanthanum silicates. The highest total conductivity value obtained at 500°C is $4.31 \cdot 10^{-4}$ $S \cdot cm^{-1}$ for $La_9NdSi_6O_{27}$. The total electrical conductivity is also a function of partial pressure of oxygen [3].

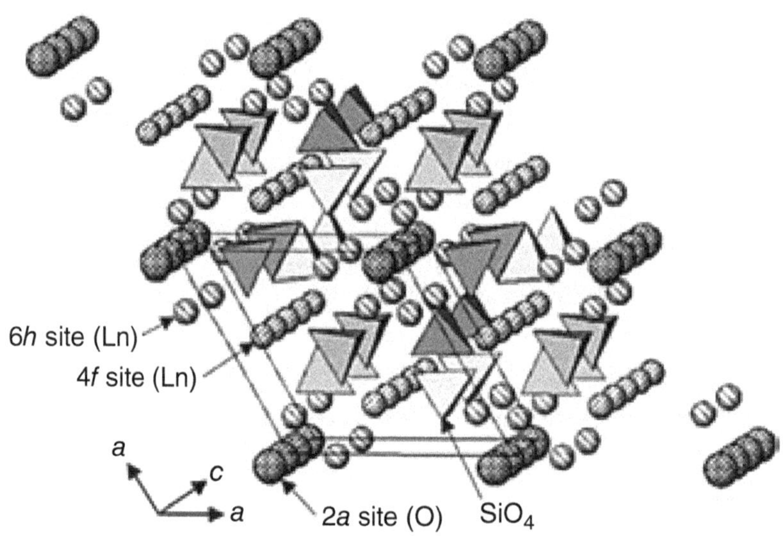

Fig. 2. Crystal structure of apatite-type rare-earth element silicate viewed along the c-axis [12].

The measurements using single crystals revealed definite anisotropy of the electrical conductivity of $Ln_{9.33}(SiO_4)_6O_2$, that is, the conductivity parallel to the c-axis is larger by one order of magnitude than that perpendicular to the c-axis. This fact clearly indicates that the channel oxide ions not bonded to silicon are the principal charge carriers in apatite-type lanthanum silicates. The structure of apatite-type rare-earth silicate is shown in **Fig. 2.** SiO_4 tetrahedra are isolated mutually, and Ln ions (REE ions, in general) at 6*h* sites (sevenfold coordinated site (x,y,¼)) [18][5] form channels, in which oxide ions at 2*a* sites are located (possess the threefold coordination with rare-earth ions at the 6*h* sites in the same plane), along the c-axis.

These mobile ions at these sites have much larger anisotropic displacement parameters in the direction of the c-axis than those in the direction of the a-axis, even at room temperature, which reflects high oxide ion conduction along the c-axis. The ninefold coordinate position (4*f* site (1/3,2/3,*z*)) is the second site for the accommodation of REE cations in the structure of apatite-type REE silicate [6],[12],[19],[20].

Since the interstitial space provided by these rare-earth ions is the smallest throughout the channel along the c-axis of the apatite structure, the migration of oxide ions through the channel will not be affected significantly even if the sizes of rare-earth ions are varied. It is therefore reasonable that the electrical conductivities of apatite-type rare-earth silicates are independent on the kind of rare-earth elements [12].

Conventional oxide ion conductors are designed on the basis of the oxygen vacancy model by the introduction of aliovalent[6] [21] cations. In $Ln_{9.33}(SiO_4)_6O_2$, however, cation vacancies are present rather than oxygen vacancies. Therefore, the introduction of cation vacancies into the structure of an oxide material may induce high oxide ion conductivity if the structure has a channel or a plane that can be a path for the migration of oxide ions [12].

Apparent exchange of O(1), O(2) and O(3) oxide ions bonded to Si was observed by ^{17}O NMR measurement on $La_{9.33}Si_6O_{26}$ by Kiyino et al [22], while it was not observed for oxide ion on the isolated site O(4). The results indicate that oxide ions bonded to Si at the position O(1), O(2) and O(3) are the main diffusion species in the oxide ion conductivity.

Trivalent and divalent dopants[7] [23] have been introduced into the $La_{9.33}(SiO_4)_6O_2$ structure according to the following nominal mechanisms [24]:

$$Si^{4+} \rightarrow M^{3+} + \frac{1}{3}La^{3+} \tag{3}$$

$$\frac{2}{3}La^{3+} \rightarrow AEE^{2+} \tag{4}$$

$$\frac{1}{3}La^{3+} \rightarrow AEE^{2+} + \frac{1}{2}O^{2-} \tag{5}$$

[5] According to the Wyckoff notation: the specification of actual coordinates of atoms within the unit or primitive cell, which can be generated by the point-group operations or may be found by reference in the International Tables for Crystallography [18].

[6] Cation with different valence. Apatite structure shows large flexibility upon the substitution of other aliovalent cations at the 'Ca' sites, pentavalent and tetravalent ions such as V^{5+}, As^{5+} and Si^{4+} at the 'P' site and halide, oxide ions at the 'OH' site [21], as was described.

[7] Dopants are also termed as doping agents. It can be defined as an impurity element added to the material structure in low concentration (usually <1 wt.% [23]) in order to alter its properties.

where M = Al, Ga, B, Co, Fe, Mn, … and AEE denotes the alkaline-earth elements (Ca, Sr and Ba). Doping with Al, Ga and B according to the formula: $La_{9.33+x/3}(SiO_4)_{6-x}(MO_4)_xO_2$, via the mechanism in **Eq. 3**, causes that bulk conductivity increases in up to two orders of magnitude in the case of Al for $x = 1 - 1.5$. If, however, the sample is stoichiometric on both cation and anion sites, as for $La_8Sr_2(SiO_4)_6O_2$, the AEE doping reduces the conductivity and increases the activation energy for the conduction compared to $La_{9.33}(SiO_4)_6O_2$.

The effect of Fe doping on the electrical properties of lanthanum silicates of the composition of $La_{10}Si_{6-x}Fe_xO_{27-x/2}$ (where $x = 0.2$, 0.4, 0.6, 0.8 and 1.0) was performed by Shi and Zhang [16] via the sol-gel process. Tetraethyl orthosilicate (TEOS), $La(NO_3)_3{\cdot}6H_2O$ and $Fe(NO_3)_3{\cdot}9H_2O$ were used as starting materials. Stoichiometric amounts of $Fe(NO_3)_3{\cdot}9H_2O$ and $La(NO_3)_3{\cdot}6H_2O$ were dissolved in the mixture of ethanol, acetic acid and distilled water. The appropriate amount of TEOS was added to the solution while continuous stirring. The solution became gradually a purple clear sol. After refluxing at 80°C for 1 – 2 h, the sol transferred to a clear gel. Then, the wet gel was dried at 100°C for 20 h. The gel was heated at 600°C for 4 h to remove water and organic components and to decompose nitrates. In order to get the desired phase, obtained precursor was then calcined at 1000°C for 4 h.

Apatite-type lanthanium silicate	Lattice parameters [Å]		V	E (600 – 800°C)	E (400 – 550°C)
	a	*c*	[Å³]	[eV]	
$La_{10}Si_{5.8}Fe_{0.2}O_{26.9}$	9.725	7.192	589.1	0.78	0.96
$La_{10}Si_{5.6}Fe_{0.4}O_{26.8}$	9.729	7.208	590.8	0.74	0.95
$La_{10}Si_{5.4}Fe_{0.6}O_{26.7}$	9.732	7.220	592.2	0.72	0.89
$La_{10}Si_{5.2}Fe_{0.8}O_{26.6}$	9.735	2.217	592.3	0.74	1.01
$La_{10}Si_5FeO_{26.5}$	9.743	7.229	593.5	0.75	1.02

Table 2. Lattice parameters of Fe-doped apatite-type lanthanium silicates [16].

All synthesized samples have hexagonal lattice structure with the space group of $P6_3/M$. The lattice parameters of prepared Fe-doped apatite-type lanthanium silicates and the activation energy of conductivities (**Eq. 1**) for different Fe contents are listed in **Table 2**. When $x = 0.6$, $La_{10}Si_{5.4}Fe_{0.6}O_{26.7}$ exhibits the lowest activation energy. The lattice parameters of $La_{10}Si_6O_{27}$ (**Table 1**) and doped specimen (**Table 2**) show that the values of ***a***, ***c*** and ***V*** increase with the content of iron. The conductivity of $La_{10}Si_{6-x}Fe_xO_{27-x/2}$ is independent of oxygen partial pressure in the range from 0 to 100 kPa, which indicates that the conductivity of all samples is mainly ionic [16].

The oxygen ionic and electronic transport in apatite ceramics with the composition of $La_{10}Si_{6-x}Fe_xO_{27-x/2}$ ($x = 1 - 2$) [25] and $La_{10-x}Si_{6-y}Al_yO_{27-3x/2-y/2}$ ($x = 0 - 33$; $y = 0.5 - 1.5$) [26],[27] was investigated by Shaula et al In both cases, the essential role of oxygen content on the ionic conductivity of apatite phase was recognized. The ion transference number[8] [28] increases with decreasing partial pressure of oxygen. Such behavior indicates that the conduction under oxidizing condition is predominantly of p-type[9] (with respect to n-type of conductivity). Similar to the foundation of Shi and Zhang [16], the conductivity of these phases is predominantly ionic and almost independent on partial pressure of oxygen. The ion transference numbers are higher than 0.99, while the p-type electronic contribution to total conductivity is about 3% (700 – 950°C, $La_{10}Si_4Fe_2O_{26}$). The oxygen ionic conductivity should increase with decreasing iron content due to higher concentration of oxygen interstitials.

Another important factor influencing the oxygen diffusion is M-site deficiency, which affects the unit cell volume and may cause the O(5) ion displacement into interstitial sites, thus creating the vacancies in the O(5) sites at fixed total oxygen content. In particular, an enhanced ionic conduction was found in the system $La_{9.33+x/3}Si_{6-x}Al_xO_{26}$, where Al doping is compensated by the A-site vacancy concentration without oxygen content variations [29],[30].

The incorporation of praseodymium in the apatite-type lattice of $La_{9.83-x}Pr_xSi_{4.5}Fe_{1.5}O_{26+\delta}$ (x = 0 – 6) decreases the unit cell volume, suppresses the Fe^{4+} formation according to Mössbauer spectroscopy[10] [31],[32],[33],[34] and increases p- and n-type electronic contributions to total conductivity under oxidizing conditions, while the level of oxygen ionic transport at temperatures above 1000 K remains unaffected [35].

Since the size of the conduction channel increases with the Mg doping, the enhancement of the ionic conductivity of lanthanum silicate-based apatites can be reached by optimizing the La content and the Mg doping level at the same time. The ionic conductivities of $La_{10}Si_{5.8}Mg_{0.2}O_{26.8}$ and $La_{9.8}Si_{5.7}Mg_{0.3}O_{26.4}$ at 800°C are 88 and 74 mS·cm^{-1} with the activation energy of 0.43 and 0.42 eV, respectively [36].

The ionic conduction in cation-deficient apatite $La_{9.33-2x/3}M_xSi_6O_{26}$, where M = Mg, Ca and Sr was investigated by Yuan et al [37]. The nature of dopant and the extent of substitution have a significant effect on the conductivity. The greatest decrease in conductivity is observed for Mg doping followed by Ca- and Sr-doped apatites. The effect is ultimately attributed to the amount of oxygen interstitials, which is affected by the crystal lattice distortion arising from the cation vacancies.

The incorporation of additional La_2O_3 into $La_{9.33}(SiO_4)_6O_2$ to form $La_{10}(SiO_4)_6O_3$ or intermediate compositions can most obviously be achieved by filling empty interstitial sites with oxygen. The only alternative scenario would involve the creation of cation vacancies on the Si sublattice, which is unlikely as Si is present as a complex anion. The incorporation of excess of La_2O_3 into $La_{9.33}(SiO_4)_6O_2$ can therefore be expressed as [10]:

[8] The fraction of total current that is transferred by a given ion is affected by its mobility. The sum of transport numbers for all ions in electrolyte is equal to one [28].

[9] The p-type carriers possess typically higher mobility [25].

[10] The technique is based on the Mössbauer effect of recoil-free nuclear resonance fluorescence [31], i.e. the phenomenon of emission or absorption of X-ray photon without the loss of energy. The Mössbauer effect has been detected in a total of 88 X-ray transitions in 72 isotopes of 42 different elements [32]. The ^{57}Fe Mössbauer isotope is the most frequently used [33]. The Mössbauer spectroscopy can be used to determine the oxidation states of iron in minerals and to identify the presence of some mineral species in samples of unknown composition [31].

$$La_{9.33}[\ ]_{0.67}(SiO_4)_6O_2 + 1/3\ La_2O_3 \rightarrow La_{10}(SiO_4)_6O_2O_i(La_{10}(SiO_4)O_3) \quad (6)$$

Thus, in the ideal pure $La_{10}(SiO_4)_6O_3$, the *4f* and *6h* sites are fully occupied by La^{3+} ions, while an extra oxygen interstitial is introduced into the lattice to maintain the electroneutrality. The oxygen interstitial may benefit the oxide ion transportation if it is located nearby the [001] direction c-axis of the conventional unit cell. From the space-filling consideration, the most appropriate sites for the oxygen occupation are in this position; however, some distortion of the O *2a* sites would be required to accommodate extra oxygen atoms. This could be achieved by decreasing the symmetry from $P6_3/M$ to $P6_3$ allowing oxygen to move from 0,0,1/4 to 0,0,x. Recent studies suggest that a range of partially occupied (0,0,x) sites may accommodate this extra interstitial oxygen. From this point of view, $La_{10}(SiO_4)_6O_3$ should exhibit higher conductivity than $La_{9.33}(SiO_4)_6O_2$ [10],[38].

Introducing Sr^{2+} cations to the La^{3+} atomic positions, as in the $La_{10}(SiO_4)_6O_3$ phase, leads to complete elimination of vacancies according to the substitution [39]:

$$La_{10}(SiO_4)_6O_3 \xrightarrow{+Sr^{2+}} La_9Sr_1(SiO_4)_6O_{2.5} \xrightarrow{+Sr^{2+}} La_8Sr_2(SiO_4)_6O_2 \quad (7)$$

The substitution of La_2O_3 by SrO, taking into account the charge balance and the oxygen content, can be represented as follows (Kröger-Vink notation[11] [40],[41]):

$$SrO + 1/2\ O_o^x + La_{La}^x \rightarrow Sr_{La}^{/} + 1/2\ V_o^{\bullet\bullet} + 1/2\ La_2O_3 \quad (8)$$

Lanthanum oxyapatite phases are substantially stable with respect to their binary oxides. The general trend in the formation enthalpies as a function of (La + Sr)/(La + Sr + Si) shows that the apatite phase becomes more energetically stable as the cation vacancy and oxygen excess concentrations decrease. The stoichiometric sample achieved by Sr^{2+} doping, with no cation vacancies or interstitial oxygen atoms, is the most stable composition. The energetics of lanthanum silicate apatite materials ($La_{9.33+x}(SiO_4)_6O_{2+3x/2}$ and $La_{10-x}Sr_x(SiO_4)_6O_{3-0.5x}$) depends on lanthanum deficiency and oxygen interstitial[12] [42],[43] concentrations, and the cation vacancy concentrations appear to be the dominant factor in energetics [39].

[11] The Kröger-Vink notation indicates the lattice position for the point defect species in the crystal and its effective electric charge relative to the perfect lattice: $\mathbf{M}_Y{}^Z$ is the atomic species **M** (or vacancy **V**) that occupies the lattice site **Y** and possesses the effective charge **Z**, where the symbols ●, ' and × are used for the effective charge +1, -1 and neutral particle, respectively) [40]. For example, $\mathbf{Al}_i{}^{\bullet\bullet\bullet}$ is Al^{3+} ion at interstitial site (*i*), $\mathbf{V}_{Al}'''$ is Al^{3+} vacancy, $V_O{}^{\bullet\bullet}$ is O^{2-} vacancy, Sr_{La}' (**Eq. 8**) means Sr^{2+} ion replacing La^{3+} at lattice site, $Ti_{Al}{}^{\bullet}$ means Ti^{4+} replacing Al^{3+} at lattice site, *e'* is electron and $h^{\bullet}$ is the hole. The equation must fulfill the following three rules: mass balance (1), electroneutrality or charge balance (2) and site ratio conservation balance (3) [41].

[12] Interstitial sites are sites between normal (equilibrium) atomic positions of ideal lattice atoms [42]. Interstitial atoms and vacancies (lattice site where atom is absent) are the simplest types of point defects in a crystal. A vacancy and interstitial atoms positioned close together are referred to as the Frenkel pair. Apart from the point defects, the line crystal defects (dislocation and disclination) are recognized [43].

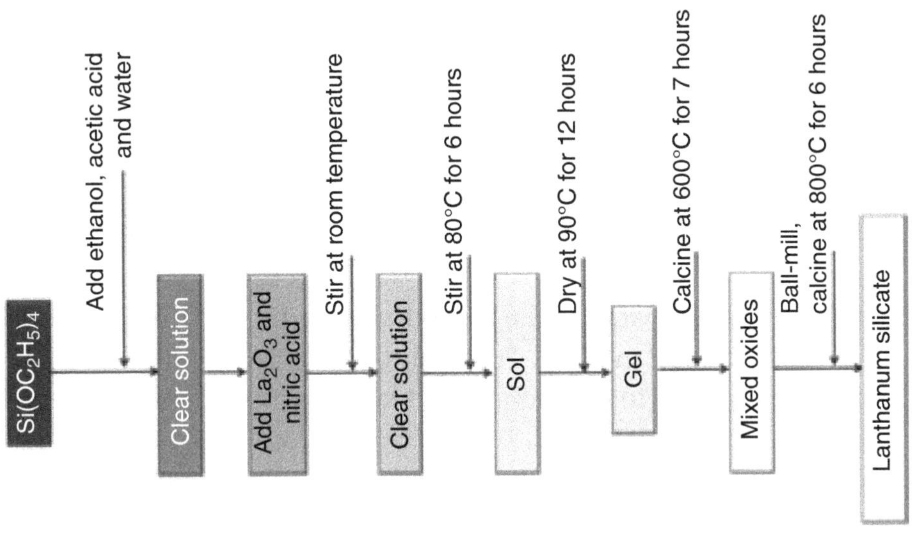

Fig. 3. Schematic diagram of preparation of lanthanum silicate by a sol-gel process [44].

The schematic diagram of the sol-gel process used by Tao and Irvine [44] for the preparation of apatite-type lanthanium silicates is shown in **Fig. 3**. The room-temperature structure is hexagonal, the space group is $P6_3$ or $P6_3/M$, with $\boldsymbol{a}$ = 9.722 and $\boldsymbol{c}$ = 7.182 Å for $La_{10}(SiO_4)_6O_3$ and $\boldsymbol{a}$ = 9.717 and $\boldsymbol{c}$ = 7.177 Å for $La_{9.33}(SiO_4)_6O_2$, i.e. the cell volume of $La_{10}(SiO_4)_6O_3$ is a little greater than that of $La_{9.33}(SiO_4)_6O_2$. Both compositions exhibit high ionic conductivity, although the grain boundary resistance is the dominant feature in the impedance spectrum of both. In general, the conductivity of $La_{10}(SiO_4)_6O_3$ is higher than that of $La_{9.33}(SiO_4)_6O_2$ and this indicates that oxygen interstitials may be introduced into the apatite lattice of $La_{10}(SiO_4)_6O_3$, which may benefit the oxygen ion transportation [44].

The $La_{10}Si_6O_{27}$ nanopowders with apatite structure were synthesized by the Li et al [45] co-precipitation method. After the calcination at 900°C and then removing of La_2O_3 by acid washing, the pure stoichiometric $La_{10}Si_6O_{27}$ nanopowders are obtained. The oxyapatite ceramics with the density higher than 95% can be obtained at rather low sintering temperature of 1300°C, and it has comparable total conductivity with the samples sintered at 1650°C from the powders prepared by solid-state reaction.

La_2O_3 and TEOS in stoichiometric amount were used by Masubuchi et al [46] as the starting materials for the preparation of both powder and film of apatite-type $La_{9.33}(SiO_4)_6O_2$ via the alkoxide hydrolysis. Lanthanum oxide was dissolved in HNO_3 (6 mol·dm^{-3}) and mixed with ethanol. Then, stoichiometric amount of TEOS in ethanol was added (La:Si = 9.33:6) to this solution. The precursor solution was obtained by refluxing for one night. This solution was heated to gelating on the hot plate followed by calcination and annealing in powder preparation. Either quartz glass or Pt foil substrate was dipped to the gelatinous solution and dried for the film preparation. It was calcined at 500°C for 1 h to remove the organic contents and then fired at 1000°C for 10 h. This preparation steps were repeated to increase the film thickness. The film showed preferred orientation of the apatite crystal in thinner film. The conductivity of sintered body was lower in about one order of magnitude than the value of single crystal perpendicular to c-axis [46].

The synthesis and the conductivities of Ti-doped apatite-type phases of the composition of $(La/Ba)_{10-x}(Si/Ge)_6O_{26+z}$, where Ti substituted at the Si/Ge site, were reported by SANSOM et al [47]. The conductivities were shown to be the highest for the samples containing either cation vacancies or oxygen excess, which is consistent with previous studies of apatite-type oxide ion conductors. However, the Ti doping was shown to generally decrease the conductivity in comparison with equivalent samples containing only Si/Ge at the tetrahedral sites, with the greatest decrease for Si-containing samples.

Vanadium-doped oxyapatite phases of the composition of $La_{10-x}V_x(SiO_4)_6O_{3+x}$ were prepared by YUAN et al [48] via the sol-gel method. The apatite phase begins to form at 800°C, which is much lower than in the case of conventional solid-state synthesis method. The best conductivity of $La_9V(SiO_4)_6O_4$ is $1.67 \cdot 10^{-2}$ S·cm^{-1}, which is significantly higher than that for lanthanum silicate oxides ($1.19 \cdot 10^{-2}$ S·cm^{-1}). The valence ion V^{5+} doped for La^{3+} does lead to the formation of hexagonal apatites even with high oxygen contents.

The phase $La_5Si_2BO_{13}$ [49],[50],[51] crystallizes with apatite-related structure (**Fig. 4**) with the space group $P6_3/M$ and the cell parameters $\boldsymbol{a}$ = 9.5587 Å, $\boldsymbol{c}$ =7.2173 Å and **Z** = 2. The composition of these apatite-like compounds can be also expressed via the general formula: $La_{9.33+x}Si_{6-2y}B_{3y}O_6$, where $0 \leq x \leq 0.67$. At limiting compositions x = 0.67, La(2) site is fully occupied, and the formula referred to the unit cell is $La_{10}Si_4B_2O_{26}$ or, more simply, $La_5Si_2BO_{13}$.

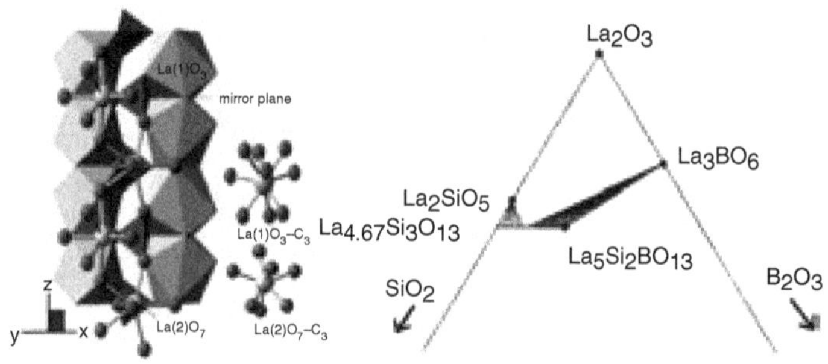

Fig. 4. The crystal structure of $La_5Si_2BO_{13}$ (a) [51] and the upper section of the ternary phase diagram La_2O_3-SiO_2-B_2O_3 at 1100°C (b) [49].

The comparison with other apatite-like structures shows lower distortion in the M(1) polyhedron and unusually short bond length from La in the M(2) site and O(4) oxygen in the column site (2.303 Å). These results can be explained in view of the presence of trivalent La and divalent O, respectively, in the M(1) and M(2) sites and in the column anion site, whereas, in apatites, these sites are occupied by divalent and monovalent ions, respectively [49].

The preparation of La-Si-O apatite-type thin films was described by VIEIRA et al [52] with Si/(La + Si) atomic ratios ranging from 0.36 to 0.43 being produced via the magnetron sputtering in reactive Ar/O discharge gas. The apatite-type lanthanum silicate phase was formed in all as-deposited films upon the annealing at 900°C for 1 h. The lanthanum silicate films obtained by annealing the as-deposited films with lower Si/(La + Si) atomic ratios have a preferential orientation with the c-axis perpendicular to the substrate, while low-intensity diffraction peaks ascribed to $La_2Si_2O_7$ phase were detected in the films deposited with higher

Si content. Preferentially oriented films have higher activation energy and lower ionic conductivity, as the ionic conductivity measurements were performed in the direction perpendicular to the c-axis. The highest ionic conductivity was obtained for the film deposited with a Si/(La + Si) atomic ratio of 0.42, with a value of 1.2×10^{-2} S·cm^{-1} at 750°C. By the incorporation of oxygen in the as-deposited films, the silicon segregation upon annealing was avoided.

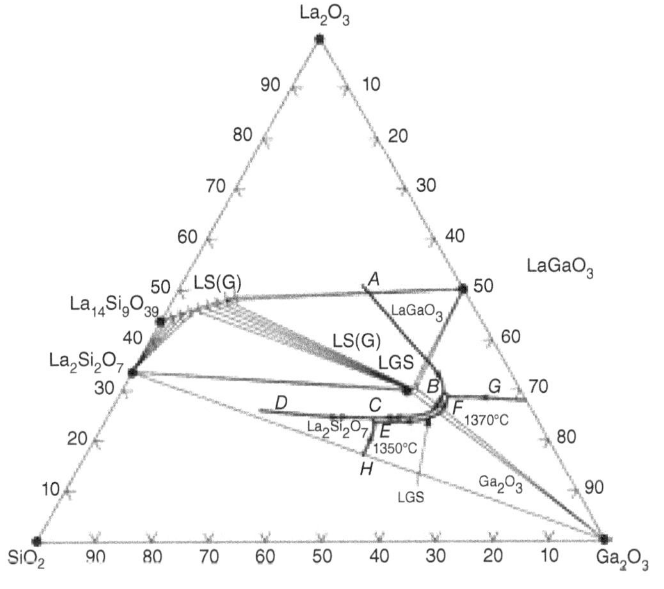

Fig. 5. Primary phase diagram of ternary system La_2O_3-Ga_2O_3-SiO_2 around LGS [53].

The formation of ternary compound with apatite structure in the system La_2O_3-Ga_2O_3-SiO_2 (**Fig. 5**) was first reported by WANG and UDA [53]. The apatite phase, which precipitates from the melt of the composition around that of stoichiometric $La_3Ga_5SiO_{14}$ (LGS), can be described by the formula: $La_{14}Ga_xSi_{9-x}O_{39-x/2}$, where $0 \leq x \leq 3.5$. Since there is a large field for the formation of solid solutions with the range extending from $La_{14}Si_9O_{39}$ to Ga_2O_3, some Si^{4+} sites are probably substituted by Ga^{3+}.

The liquidus surface of LS(G) was determined to be the field on the Ga_2O_3-poor side of boundary curve ABCD. The liquidus surface of LS(G) covers the stoichiometric composition of LGS. In this field, the crystallization of LS(G) aciculae was observed in all samples that were heated to temperatures above 1500°C. The liquidus volume of LGS is denoted by the field BCEF. It seems to be a narrow field in the composition between the liquidus surfaces of LS(G) and Ga_2O_3. E and F are eutectic points, where LGS + $LaGaO_3$ + Ga_2O_3 + liquid and LGS + Ga_2O_3 + $La_2Si_2O_7$ + liquid were found, respectively [53].

The CaO-La_2O_3-SiO_2-P_2O_5 phase diagram was investigated by EL OUENZERFI et al [54] in order to determine a domain inside which all points correspond to pure apatitic oxyphosphosilicates with the general formula: $Ca_xLa_y(SiO_4)_{6-u}(PO_4)_uO_t$. The defined domain (**Fig. 6**) is only a part of the whole existence domain of the apatitic structure, but it allows to prepare pure apatitic samples with well-controlled composition.

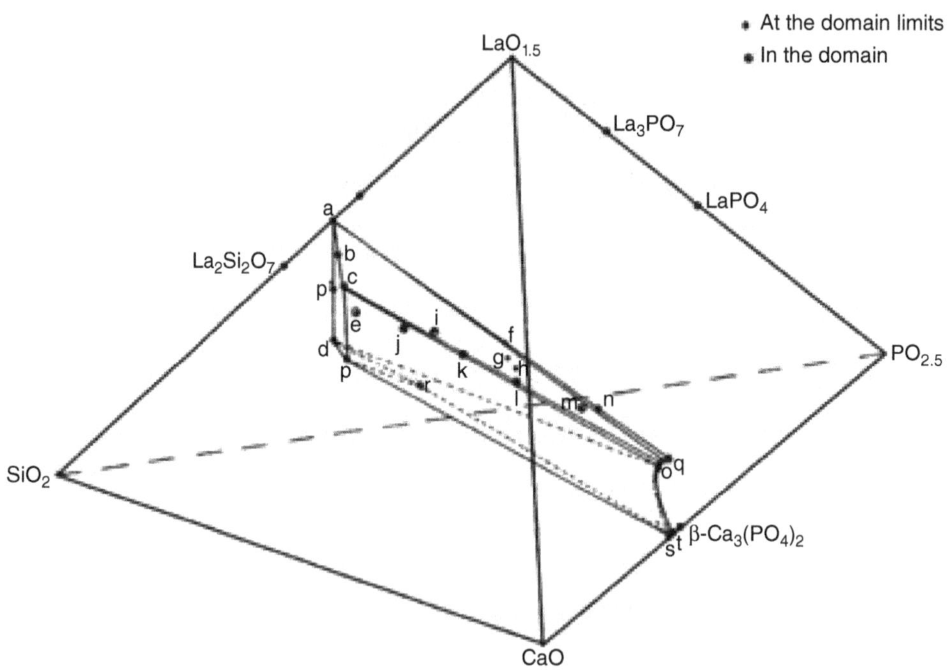

Fig. 6. Britholite stability domain in the CaO-SiO_2-La_2O_3-P_2O_5 quaternary system [54].

For these samples, the continuous change of the stoichiometry of each element proves that it exists as a solid solution including the oxygen content. This observation completes the literature data where britholites are presented as limited to three series corresponding to the stoichiometry [54],[55]:

1. $Sr_{2+x}La_{8-x}(SiO_4)_{6-x}(PO_4)_xO_2$ with $0 \leq x \leq 6$ [13];

2. $Sr_{3+x}La_{6-x}(SiO_4)_{6-x}(PO_4)_x$ with $0 \leq x \leq 1.5$;

3. $Sr_{4+x}La_{6-x}(SiO_4)_{6-x}(PO_4)_xO$ with $0 \leq x \leq 6$.

Inside this domain, the solid solution continuously varies between pure phosphate apatites $Ca_xLa_y(PO_4)_6O_t$ and pure silicate apatites $Ca_xLa_y(SiO_4)_6O_t$ and also between oxyapatites $Ca_xLa_y(SiO_4)_{6-u}(PO_4)_uO_t$ and nonoxyapatites $Ca_xLa_y(SiO_4)_{6-u}(PO_4)_u$.

During the investigation of the kinetics of solid-state sintering[14] of strontium-doped apatite-type lanthanum silicates ($Sr_xLa_{10-x}Si_6O_{27-x/2}$) under isothermal conditions (1250 – 1550°C), Bonhomme et al [56] recognized that the densification mechanism of the apatite ceramics was

[13] Wanmaker et al [55] reported the synthesis of apatite-type compounds of the composition:

(a) $M^{2+}(II)M(III)_{8-x}(SiO_4)_{6-x}(PO_4)_xO_2$, where $0 \leq x \leq 6$;

(b) $M(II)_{3+x}M(III)_{6-x}(SiO_4)_{6-x}(PO_4)_x$, where $0 \leq x \leq 1.4$;

(c) $M(II)_{4+x}M(III)_{6-x}(SiO_4)_{6-x}(PO_4)_xO$, where $0 \leq x \leq 6$.

where M(II) = Ca, Sr, Ba, Mg, Zn or Cd and M(III) Y or La. The paper also contains structural data for several other newly prepared oxy-britholites, including $Zn_2La_8(SiO_4)_6O_2$, $BaMgY_8(SiO_4)_6O_2$, $Zn_2Y_8(SiO_4)_6O_2$, $Cd_2Y_8(SiO_4)_6O_2$, $Ca_4La_5(SiO_4)_5(PO_4)$ and $Ba_4La_5(SiO_4)_5(PO_4)$.

controlled by the diffusion of rare-earth element (La) at the grain boundaries. This process showed the activation energy of 470 kJ·mol^{-1}.

The ternary phase diagram Al_2O_3-SiO_2-La_2O_3 at 1300°C (**Fig.** 7) was investigated by Mazza and Ronchetti [57]. $La_{14}Si_9O_{39}$ was described by Kuz'min and Belov [58] as an apatite-like structure of hexagonal symmetry (space group $P6_3/M$). Isomorphous compounds were also reported for Nd [59], Ce [60] and Sm [58]. The $La_{14}Si_9O_{39}$ compound extends its stability range in the interior of the phase diagram, forming the solid solution of the type $La_{14+1x/3}Si_{9-x}Al_xO_{39}$, which is stable from $x = 0$ to $x = 1.5$. This substitution stoichiometry (Al + 1/3La ↔ Si) can be described as a tetrahedral Al for Si substitution on the *6h* position and contemporary occupation of vacant La sites [57].

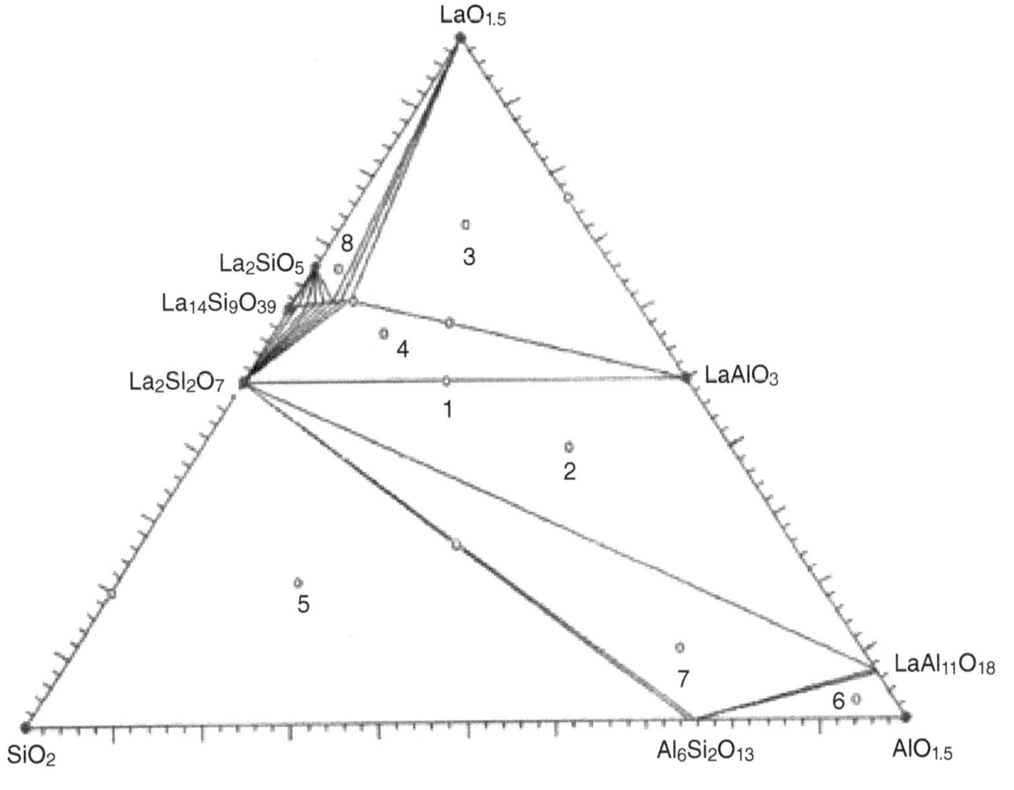

Fig. 7. Ternary phase diagram at 1300°C in air [57].

[14] The densification rate is considered as the function of temperature (*T*) and mean grain size (D_m). Constant *A* depends on the surface energy (γ_{sg}) of grains, on the apatite molar volume (Ω) and on average coefficient of diffusion of limiting species *D*. This relationship can be written as follows [56]:

(a) $\frac{d\rho}{dt} = \frac{AD}{TD_m^n}$

The coefficient of diffusion *D* is thermally activated:

(b) $D = D_0 \exp\left(-\frac{E_a}{RT}\right)$

where D_0 is the pre-exponential coefficient of diffusion, *R* is the universal gas constant and E_a is the apparent activation energy of diffusion of the rate limiting process. Exponent *n* in **Eq. (a)** depends on the mechanism of transport of the limiting species governing the kinetics of densification.

Apatite-type lanthanium germanates

Lanthanum germanate and silicate apatite-based materials, both undoped and with partial substitution of, for example, Al, B instead of Si and Sr in place of La, are promising oxide ion conductors with potential applications as high-temperature solid electrolytes. Considerable uncertainties remain over the stoichiometry, the defect structure and the conductivity variations within various apatite systems, partly caused by the fact that the La:(Ge, Si) ratio is variable, giving rise to the solid solutions in the undoped systems as well as to the solid solutions formed by partial replacement of La and/or (Si, Ge) together with, depending on the solid solution mechanism, variations in oxygen content [24].

The preparation of single crystal of apatite-type lanthanium germanate of the composition of $La_{9.33}Ge_6O_{26}$ was reported by Nakayama and Sakamoto [61]. The mixtures of La_2O_3 and GeO_2 were well mixed in ethanol under an atomic ratio of La:Ge = 9.33:6 using a ball mill. The mixture was dried and then calcined in air at 1000°C for 2 h. The resulting $La_{9.33}Ge_6O_{26}$ powders were further ball milled into finer powders. After the pre-sintering, the closely packed $La_{9.33}Ge_6O_{26}$ powders were heated at 1300°C for 2 h in air, and the surface of the specimen was mirror polished. The polished surface of polycrystalline $La_{9.33}Ge_6O_{26}$ ceramics was then bonded to a <001> face of $La_{9.33}Ge_6O_{26}$ seed crystal prepared by the Czochralsky (**Section**) method. On heating of the bonded sample at 1525 – 1550°C, continuous grain growth of polycrystalline $La_{9.33}Ge_6O_{26}$ occurred and the single crystal was gradually grown from the seed crystal into the polycrystalline region [61].

Apatite-type lanthanium germanate possesses hexagonal structure with the space group $P6_3/m$ and the lattice parameters: $a = 9.9256$ and $c = 7.2900$ Å, $V = 621.97$ Å^3 and $Z = 2$. The calculated density of the phase is 2.148 g·cm^{-3}. Similar to apatite-type lanthanium silicate ($La_{9.33}Si_6O_{26}$) described above, the structure of $La_{9.33}Ge_6O_{26}$ (**Fig. 8**) contains two different sites for atoms of La. The La(1) and La(2) sites are located at *4f* and *6h*, respectively. While the $La_{9.33}Ge_6O_{26}$ single crystal showed little anisotropy in conductivity, the conductivity of $La_{9.33}Si_6O_{26}$ single crystal gave 100 times higher value parallel to the c-axis than that perpendicular to the c-axis at each temperature (**Section**) [61].

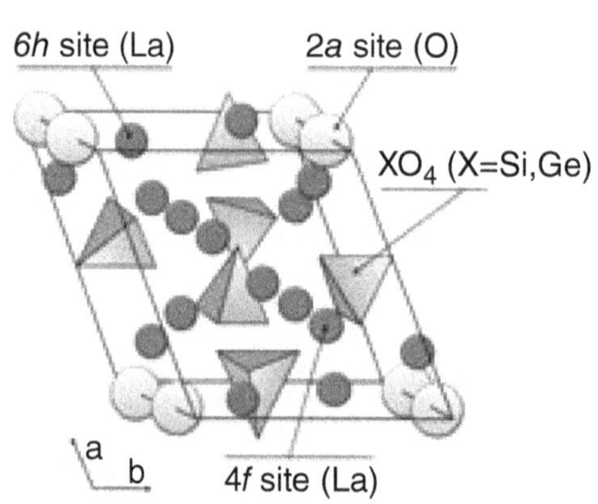

Fig. 8. Hexagonal structure proposed for the apatite-type phase of $La_{9.33}X_6O_{26}$, where X = Si and Ge [61].

The selective doping of $La_{9.33+x}(GeO_4)_6O_{2+3x/2}$ with Y leads to the stabilization of hexagonal lattice, even at high oxygen contents. Furthermore, this has the effect of enhancing the low-temperature conductivities [62]. Depending on the composition, the cell can be either hexagonal or triclinic, with the evidence of reduced low-temperature conductivities for the latter, attributed to increased defect trapping in this lower symmetry cell. In summary, it was shown that the series $La_8Y_2(GeO_4)_{6-x}(GaO_4)_xO_{3-x/2}$ can be prepared for $0 \leq x \leq 2$ with all samples showing the hexagonal symmetry, compared to the series without Y co-doping, $La_{10}(GeO_4)_{6-x}(GaO_4)_xO_{3-x/2}$, for which all compositions display the triclinic symmetry [24],[62], [63].

The effect of Ga doping of the oxygen stoichiometric series containing the cation vacancies, $La_{7.33+y/3}Y_2(GeO_4)_{6-y}(GaO_4)_yO_2$ ($0 \leq y \leq 2$), single-phase samples was obtained for $y \geq 1.0$, with small impurities observed at lower Ga contents. The conductivities were shown to increase with increasing cation vacancy content, reaching the values of ≈0.02 $S \cdot cm^{-1}$ at 800°C, which are similar to the oxygen excess series. These results are in agreement with previous reports on the apatite systems, which showed that the oxide ion conductivity was maximized in samples containing the oxygen excess and/or the cation vacancies [24], [62],[63].

The series of apatite-type silicates/germanates of the composition of $La_{8+x}Sr_{2-x}Si_6O_{26+x/2}$ ($0 \leq x \leq 1$) and $La_{8+x}Sr_{2-x}Ge_6O_{26+x/2}$ ($0 \leq x \leq 2$) were prepared from high-purity La_2O_3, $SrCO_3$, SiO_2 and GeO_2 by Orera et al [64] via the thermal treatment of these components mixed in the stoichiometric ratio.

The extent of, and the structural changes within, the apatite domain in the $LaO_{1.5}$-GeO_2-SrO ternary system at 1100°C was studied and the single-phase samples were obtained for $La_{9.33+x-2y/3}Sr_y(GeO_4)_6O_{2+1.5x}$ with x = 0.17 and 0.34. The hexagonal to triclinic transition is clearly associated with increasing oxygen content rather than with filling the La sites by the addition/substitution of Sr into the structure. The limits of undoped solid solution are ~$0.17 \leq x \leq 0.5$ at 1100°C [24].

The hydrothermal synthesis of apatite-type compound $NaRE_9(GeO_4)_6O_2$ (RE = Nd, Pr) with the hexagonal structure of the space group of $P6_3/M$ was described by Emirdag-Eanes et al [65]. The structure is composed of REO_7 and REO_9 polyhedra as well as GeO_4 tetrahedra (**Fig. 10**). The unit cell dimensions are: a = 9.782(1) Å, c = 7.083(1)Å (T = 293 K) and V = 587.0(2)Å3 for REE = Nd and a = 9.802(1) Å, c = 7.116(1)Å (T = 293 K) and V = 592.1(2)Å3 for REE = Pr.

The high-temperature flux method for the preparation of single crystal of hexagonal $NaLa_9Ge_6O_{26}$ apatite-type germanate (space group $P6_3/M$, $\boldsymbol{a}$ = 9.883, $\boldsymbol{c}$ = 7.267 Å and $\boldsymbol{Z}$ = 1) was used by Takahashi et al [66]. The crystal structure (**Fig. 11**) was found to be similar to that of silicate oxyapatite $NaY_9Si_6O_{26}$. The 4f cation sites are occupied disorderedly by La and Na. On the other hand, the 6h cation sites are occupied by La only. This compound constitutes a new member of the oxyapatite-type structure family with the composition given by general formula: $A_xLn_{10-x}B_6O_{24}O_{3-x}$.

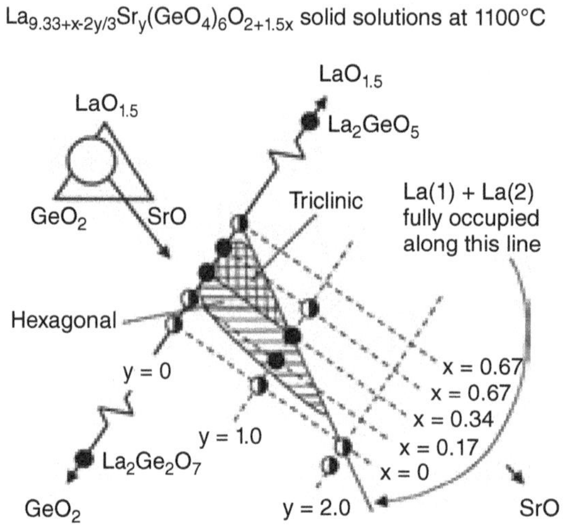

Fig. 9. Apatite solid solutions in the $LaO_{1.5}$-GeO_2-SrO ternary system [24]: pure phases are indicated by filled circles and the presence of secondary phases is shown by half-filled circles.

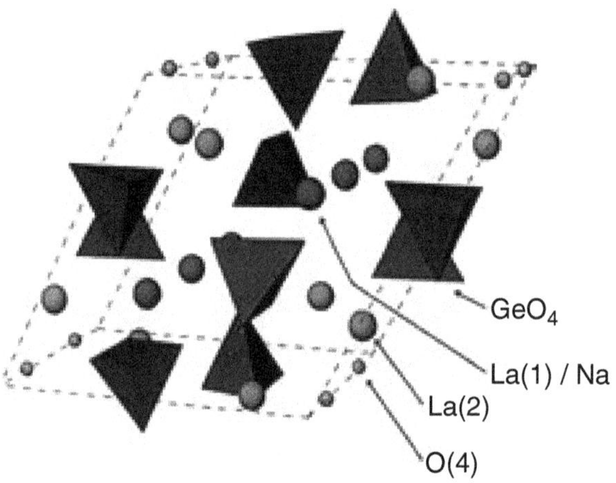

Fig. 11. The structure of sodium lanthanum germanate $NaLa_9Ge_6O_{26}$ [66].

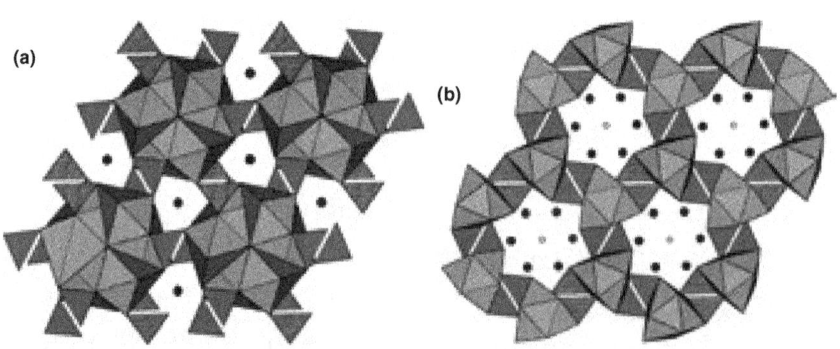

Fig. 10. The unit cell view with GeO_4 and NdO_7 (a) and GeO_4 and NdO_9 down the c-axis. NdO_7 and NdO_9 are dotted polyhedra, and GeO_4 are lined polyhedra [65].

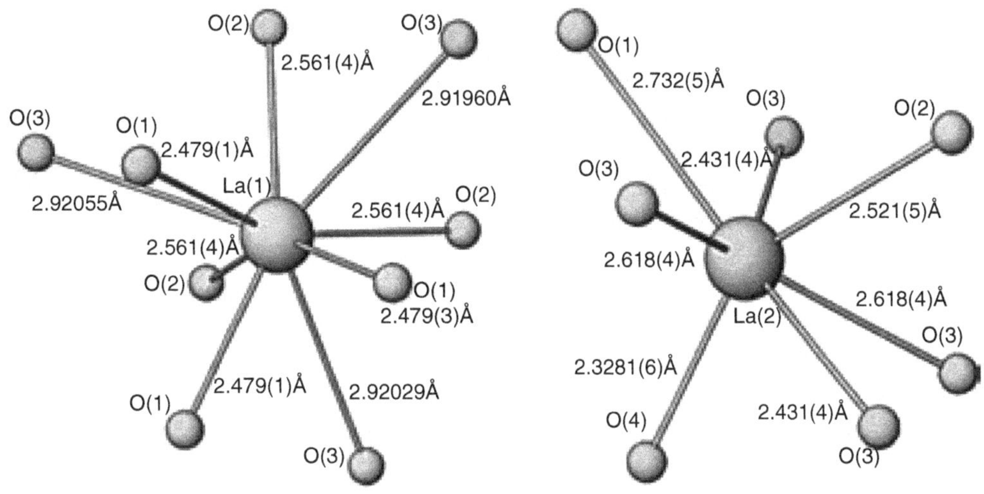

Fig. 12. Coordination environment of La atoms [66].

The coordination environments of La atoms by O atoms are shown in **Fig. 12**. La(1) atom on the 4*f* site is coordinated by nine O atoms. It is linked to three O(1) atoms in a distance of about 0.2479 nm, to three O(2) atoms in a distance of about 0.2561 nm and to three O(3) atoms in a distance of about 0.2924 nm. Because the distance between La(1) and O(3) is relatively large, La(1) atom can be also regarded to be in sixfold coordination, the environment of which is fairly distorted from an ideal octahedron. On the other hand, La(2) atom, which occupies the 6h position, is coordinated by seven O atoms, that is, O(1), O(2), O(4) and O(3). The distances of those two types of bonds between La(2) and O(3) atoms are 0.2618 nm×2 and 0.2431 nm×2, respectively. Those between La(2) and O(1), O(2) and O(4) are 0.2732, 0.2521 and 0.23281 nm, respectively [66].

Apatite-type borates

Two high terbium content apatites $Tb_5Si_2BO_{13}$ (a = 9.2569 Å, c = 6.8297 Å, V = 506.83 Å^3 and **Z** = 2) and $Tb_{4.66}Si_3O_{13}$ (a = 9.493 Å, c = 6.852 Å, V = 534.70 Å^3 and **Z** = 2) were prepared by CHEN and LI [67] via spontaneous crystallization and synthesized with high purities and excellent crystallinities by the sol-gel process. Both compounds are isostructural with $P6_3/M$ space group and exhibit paramagnetic behavior down to 2 K. Owing to high Tb^{3+} ion concentrations and good transmittance in the range from 500 to 1500 nm, $Tb_5Si_2BO_{13}$ and $Tb_{4.66}Si_3O_{13}$ may be promising magneto-optical materials in the visible-near-IR range.

Both $Tb_5Si_2BO_{13}$ and $Tb_{4.66}Si_3O_{13}$ contain two distinct sites for Tb^{3+} cations, which are depicted in **Fig. 13**. Tb(2) is at the 4*f* site, which is on a threefold axis and coordinated by nine oxygen ions. However, for $Tb_{4.66}Si_3O_{13}$, the Tb(2) site is not fully occupied but leaves one of six Tb(2) positions randomly vacant while fully occupied in $Tb_5Si_2BO_{13}$. In contrast, Tb(1) at the 6*h* site is fully occupied and sevenfold coordinated in both $Tb_5Si_2BO_{13}$ and $Tb_{4.66}Si_3O_{13}$. Moreover, in $Tb_5Si_2BO_{13}$, one third of Si is disorderly occupied by B, which gives rise to extra 1/3 Tb^{3+} ion for the charge balance. The $Tb(2)O_9$ polyhedron consists of Tb(2) and nine oxygens along the c-

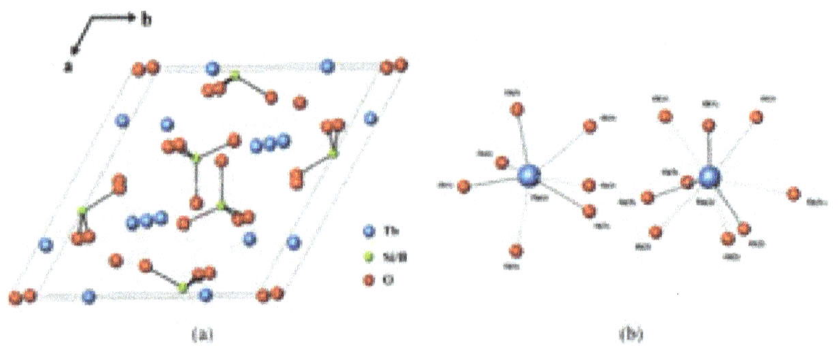

Fig. 13. Unit cell of $Tb_5Si_2BO_{13}$ (a) and coordination environment of kinds of Tb cations (b) [67].

axis. Six Tb(1) comprise a sixfold channel parallel to the c-axis. It is worth to note that the channel is considered to play an extremely important role in oxide ion conductivity [67].

The structure and optical properties of noncentrosymmetric borate $RbSr_4(BO_3)_3$ (RSBO) was described by XIA and LI [68]. RSBO can be viewed as a derivative of the apatite-like structure. Based on the anionic group approximation, the optical properties of the compound are compared to those of the structure-related apatite-like compounds with the formula "$A_5(TO_n)_3X$". When the structures of all apatite-like crystals are presented in orthorhombic unit cell, the arrangements of planar anionic BO_3 groups are all similar to one-third of the BO_3 groups aligned perfectly parallel at corner- and face-centered locations, whereas the other two-thirds of BO_3 groups are distributed differently.

Europium borate fluoride, $Eu_5(BO_3)_3F$ with an apatite-type structure, was synthesized by KAZMIERCZAK and HÖPPE [69] as single-phase crystalline powder starting from europium oxide, europium fluoride and boron oxide at 1370 K. $Eu_5(BO_3)_3F$ crystallizes in the space group PNMA.

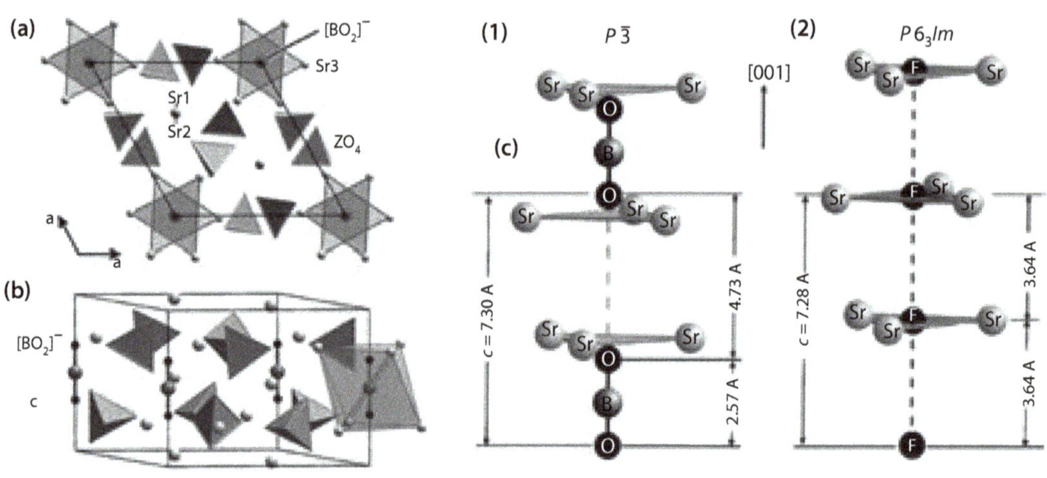

Fig. 14. The crystal structure of $Sr_{10}[(PO_4)_{5.5}(BO_4)_{0.5}](BO_2)$ (1): (a) the projection along [001] showing the channels formed by Sr3 (gray triangles) and the positions of the XO_4 tetrahedra (gray; Z = 11/12 P + 1/12B) and (b) the side view with emphasized $[BO_2]^-$ groups and the coordinating trigonal antiprism formed by Sr3. The comparison of the arrangement of $[BO_2]^-$ and F- ions within the Sr channels of $Sr_{10}(PO_4)_{5.5}(BO_4)_{0.5}(BO_2)$ (c(1)) and $Sr_{10}(PO_4)_6F_2$ (c(2)) [70].

The structure of single crystal of strontium phosphate orthoborate metaborate ($Sr_{10}[(PO_4)_{5.5}(BO_4)_{0.5}](BO_2)$) that was grown from the melt by CHEN et al [70] is shown in **Fig. 14**(a). The phase crystallizes in the space group P3 with the cell parameters $\boldsymbol{a}$ = 9.7973 Å, c = 7.3056 Å, $\boldsymbol{V}$ = 607.19 Å^3 and $\boldsymbol{Z}$ = 1. The crystal structure is closely related to apatite and contains linear metaborate groups, $[BO_2]^-$ (point group $D_{\infty h}$, B-O = 1.284 Å), taking positions within the channels running along the threefold inversion axis. Strontium sites are found to be fully occupied, while $[PO_4]^{3-}$ tetrahedra are partially replaced by$[BO_4]^{5-}$ groups.

The comparison of the nearest neighbors around $[BO_2]^-$ and F^- located within the channels is shown in **Fig. 14(b,c)**. F^- ions (0,0,1/4) are situated on the mirror plane in the center of Sr triangle. As a result, constant F…F distances of 3.64 Å (a/2) are observed along [001] (**b**). In $Sr_{10}[(PO_4)_{5.5}(BO_4)_{0.5}](BO_2)$, the incorporation of boron atoms between two O atoms draws these atoms closer (d(O-B-O) = 2.57 Å) and at the same time increases the gaps between two neighboring $[BO_2]^-$ units (d(O…O) = 4.73 Å), which results in alternating O…O distances along the c-axis (**a**) [70].

By replacing Mn in $YCa_3(MnO)_3(BO_3)_4$ with trivalent Al and Ga, two new borates with the compositions of $YCa_3(MO)_3(BO_3)_4$ (M = Al, Ga) were prepared via the solid-state reaction by YU et al [71]. The phases are isostructural to gaudefroyite with the hexagonal space group $P6_3/$M. The cell parameters of $\boldsymbol{a}$ = 10.38775 Å, $\boldsymbol{c}$ = 5.69198Å for the Al-containing compound and $\boldsymbol{a}$ = 10:5167 Å, $\boldsymbol{c}$ = 5:8146 Å for the Ga analogue were obtained from the refinements. The structure is constituted of AlO_6 or GaO_6 octahedral chains interconnected by BO_3 groups in the ***ab*** plane to form a Kagomé-type lattice[15] [72],[73],[74], leaving trigonal and apatite-like tunnels. It was found that most rare-earth and Cr, Mn ions can be substituted into the Y^{3+} and M^{3+} sites, respectively, and the preference of rare-earth ions to be located in the trigonal tunnel is correlated to the sizes of the M^{3+} ions.

Other apatite-type REE silicates

Hexagonal apatite-type phase of the composition of $Pr_9K(SiO_4)_6O_2$ (space group $P6_3$/M, a = 9.6466 Å and c = 1136 Å, V = 573.28 Å^3, ρ_{calc} = 5.48 g·cm^{-3} and Z = 1) was synthesized by WERNER AND KUBEL [75] in a potassium fluoride flux. Potassium fills one (4f) of two metal positions present in the structure (**Fig. 15**) with the occupancy factor of 25%. The remaining positions of this site (Pr2/K2) are occupied by praseodymium.

---

15 The Kagomé lattice (**d**) is one of the most interesting lattices in 2D, especially in materials in which the Kagome lattice is built from magnetic ions. Each of its vertices touches a triangle, hexagon, triangle and hexagon (the planes of corner-sharing equilateral triangles). The vertices correspond to the edges of the hexagonal (honeycomb) lattice (**c**), which in turn is the dual of triangular lattice (can be derived from triangular lattice by periodical removal of ¼ sites) (**b**). Since it has the same coordination number (z = 4), the Kagomé lattice is also related to the square lattice (**a**) [72], [73]. Numerous Kagomé compounds built from stacked Kagomé layers were found in Alunite (Jarosite, $KFe^{3+}{}_3(SO_4)_2(OH)_6$, (**e**)) family of minerals [74].

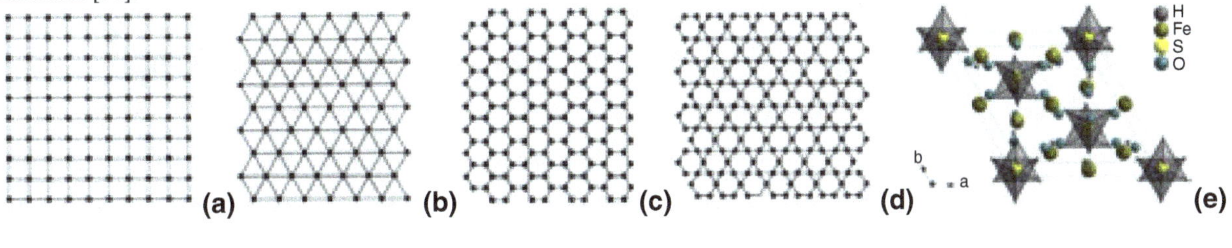

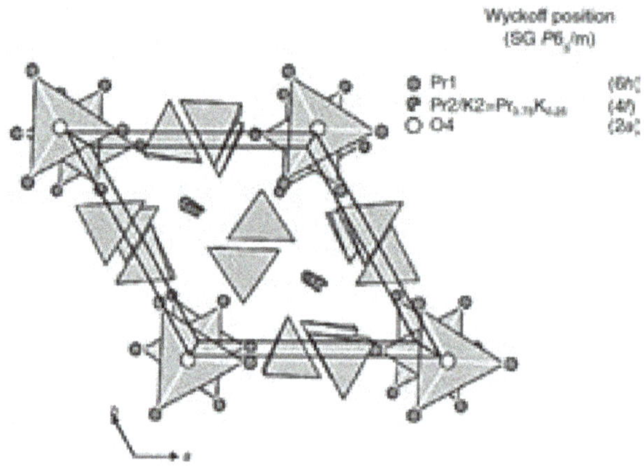

Fig. 15. The perspective view of $Pr_9K(SiO_4)_6O_2$ along the c-axis: regular planar coordination of the channel oxygen O(4) and the SiO_4 tetrahedra are marked in gray, and the unit cell is outlined [75].

Oxygen from the silicate groups forms a coordination polyhedron (ninefold) in the shape of a distorted threefold capped trigonal prism. These face sharing [(Pr2/K2)O_9]-polyhedra build up chains, which are interconnected via the SiO_4 groups. The resulting channel framework accommodates sevenfold oxygen-coordinated praseodymium (Pr1), attached to the inside of the tubes that are aligned parallel to the c-axis. Oxide ions O4, located on the longitudinal axis of the channels, exhibit anomalously high atomic displacement parameters along the c-direction [75].

Single crystals of apatite-type $Nd_{9.33}(SiO_4)_6O_2$, $Pr_{9.33}(SiO_4)_6O_2$ and $Sm_{9.33}(SiO_4)_6O_2$ were described in **Section**. The structure of samarium orthosilicate oxyapatite ($Sm_5(SiO_4)_3O$, **Fig. 16**) was resolved by Morgan et al [76]. The phase crystallizes in hexagonal system with the space group P6$_3$/M and the cell parameters: $\boldsymbol{a}$ = 9.4959 Å, $\boldsymbol{c}$ = 7.0361 Å, $\boldsymbol{c}$:$\boldsymbol{a}$ = 0.7410 and $\boldsymbol{V}$ = 549.46 Å^3. The preparation and the structure of single crystal of strontium tetrapraseodymium tris(silicate) oxide ($SrPr_4(SiO_4)_3O$), which was grown by the self-flux method using $SrCl_2$, was described by Sakakura et al [77].

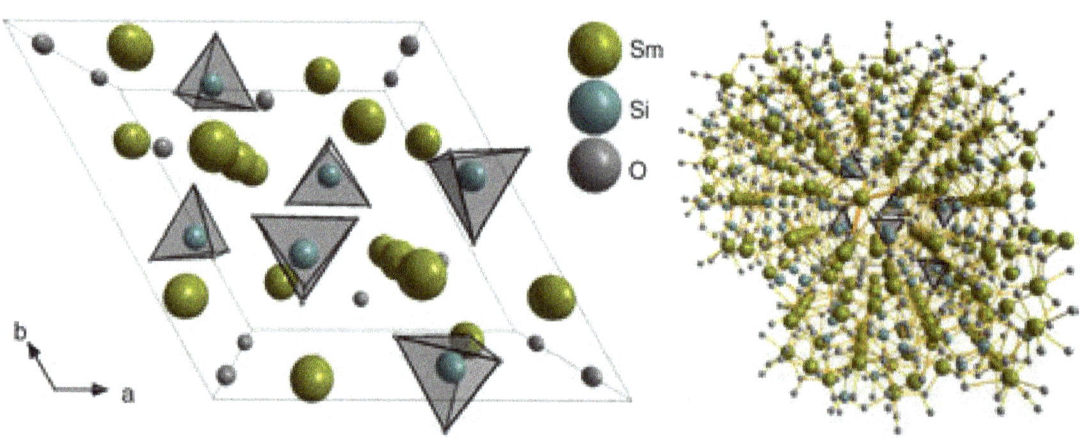

Fig. 16. The structure of $Sm_5(SiO_4)_3O$ viewed along the c-axis [76].

The M(2) sites are almost exclusively occupied by praseodymium. The complete series of apatite-like compounds $REE_{9.33}\square_{0.67}[SiO_4]_6O_2$, $LiREE_9[SiO_4]_6O_2$ and $NaREE_9[SiO_4]_6O_2$ were synthesized by FELSCHE [78] with REE: La → Lu. Apatite-type neodymium silicates doped with various cations at the Si site, $Nd_{10}Si_5BO_{27-\delta}$ (B = Mg, Al, Fe, Si), were synthesized by XIANG et al [79] via the solid-state reaction.

The crystal growth and the structure of three new neodymium-containing silicates, $Na_{0.50}Nd_{4.50}(SiO_4)_3O$, $Na_{0.63}Nd_{4.37}(SiO_4)_3O_{0.74}F_{0.26}$ and $Na_{4.74}Nd_{4.26}(O_{0.52}F_{0.48})[SiO_4]_4$, prepared using the eutectic mixture of KF/NaF were investigated by LATSHAW et al [80]. $Na_{0.50}Nd_{4.50}(SiO_4)_3O$ and $Na_{0.63}Nd_{4.37}(SiO_4)_3O_{0.74}F_{0.26}$ adopt the apatite structure and crystallize in hexagonal space group P6$_3$/M, while $Na_{4.74}Nd_{4.26}(O_{0.52}F_{0.48})[SiO_4]_4$ crystallizes in tetragonal space group I-4 and exhibits rare-earth mixing on the sodium site. The unit cell parameters of the crystals are:

1. $Na_{0.50}Nd_{4.50}(SiO_4)_3O$: $a = 9.5400$ Å and $c = 7.033$ Å;
2. $Na_{0.63}Nd_{4.37}(SiO_4)_3O_{0.74}F_{0.26}$: $a = 9.5533$ Å and $c = 7.0510$ Å;
3. $Na_{4.74}Nd_{4.26}(O_{0.52}F_{0.48})[SiO_4]_4$: $a = 12.1255$ Å and $c = 5.4656$ Å.

Double REE silicate $Gd_{4.33}Ho_{4.33}(SiO_4)_6(OH)_2$ with the hydroxylapatite structure was synthesized by WANG et al [81] using the piston-cylinder high-pressure apparatus at the pressure of 2.0 GPa and the temperature of 1450°C. Since they have nearly identical chemical character, two REE cations (Ho and Gd) are distributed randomly among the M(1) and M(2) sites, and the charge balance is maintained by the cation vacancies in M(1). The presence of two different REE cations in the same compound might promote better understanding of the cooperative effects of ions under the solid-state conditions (**Fig. 17**).

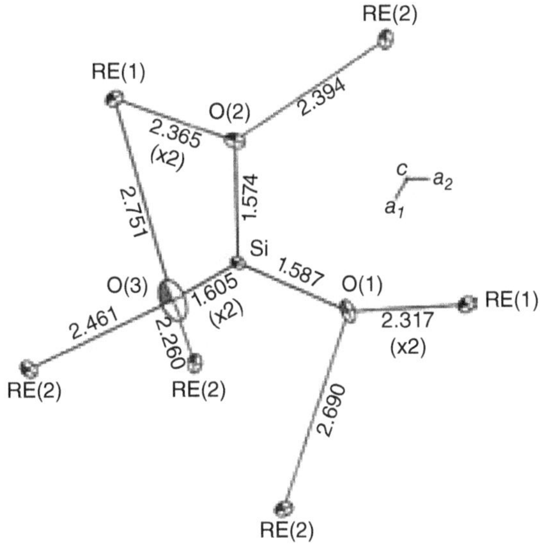

Fig. 17. Bond distances (Å) and anisotropic displacement in $Gd_{4.33}Ho_{4.33}(SiO_4)_6(OH)_2$: note that the exaggerated anisotropic displacement of O(3) is attributable to the high proportion of vacancies in REE(1) and near-equatorial distribution of strong bonds to Si and REE(2) [81].

The structure analysis reveals that the hexagonal compound crystallizes in usual apatite space group $P6_3/M$ with lattice parameters $\boldsymbol{a}$ = 9.3142 Å and $\boldsymbol{c}$ = 6.7010 Å. REE(1) atoms are connected to nine oxygen atoms with the REE(1)-O bond distances ranging from 2.317 to 2.751 Å (mean 2.478 Å) and REE(2) atoms are connected to seven oxygen atoms with the RE(2)-O bond distances ranging from 2.223 to 2.690 Å (mean 2.393 Å). Oxygen anion O(4) in the apatite channel is located on the 6_3 axis and coordinated with three REE(2) cations arranged in a tricluster perpendicular to the c-axis. An isotropic displacement parameter was used for O(4), and H atom was assumed to ride on it. OH^- anions are stacked in regular column in the apatite channel, and in locally ordered structure, their polar direction is flipped in neighboring channel [81].

The formation of apatite-type phases of the composition of $KNd_6(SiO_4)_6O_2$ from the glass precursor ($4K_2O$-Nd_2O_3-$17SiO_2$) during the hydrothermal experiments (500°C and 825 bar) carried out at KOH molarities of 6 or greater was reported by Haile et al [82]. High temperatures, high pressures and long times tended to favor the synthesis of this apatite-type phase over the $K_8Nd_3Si_{12}O_{32}OH$ phase. In comparison with the potassium system, the concentration of NaOH required for the synthesis of $NaNd_6(SiO_4)_6O_2$ phase (system $4Na_2O$-M_2O_3-$17SiO_2$, where M = Nd and Y) is very low. The formation of apatite-type phases in the $4Na_2O$-Y_2O_3-$17SiO_2$ system was not observed [83].

Two series of strontium-lanthanum apatites, $Sr_{10-x}La_x(PO_4)_{6-x}(SiO_4)_xF_2$ and $Sr_{10-x}La_x(PO_4)_{6-x}(SiO_4)_xO$ with $0 \le x \le 6$, were synthesized by Boughzala et al [84] via the solid-state reaction in the temperature range of 1200 – 1400°C:

$$\begin{aligned} &3\ SrCO_3 + (6-x)/2\ Sr_2P_2O_7 + SrF_2 + x\ SiO_2 + x\ La(OH)_3 \rightarrow \\ &Sr_{10-x}La_x(PO_4)_{6-x}(SiO_4)_x F_2 + 3\ CO_2 + 3x/2\ H_2O \end{aligned} \tag{9}$$

and

$$\begin{aligned} &4\ SrCO_3 + (6-x)/2\ Sr_2P_2O_7 + x\ SiO_2 + x\ La(OH)_3 \rightarrow \\ &Sr_{10-x}La_x(PO_4)_{6-x}(SiO_4)_x O + 4\ CO_2 + 3x/2\ H_2O \end{aligned} \tag{10}$$

where x = 0, 1, 2, 4 and 6. $Sr_2P_2O_7$ was synthesized by the following reaction at 900°C:

$$2\ SrCO_3 + 2\ (NH_4)_2HPO_4 \rightarrow Sr_2P_2O_7 + 2\ CO_2 + 3\ H_2O + 4\ NH_3 \tag{11}$$

The raw meal was prepared via mixing $SrCO_3$, La_2O_3, SiO_2, SrF_2 and $(NH_4)_2HPO_4$ in required stoichiometric amounts ($0 \le x \le 6$). The mixture was ground in an agate mortar, pressed to pellets and calcined at the temperature of 900°C for 12 h under the flow of argon ($Sr_{10-x}La_x(PO_4)_{6-x}(SiO_4)_xF_2$) and oxygen ($Sr_{10-x}La_x(PO_4)_{6-x}(SiO_4)_xO$). The product was ground and pressed again in order to improve its homogeneity. Next, thermal treatment was performed at the temperature of 1200 and 1400°C (depending on the content of SiO_2) for 12 h. The samples were heated and cooled with the rate of 10°C·min^{-1}. The incorporation of La^{3+} and SiO_4^{4-} ions into the apatite structures, i.e. the substitution of the pair La^{3+} and SiO_4^{4-} for Sr^{2+} and PO_4^{3-}, induced an increase of parameter ***a*** and decrease of parameter ***c*** (**Fig. 18**) [84].

The formation of nanocrystalline Ce-Yb mixed silicate-type oxyapatite of the composition of $Yb_yCe_{9.33-y}(SiO_4)_6O_2$ via the solid-state synthesis was described by MAŁECKA and KĘPIŃSKI [85]. The phase was identified as an intermediate formed during the synthesis of Ce-Yb silicates.

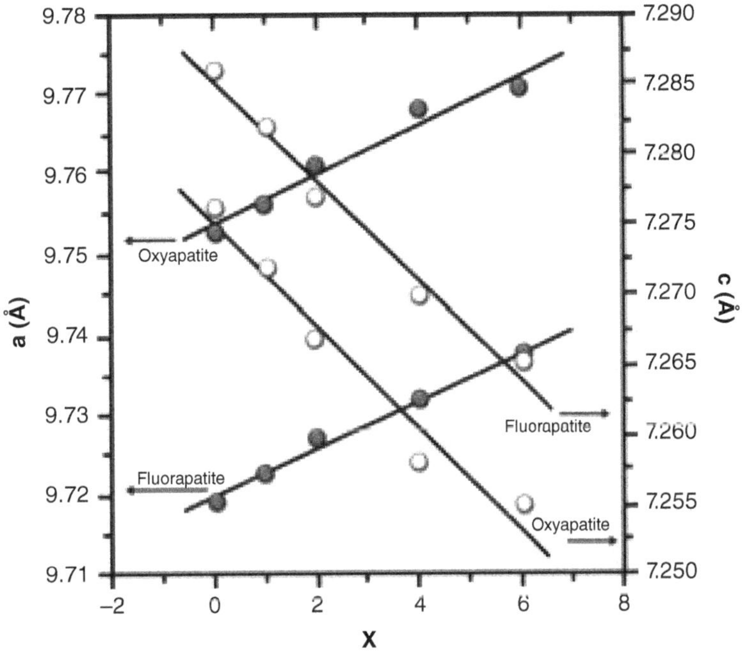

Fig. 18. The variation of *a* and *c* parameters of $Sr_{10-x}La_x(PO_4)_{6-x}(SiO_4)_xF_2$ and $Sr_{10-x}La_x(PO_4)_{6-x}(SiO_4)_xO$ phases with the value of *x* [84].

Different compositions of apatite-type $La_{10}Si_{6-x}W_xO_{27+\delta}$ ceramics were prepared successfully by XIANG et al [86] via the high-temperature solid-state reaction route. Doping with W^{6+} is beneficial to the removal of La_2SiO_5 impurity phase. When the doping content of W^{6+} is more than 0.1, the rod-like grains of $La_{10}Si_{6-x}W_xO_{27+\delta}$ ceramics are replaced gradually by equiaxed apatite-type grains, and randomly shaped convex $La_6W_2O_{15}$ particles appear at the grain boundaries. While doping with Nb^{5+} leads to the hexagonal-phase $La_{9.5}Ge_{5.5}Nb_{0.5}O_{26.5}$, the addition of MO^{6+} leads to the compound $La_{9.5}Ge_{5.5}Mo_{0.5}O_{26.75}$ with triclinic symmetry [87].

Apatite-type yttrium silicates

Yttrium silicates

The formation of the phase with the composition ($Y_4Si_3O_{12}$, $Y_4(SiO_4)_3$ or $2Y_2O_3{\cdot}3SiO_2$ [88]), which is stable between 1650 and 1950°C, was reported by TOROPOV and BONDAR [89] in the binary system of Y_2O_3-SiO_2 and by TOPOROV and FEDOROV [90] in the ternary system of CaO-Y_2O_3-SiO_2

(**Fig. 19(a)**). This was the first reported occurrence of the phase with the composition between yttrium orthosilicate (Y_2SiO_5, oxyorthodsilicate, YSO, Y_2O_3:SiO_2 = 1:1 [91]) and yttrium disilicate $Y_2Si_2O_7$ (yttrium pyrosilicate, YPS, 1:3). The structure of this phase was described as the garnet type [88].

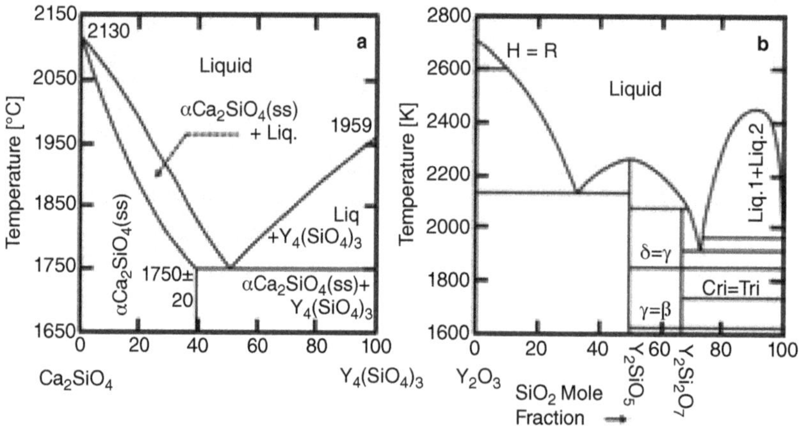

Fig. 19. Phase equilibrium in the system Ca_2SiO_4-$Y_4(SiO_4)_3$ according to Toporov and Fedorov [90] and Y_2O_3-SiO_2 phase diagram [92].

Since then, authors have disagreed about the existence of such a phase because the attempts to make it starting with yttria and silica powders resulted in the formation of only Y_2SiO_5 and $Y_2Si_2O_7$ [93]. This phase was not reported either by other studies of Y_2O_3-SiO_2 system [92],[94],[95], which contains two compounds. Y_2SiO_5 and$Y_2Si_2O_7$ were found, with two (A and B) and five (*y*, α, β, γ and δ, also called *y*, B, C, D and E [96]) polymorphs, respectively. The first has a congruent melting, whereas the second has an incongruent one (**Fig. 19(b)**).

Nevertheless, the formation of oxyapatite phase of the composition of $Y_{4.67}\square_{0.33}(SiO_4)_3O$ (7:9) prepared via the oxidation of nitrogen apatite $Y_5(SiO_4)_3N$ was reported by other authors [7],[97]. The apatite-like phase $Y_{4.67}(SiO_4)_3O$ possesses hexagonal structure with the space group $P6_3/M$, $a = 9.368$ and $c = 6.735$ Å [78],[98]. The specific gravity of the phase is 4.39 g·cm^{-3} and the hardness on the Mohs scale is 5 – 7 [99].

Since the structure of YSO containing two different types of anions includes the $[SiO_4]^{4-}$ complex ion and an additional non-silicon-bonded oxygen ion (NBO), it could be written as $Y_2(SiO_4)O$. This compound also displays two structure types of monoclinic symmetry with different linking of O-Y_4 tetrahedra. Low-temperature X_1 phase and high-temperature X_2 phase belong to the space groups of $P2_1/c$ (Z = 2) and C2/c (Z = 8), respectively [91].

The samples of the composition of $Y_4(SiO_4)_3$, and similar ones containing small amount of iron oxide, corresponding to an overall composition of $Fe_{0.2}Y_4(SiO_4)_3O_{0.2}$, were produced by the mixed powder method and by the sol-gel route using yttrium nitrate ($Y(NO_3)_3 \cdot 5H_2O$), TEOS (tetraethylorthosilicate) and iron nitrate ($Fe(NO_3)_3 \cdot 9H_2O$) by Parmentier et al [7]. Nitrate was dissolved in ethanol/water mixture (volume ratio 7:3), the amount of the latter being controlled to give final Si concentration. Iron nitrate was added at this stage in calculated amounts corresponding to the final iron-doped apatite composition. The solution was stirred for a few hours and TEOS was added to give the appropriate silicon content and then the solution was placed in an oven at 60°C until the gelation occurred. The gel was dried at 80°C and calcined at 600°C for 1 h.

Powders prepared by the two routes were uniaxially pressed into pellets and treated to temperatures up to 1650°C in air in a Pt crucible, or for the heat treatments at 1700°C, carbon element furnace was used, and the samples were heated in a BN-lined crucible in nitrogen atmosphere. Iron appears to have two roles depending on the temperature; it stabilizes the apatite phase at high temperatures when produced by the sol-gel route and catalyzes the decomposition of sol-gel-derived apatite at low temperatures [7].

A new phase of yttrium magnesium silicate having the apatite structure was prepared by SUWA et al [100] at 1500°C in air. Its chemical composition can vary from $(Y_4Mg)Si_3O_{13}$ to $(Y_{4.33}Mg_{1.13})Si_{3.34}O_{13}$. The hexagonal unit cell dimensions $\boldsymbol{a}_0$ and $\boldsymbol{c}_0$ of $(Y_4Mg)Si_3O_{13}$ are 9.298 ± 0.002A and 6.635 ± 0.001A, respectively, and its axial ratio c/a is 0.714. It is optically uniaxial negative with $\varepsilon = 1.810 \pm 0.005$, $\omega = 1.820 \pm 0.005$ and $\omega - \varepsilon = 0.010$. The cleavages parallel and perpendicular to the c-axis were recognized. The formation of the apatite-type phase of the composition of $NaY_9Si_6O_{26}$ in the ternary system Na_2O-Y_2O_3-SiO_2 was also reported by LEE et al [101].

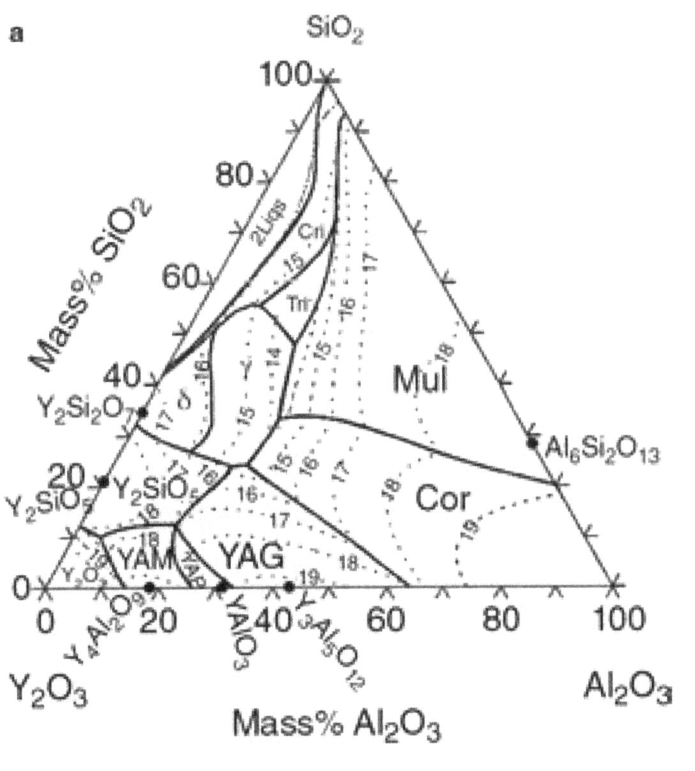

Fig. 20. Calculated liquidus surface of the Y_2O_3-Al_2O_3-SiO_2 system: three-phase equilibria with liquid phase (thick lines), liquidus surfaces for various solids (labeled area) and isothermal section (dotted lines, temperature in hundreds °C) [92].

The phase diagram of Al_2O_3-SiO_2-La_2O_3 system (**Fig. 20**) can be compared with the Y_2O_3-Al_2O_3-SiO_2 ternary diagram examined by BONDAR and GALAKHOV in 1964 [102]. The latter represents the only other example of REE_2O_3-Al_2O_3-SiO_2 phase diagram found so far in the literature. The authors identified the liquidus surface of the whole ternary field, but they failed to elucidate the subsolidus phase relationships among different binary compounds. Due to the much smaller ionic size of Y^{3+} ion with respect to La^{3+} ion (1.18 and 1.015 Å for the eightfold coordination, respectively [103]) and lower bond-valence parameter (2.019 and 2.172 Å [104]), the stability of the binary compositions is substantially altered [57].

The β-alumina-like phase $LaAl_{11}O_{18}$ is no longer stable, while the garnet-like phase $Y_3Al_5O_{18}$ and $Y_4Al_2O_9$ monoclinic compound exist. The lacunar apatite-like phase $Y_{14}Si_9O_{39}$ reported by Wills et al [105] does not appear in the Y_2O_3-Al_2O_3-SiO_2 ternary diagram; however, a compound with similar Y/Si atomic ratio, namely $Y_4Si_3O_{12}$, also reported by Wills et al [105], does. Since the formation of $Y_4Si_3O_{12}$ phase was not confirmed, it may be stabilized by impurities [57],[92].

AM and AEE-yttrium orthosilicate oxyapatites

Alkaline metals (AM) and alkaline-earth element oxyapatites (oxybritholites) are described in this chapter. Phosphate minerals of the apatite supergroup possess strong affinity for strontium [106]. The apatite-type phase of the composition of $Sr_2Y_8Si_6O_{26}$ ($Sr_2Y_8(SiO_4)_6O_2$) was prepared by Zuev et al [107] via the two-stage calcination of mixture of $SrCO_3$, Y_2O_3, Eu_2O_3 and SiO_2 in air in order to investigate the spectral characteristics of $Sr_2Y_8(SiO_4)_6O_2$:Eu polycrystals. The structure of apatite phase is shown in **Fig. 21**. The sol-gel synthesis and the characterization of $Sr_2Y_{8(1-x)}Eu_{8x}Si_6O_{26}$ solid solution doped by Eu, where $x = 0.01 - 0.4$, was described by Karpov and Zuev [108]. The formation of calcium analogue ($Ca_2Y_8(SiO_4)_6O_2$, calcium-yttrium-silicate oxyapatite) was observed during the crystallization of SiO_2-Al_2O_3-CaO-Na_2O-K_2O-F-Y_2O_3 glass ceramics [109] and during the degradation of advanced environmental barrier coatings [110]. The phase $Ca_4Y_6(SiO_4)_6O_2$ was often prepared in order to investigate its luminescent properties [111],[112].

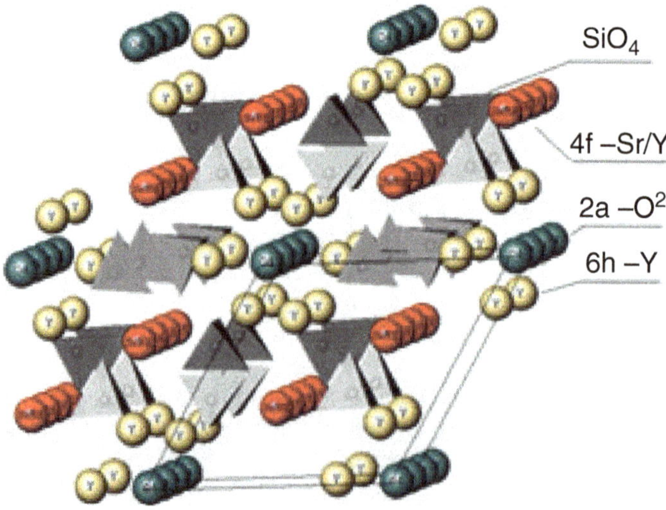

Fig. 21. The structure of $Sr_2Y_8(SiO_4)_6O_2$ [107].

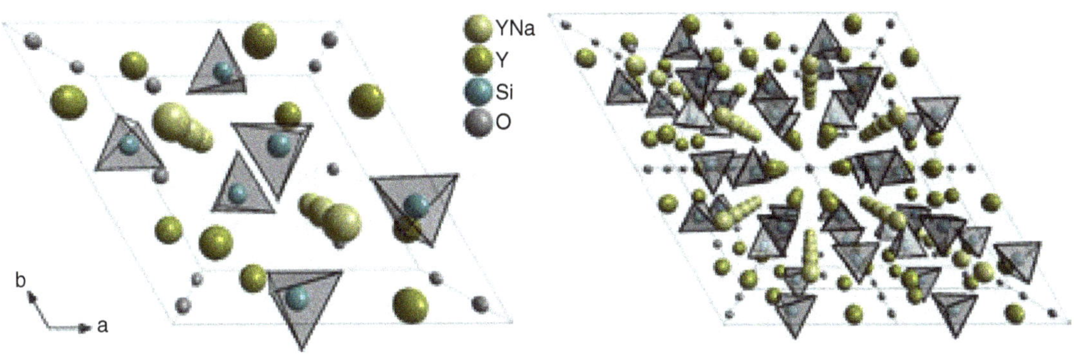

Fig. 22. The structure of $NaY_9(SiO_4)_6O_2$ oxyapatite (the perspective view along the c-axis) [114].

The precipitation of $NaY_9(SiO_4)_6O_2$ apatite-type compound (sodium nonayttrium hexakis(silicate) dioxide) in the SiO_2-B_2O_3-Al_2O_3-Y_2O_3-CaO-Na_2O-K_2O-F glass-ceramics system (**Section**) was described by VAN'T HOEN et al [113]. The hexagonal structure of $NaY_9(SiO_4)_6O_2$ oxyapatite (**Fig. 22**) was resolved by GUNAWARDANE et al [114]. The phase crystallizes in $P6_3/M$ space group with the cell parameters $\boldsymbol{a}$ = 9.334, $\boldsymbol{c}$ = 6.759 Å, $\boldsymbol{c}$:$\boldsymbol{a}$ = 0.7241, V = 509.97 Å and **Z** = 1).

Lithium yttrium orthosilicate ($LiY_9(SiO_4)_6O_2$, lithium nonayttrium hexakis(silicate) dioxide) crystallizes in centrosymmetric space group $P6_3/M$. The structure closely resembles those of fluorine apatite. There are two different crystallographic sites for Y^{3+} ion, which are coordinated by seven and nine O atoms. One-fourth of the nine-coordinated site is occupied by Li atoms, thus maintaining the charge balance. Si atom occupies the tetrahedral site [115].

The preparation, the properties and the effect of sintering additives of hexagonal ($P6_3/M$) strontium-yttrate-silicate oxyapatite (oxybritholite2) with the composition of $SrY_4(SiO_4)_3O$ as the main product of sinter-crystallization process, in which the non-equilibrium melt was formed in the temperature interval from 1300 to 1550°C in the SrO-Y_2O_3-SiO_2 system, was described by PTÁČEK et al [116]. The formation of non-equilibrium melt is facilitated by borate fluxes, alkaline fluxes and talc. The apparent activation energy and the frequency factor of the sinter-crystallization process were determined to be 1525 kJ mol^{-1} and $1.04 \cdot 10^{45}$ s^{-1}, respectively. The material shows low value of linear thermal expansion coefficient of $(1.1 \pm 0.1) \cdot 10^{-6}$°$C^{-1}$ in the temperature range from 25 to 850°C.

The course of synthesis can be expressed by the following reaction formula [116]:

$$AEECO_3 + 2\ Y_2O_3 + 3\ SiO_2 \rightarrow AEE\ \ Y_4\left[SiO_4\right]_3 O + CO_2(g) \tag{12}$$

This **reaction 12** is too general to describe formed intermediates ($SrSiO_3$, Sr_2SiO_4, SrY_2O_4, …[16]) and the process of sinter-crystallization of apatite:

[16] Detailed description of formed intermediates can be found in work [116].

$$\begin{aligned}&AEECO_3 + 2\ Y_2O_3 + 3\ SiO_2 \rightarrow \text{intermediates} \rightarrow \\ &\text{non-equilibrium melt} \rightarrow AEE\ Y_4[SiO_4]_3O + CO_2(g)\end{aligned} \tag{13}$$

Since the formation of $SrY_4(SiO_4)_3O$ proceeds thorough non-equilibrium melt phase, the effect of sintering additives such as borate fluxes, fluorides and carbonates of alkaline metals as well as talc was investigated. Sintering additives facilitate the formation of melt phase and increase the length of sinter-crystallization interval. The expansion after the thermal decomposition of strontium carbonate is reduced as well. Calcinate, treated to the temperature lower than the temperatures of sinter-crystallization interval, has hydraulic activity. Therefore, it can be applied in special composite cements as an activator for latent hydraulic and pozzolanic materials.

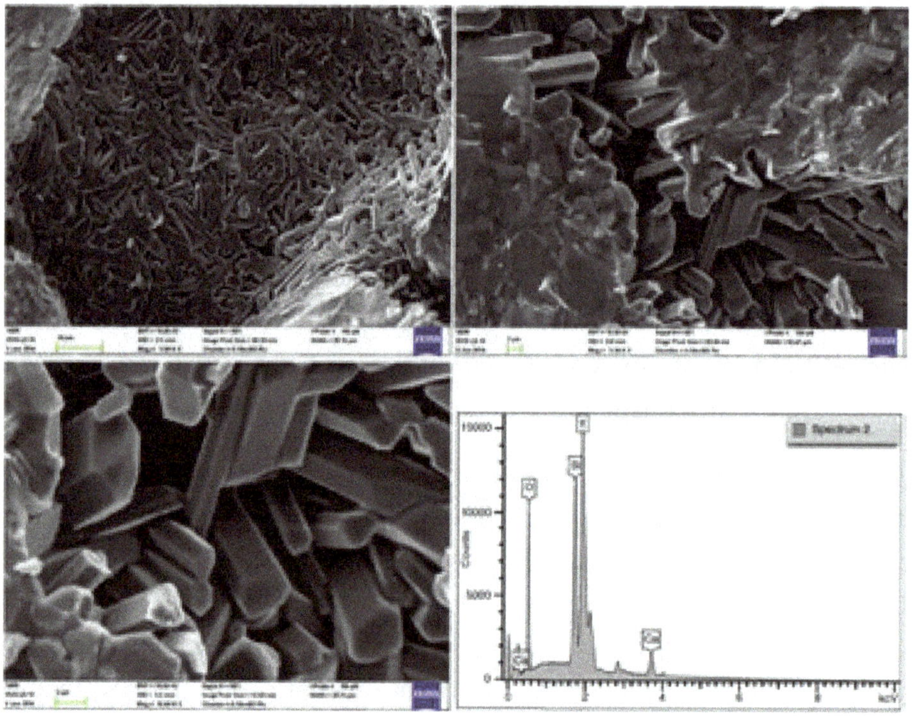

Fig. 23. SEM picture and WDX analysis of hexagonal crystal of $CaY_4(SiO_4)_3O$ apatite phase.

Effect on sinter-crystallization process	Decreasing intensity or temperature of effect →
Expansion before sinter-crystallization	Pure >> NaF >> Talc ≈ Li_2CO_3 ≈ $Li_2B_4O_7$ >> Na_2CO_3 > $LiBO_2$ > K_2CO_3
Firing shrinkage (sample treated to 1600°C)	$LiBO_2$ > $Li_2B_4O_7$ ≈ Pure > NaF > K_2CO_3 ≈ Na_2CO_3 > Talc > Li_2CO_3
Initial temperature of sinter-crystallization	Pure ≈ Na_2CO_3 ≈ Talc > $Li_2B_4O_7$ > K_2CO_3 ≈ NaF > $LiBO_2$ > Li_2CO_3
Maximum rate of sinter-crystallization	Li_2CO_3 > Pure > Li_2B4O_7 ≈ Talc ≈ NaF ≈ K_2CO_3 ≈ Na_2CO_3 > $LiBO_2$
Length of interval of sinter-crystallization	$LiBO_2$ >> $Li_2B_4O_7$ ≈ NaF > Li_2CO_3 > pure > Talc > Na_2CO_3 > K_2CO_3

Table 3. The influence of sintering additive on the behavior during thermal treatment [116].

After the process of sinter-crystallization, the reactivity of glassy phase with water drops. A significant benefit of talc is the fact that the glassy phase surrounding the crystals of apatite phase becomes resistant against the influence of water with this sintering additive. Furthermore, magnesium is not being incorporated into the structure of apatite phase during the crystallization of $SrY_4(SiO_4)_3O$ from non-equilibrium melt. The influence of sintering additives on the behavior during the thermal treatment is summarized in **Table 3** [116].

The important feature of this compound is the formation of colored center after the exposition to X-ray radiation (**Fig. 24**); hence, the prepared material is an important candidate for optical applications, sensors and dosimeters.

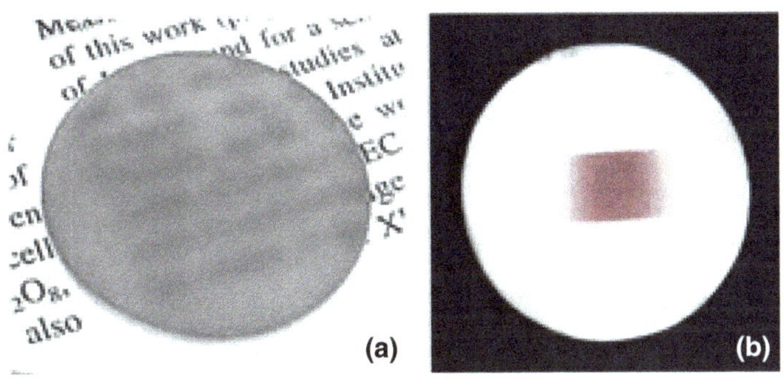

Fig. 24. Semitransparent disc of $SrY_4(SiO_3)_4O$ sintered specimen (a) and formation of colored center on X-ray irradiated area (Cu(Ka), 40 kV and 30 mA) [116].

On the other hand, this reaction also indicates that the synthesis of individual apatite analogues ($AEEY_4(SiO_4)_3O$, where AEE = Ca, Sr and Ba) and their solid solutions proceeds via similar ical intermediates formed in the temperature range, which is affected by the thermal stability of AEE carbonates that increases in the order: $CaCO_3$, $SrCO_3$ and $BaCO_3$.

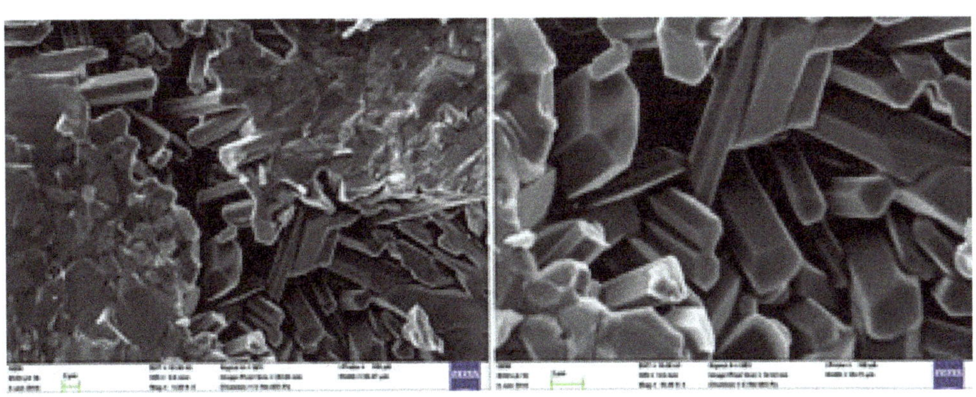

Fig. 25. SEM picture of hexagonal crystal of $CaY_4[SiO_4]_3O$ apatite phase.

While the synthesis of $CaY_4(SiO_4)_3O$ leads to well-developed hexagonal crystals (**Fig. 25**), the attempts for the preparation of $BaY_4(SiO_4)_3O$ phase were not successful. This synthesis leads to well-developed crystals of yttrium orthosilicate (Y_2SiO_5) surrounded by BaO-Y_2O_3-SiO_2 glassy phase (**Fig. 26**).

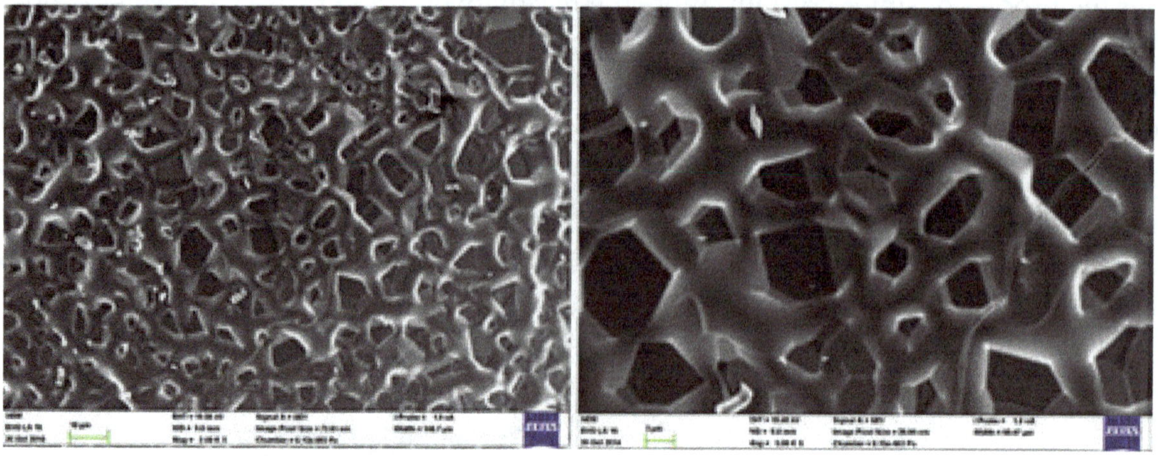

Fig. 26. SEM of Y_2SiO_5 crystal surrounded by barium-containing glassy phase.

The investigation of this system leads to the conclusion that non-limited $Ca^{2+} \leftrightarrow Sr^{2+}$ substitution can be performed in the binary system of (Ca-Sr)$Y_4[SiO_4]_3O$. On the contrary, the $BaY_4[SiO_4]_3O$ analogue of $CaY_4[SiO_4]_3O$ and $SrY_4[SiO_4]_3O$ apatite cannot be prepared; therefore, the extent of $Ca^{2+} \leftrightarrow Ba^{2+}$ and $Sr^{2+} \leftrightarrow Ba^{2+}$ substitutions is limited to 28 ± 4 and 38 ± 4%, respectively. The field of ternary solid solutions in the AEE$Y_4[SiO_4]_3O$ system, where AEE = Ca, Sr and Ba, is shown in **Fig. 27**.

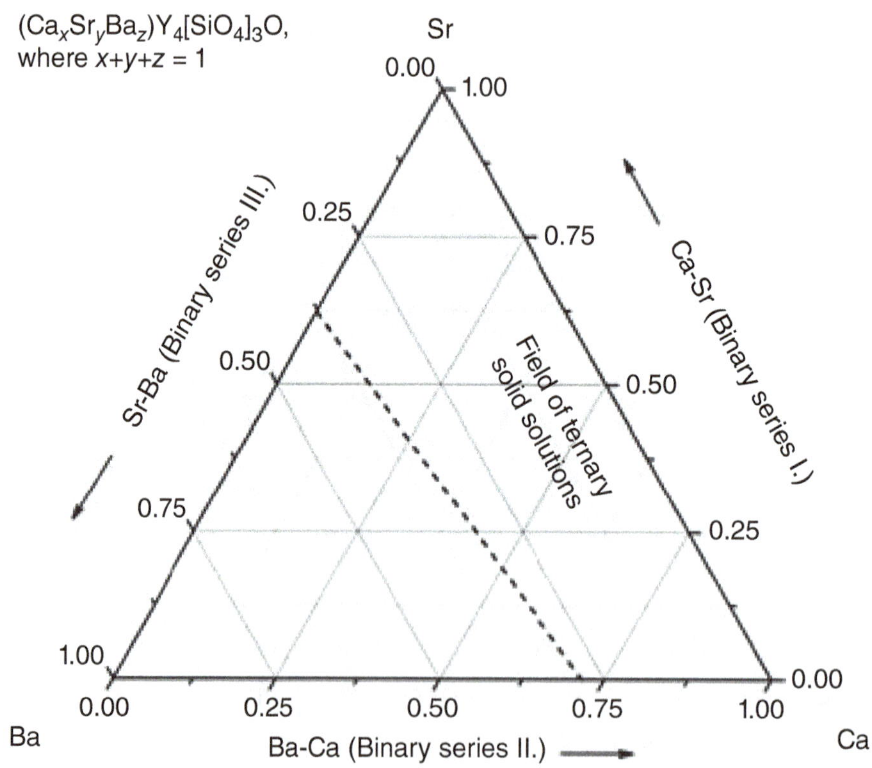

Fig. 27. The miscibility in the AEE $Y_4(SiO_4)_3O$ system.

N-apatite

The main secondary phases in Jänecke prism[17] [117] for Si_3N_4-Al_2O_3-SiO_2-Y_2O_3-YN-AlN system[18] are shown in **Fig. 28**. The formal exchange of oxygen by nitrogen leads to the compounds of N-apatite ($Y_{10}(SiO_4)_6N_2$, H-phase[19]), N-melilite[20] [118] ($Y_2Si_3O_3N_4$, M-phase), N-wollastonite ($YSiON_2$, K-phase) and N-woehlerite ($Y_4Si_2O_7N_2$, J-phase). The latter one forms a complete solid solution with $Y_4Al_2O_9$ (YAM) of the composition of $Y_4Si_{2-x}Al_xO_{7+x}N_{2-x}$ (Jss-phase) [119],[120], [121],[122],[123],[124].

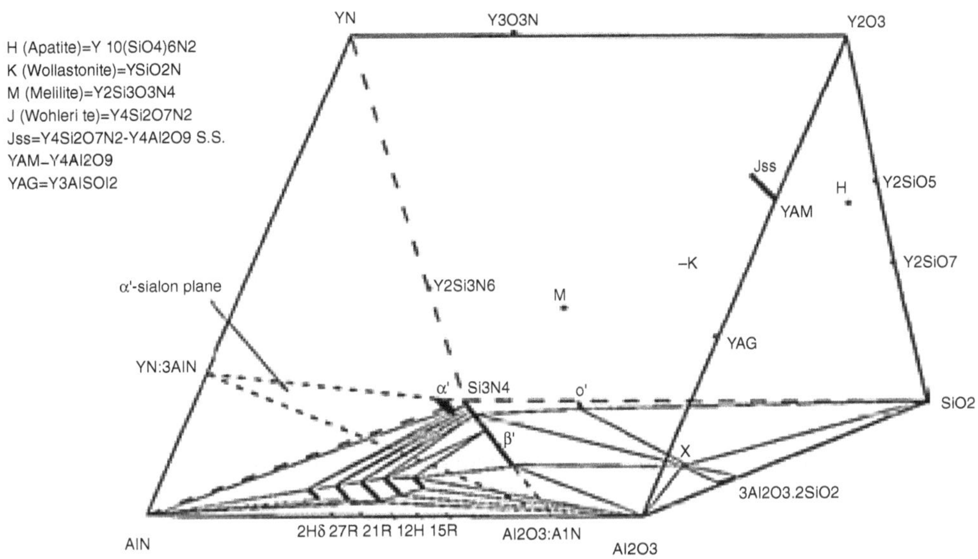

Fig. 28. Representation of the Si_3N_4-Al_2O_3-SiO_2-Y_2O_3-YN-AlN phase equilibrium.

$Y_{10}(SiO_4)_6N_2$ (N-apatite, H-phase, $(Y,Si,\square)_{10}[Si(O,N)_4]_6(O,N,\square)_2$ [125], silicon-yttrium oxynitride) was first identified by Rae et al [126] as a compound with the compositional mixture of $10Y_2O_3{\cdot}9SiO_2{\cdot}Si_3N_4$ that was stable up to 1750°C. There were other suggested compositions, such as $Y_{10}Si_7O_{23}N_4$ [127]. Later work by Gauckler et al [128] established N-apatite as a stoichiometric compound with the formula unit of $Y_{10}[SiO_4]_6N_2$ and the apatite structure (space group $P6_3/M$). The lattice constants of the hexagonal cell were reported to be $\boldsymbol{a} = 9.638$

[17] Jänecke prism is used to visualize the phase relationships among α-sialon, β-sialon and other phases in the M-Si-Al-O-N system. α- and β-sialons are isostructural with α- and β-Si_3N_4, respectively. The substitution of Al-O for Si-N in β-Si_3N_4 yields β-sialon with general formula: $Si_{6-x}Al_zO_zN_{8-z}$ $(0 < z < 4.2)$. The structure is built up by Si and Al tetrahedra coordinated with oxygen and nitrogen. The unit cell contains two Si_3N_4 units. α-sialons are solid solutions based on the α-Si_3N_4 structure, with the general formula: $M^{p+}{}_xSi_{12-(m+n)}Al_{(m+n)}O_nN_{16-n}$, where **M** is metal ion such as Li, Ca, Ba, Y and RE with a valence of **p+** and index **m** = **px** [117].

[18] Yttria is often used additive to improve the sintering behavior of Si_3N_4 [124].

[19] In dependence on the system composition, the general composition of H-phase (N-apatite) can be written as $(M,REE)_{10}(SiO_4)_6N_2$. The specification of cations then leads to the names such as Mg-Nd-N-apatite [123].

[20] The melilite-type structure (tetragonal mineral melilite $((Ca,Na)_2(Al,Mg,Fe^{2+})(Si,Al)_2O_7))$ is sorosilicate from the group of melilite, first described (Capo di Bove, near Rome, Italy) in 1976 and named from the Greek words *meli* "honey" and *lithos* "stone"). $Y_2Si_3O_3N_4$ was described by Fang et al [118]. N atoms fully occupy the bridging site (2*c*) and O atoms fully occupy the terminal site (4*e*) with 2 O and 6 N atoms at the bridging 8*f* site. The preferential distribution of O and N atoms at the 8*f* site results in two different local coordinations of Y and three different types of Si atoms.

Å and c = 6.355 Å. The electronic structure and bonding of the complex ceramic crystal $Y_{10}(SiO_4)_6N_2$ was studied by CHING et al [121]. This crystal is an insulator with direct band gap of 1.3 eV. It has some unique properties related to one-dimensional chain structure in the c-direction and planar N-Y bonding in the x-y plane.

The ternary phase diagrams of the Si_3N_4-Y_2O_3-SiO_2 [123] and Si_3N_4-La_2O_3-SiO_2 systems [129] are shown in **Fig. 29(a)** and (**b**). The apatite phase is able to form various solid solutions that may influence the development of strength in silicon nitride densified by yttria [130].

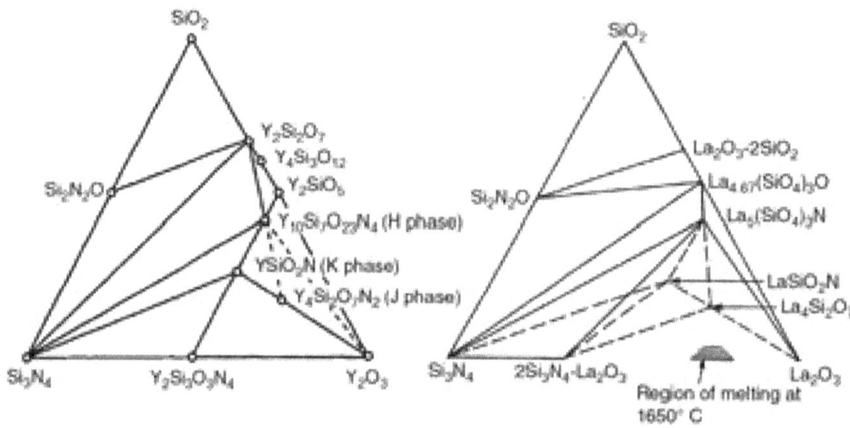

Fig. 29. Phase relationships in the systems Si_3N_4-SiO_2-Y_2O_3 [123] (a) and Si_3N_4-SiO_2-La_2O_3 at 1700°C and 1550°C (dashed lines) [129] (b).

The hexagonal lanthanum N-apatite phase of the composition of $La_5(SiO_4)_3N$ (isostructural with apatite) can be prepared from the mixture of La_2O_3 and Si powder sintered at temperatures in the range from 900 to 1200°C under the flow of nitrogen. The melting temperature of this phase was determined to be ~1600°C. It was observed that continuous heating and addition of Pd into the reaction mixture favored the formation of $La_5(SiO_4)_3N$. Prolonged heating of this compound yields $La_{4.67}(SiO_4)_3O$ [129],[131],[132],[133],[134],[135]. The absorption bands observed in infrared spectrum of lanthanum oxynitrides are introduced in **Table 4**.

Wavenumber [cm^{-1}]	Mode	Frequency [cm^{-1}]	[cm^{-1}]
225	δ (Si-O) (A_1)	730	ν (Si-N)
270	Si-O_2	800	SiO_2
327	δ (Si-O) (B_2)	840	Si-N
337	Si-O-Si (A_1)	872	SiO_4 (ν_1)
380	SiO_4 (ν_4)	909	ν (Si-N) vs
376 – 385	δ (Si-N) (sh)	930	Si-O-N_x
396	SiO_4 (ν_4)	940	ν Si-N vs
432	δ (Si-N) (sh)	960	
448	Si-N s or O-Si-O bend.	905	

Wavenumber [cm^{-1}]	Mode	Frequency [cm^{-1}]	[cm^{-1}]
462	SiO_4 (ν_4)	934	SiO_4 (s) (ν_3)
475		978	
490	ν Si-O (A_1)	980	ν Si-N (s)
542	ν Si-N (s)	1060	ν Si-O (ν_3)
600	ν Si-O	1090	ν Si-N (sh)
648	ν Si-O (w)	1130	ν Si-O (B_1)
679	ν Si-O m (B_2)	—	—

Table 4. The absorption bands observed in infrared spectrum of lanthanum oxynitrides [131].

Mitomo et al [129] used the mixture of Si_3N_4 and La_2O_3 powder heated in a 10 mm diameter carbon die using a hot-pressing apparatus. In some compositions and at temperatures above 1600°C, liquid phases were squeezed out of the die by applied pressure resulting in a change in the overall composition. The compacted powder was therefore heated at the pressure of 20 MPa up to 1400°C and the temperature was then raised to 1700°C without the pressure. The specimen was kept at 1700°C for 1 h and then allowed to cool.

The sintering temperature of Si_3N_4 with La_2O_3 additions is 1700 to 1800°C. Heating of powder mixture of various Si_3N_4/L_2O_3 ratios at 1700°C results in the formation of $2Si_3N_4{\cdot}La_2O_3$, $La_5(SiO_4)_3N$[21] and β-Si_3N_4 or a glass. The reactions occurring during heating were determined as follows [129]:

$$\begin{aligned} Si_3N_4 + La_2O_3 &\xrightarrow{1200\text{–}1250°C} Si_3N_4 + \left(La_4Si_2O_7N_2 + LaSiO_2N\right) \\ &\xrightarrow{1400\text{–}1500°C} LaSiO_2N + Si_3N_4 \\ &\xrightarrow{1650\text{–}1750°C} 2\,Si_3N_4 \cdot La_2O_3 + \text{liquid} \\ &\xrightarrow{20°C} 2\,Si_3N_4 \cdot La_2O_3 + La_5(SiO_4)_3N + \text{glass} \end{aligned} \tag{14}$$

The results of Sakai et al [136] indicate that N-apatite and N-diopside containing grain boundary phase may improve the oxidation resistance of silicon nitride. Since the oxidation of Si_3N_4 leads to the formation of protective SiO_2 layer on the surface:

$$Si_3N_4 + 3\ O_2 \rightarrow 3\ SiO_2 + 2\ N_2 \tag{15}$$

[21] The presence of $La_5(SiO_4)_3N$ is inevitable in the production of high-density materials by liquid-phase sintering; therefore, the amount of $La_5(SiO_4)_3N$ and glassy phase must be minimized to obtain materials with good high-temperature strengths [129].

the silicon nitride shows excellent oxidation resistance. Formed SiO_2 then reacts with the grain boundary constituents to form silicates:

$$SiO_2 + MeO_x \rightarrow MeSiO_{2+x} \tag{16}$$

but in the case of MgO, the formed layer did not act as protection [136]. $Y_{4.67}(SiO_4)_3O$ apatite (britholite phase[2]) is formed as the oxidation product of silicon yttrium oxynitride (H-phase) in the temperature range from 700 to 1400°C [137],[138].

REE vanadocalcic apatite

The synthesis and physicochemical study of rare-earth-containing vanadocalcic oxyapatites where the pair Ca^{2+} and □ was substituted by Ln^{3+} and $1/2O^{2-}$ was described by Benmoussa et al [139]. This substitution leads to lanthanum and praseodymium dioxyapatites $Ca_8Ln_2(VO_4)_6O_2$, where Ln = La and Pr. Regarding rare earths such as neodymium, samarium, europium, gadolinium and terbium, the Ln^{3+} content limit varies from one REE to another. It decreases when the REE ionic radius declines.

Apatite-type yttrium phosphates

The following compositions having the apatite structure were prepared by Wanmaker et al [140]:

a. $Me(II)_{2+x}Me(III)_{8-x}(SiO_4)_{6-x}(PO_4)_xO_2$, where $0 \leq x \leq 6$;

b. $Me(II)_{3+x}MeIII_{6-x}(SiO_4)_{6-x}(PO_4)_x$, where $0 \leq x \leq 1.5$;

c. $Me(II)_{4+x}Me(III)_{6-x}(SiO_4)_{6-x}(PO_4)_xO$, where $0 \leq x \leq 6$.

with Me(II) = Ca, Sr, Ba, Mg, Zn or Cd and Me(III) = Y or La. Among these, there are several new compounds, e.g. $Zn_2La_8(SiO_4)_6O_2$, $BaMgY_8(SiO_4)_6O_2$, $Zn_2Y_8(SiO_4)_6O_2$, $Cd_2Y_8(SiO_4)_6O_2$, $Ca_4La_5(SiO_4)_5(PO_4)$ and $Ba_4La_5(SiO_4)_5(PO_4)$. The crystallographic parameters were determined and their luminescence was studied. The most efficient activator proved to be trivalent antimony, especially in the compositions of type I. At 77°K, an emission band at about 400 nm was observed in many of these apatites.

The humidity-sensitivity of yttrium-substituted calcium oxyhydroxyapatites was studied by Owada et al [141]. The logarithm of the electrical resistance of present sensors decreased linearly with increasing relative humidity (RH) from 30 to 65%. The resistance of $[Ca_{9.0}Y_{1.0}](PO_4)_6[O_{1.5}\square_{0.5}]$ with the largest OH vacancy content was about one order of magnitude lower than that of calcium hydroxyapatite. It was found that the larger the ratio of surface hydroxyl groups per unit surface area in the sample, the lower the resistance and the higher the amount of OH vacancies.

A ceramic proton conductor was obtained in the solid solutions of yttrium-substituted oxyhydroxyapatite $(Ca_{10-x}Y_x)(PO_4)_6((OH)_{2-x-2y}O_{x+y}\square_y)$ [142]. Using the hydrogen concentration cells, it was confirmed that the specimens with the composition of $x \leq 0.65$ have the protonic transference number (t_i) is equal to one, while the values of t_i of specimens with $0.65 < x < 1$ were smaller than one. The conduction properties were also dependent on the composition of apatites. At $x = 0.65$, the conductivity (σ) showed the maximum value ($5{\cdot}10^{-4}S{\cdot}cm^{-1}$ at 800°C) in the relationship between σ and x, while the activation energy was the lowest (about 1.0 eV) at corresponding x. The applicability of proton conductive apatite for a fuel cell was discussed in **Section**.

Author details

Petr Ptáček

Brno University of Technology, Czech Republic

References

[1] Richards R. Surface and Nanomolecular Catalysis. CRC Press, 2006. ISBN: 978-1420015751

[2] Kendrick E, Slater PR. Synthesis of Ga-doped Ge-based apatites: Effect of dopant and cell symmetry on oxide ion conductivity. Materials Research Bulletin 2008;43(12) 3627–3632.

[3] Xiang J, Liu Zhan-G, Ouyang Jia-H, Yan Fu-Y. Synthesis, structure and electrical properties of rare-earth doped apatite-type lanthanum silicates. Electrochimica Acta 2012;65 251–256.

[4] Fuentes AF, Rodríguez-Reyna E, Martínez-González LG, Maczka M, Hanuza J, Amador U. Room-temperature synthesis of apatite-type lanthanum silicates by mechanically milling constituent oxides. Solid State Ionics 2006;177(19-25) 1869–1873.

[5] Tao Sh, Irvine JTS. Preparation and characterisation of apatite-type lanthanum silicates by a sol-gel proces. Materials Research Bulletin 2001;36(7-8) 1245–1258.

[6] Martínez-González LG, Rodríguez-Reyna E, Moreno KJ, Escalante-García JI, Fuentes AF. Ionic conductivity of apatite-type rare-earth silicates prepared by mechanical milling. Journal of Alloys and Compounds 2009; 476(1-2) 710–714.

[7] Parmentier J, Liddell K, Thompson DP, Lemercier H, Schneider N, Hampshire S, Bodart PR, Harris RK. Influence of iron on the synthesis and stability of yttrium silicate apatite. Solid State Sciences 2001;3(4) 495–502.

[8] Hamon C, Marchand R, Maunaye M, Gaudé J, Guyader J. Revue de chimie minérale 1975;12 259.

[9] Kuzmin EA, Belov NV. Doklady Akademii Nauk SSSR 1965;165 88.

[10] Felsche J. Naturwissenschaften 1967;29 325.

[11] McCarthy GJ, White WB, Roy RJ. Preparation of $Sm_4(SiO_4)_3$. Journal of Inorganic and Nuclear Chemistry 1967;29 253–254.

[12] Higuchi M, Masubuchi Y, Nakayama S, Kikkawa S, Kodaira K. Single crystal growth and oxide ion conductivity of apatite-type rare-earth silicates. Solid State Ionics 2004;174(1-4) 73–80.

[13] Arrhenius SA. Über die Dissoziationswärme und den Einfluss der Temperatur auf den Dissociationsgrad der Elektrolyte. Zeitschrift für Physikalische Chemie 1889;4 96–116.

[14] Arrhenius , SA. Über die Reaktionsgeschwindigkeit bei der Inversion von Rohrzucker durch Säuren. Zeitschrift für Physikalische Chemie 1889;4: 226–248.

[15] Clugston M, Flemming R. Advanced Chemistry. A New Mainstream Text for the New Specifications, Advanced Science. OUP Oxford, 2000. ISBN: 978-0199146338

[16] Shi Q, Zhang H. Electrical properties of iron doped apatite-type lanthanum silicates. Journal of Rare Earths 2012;30(12) 1235–1239.

[17] Takeda N, Itagaki Y, Aono H, Sadaoka Y. Preparation and characterization of $Ln_{9.33+x/3}Si_{6-x}Al_xO_{26}$ (Ln = La, Nd and Sm) with apatite-type structure and its application to a potentiometric O_2 gas sensor. Sensors and Actuators B: Chemical 2006;115(1) 455–459.

[18] El-Batanouny M, Wooten F. Symmetry and Condensed Matter Physics: A Computational Approach. Cambridge University Press, 2008. ISBN: 978-1139469517

[19] Lambert S, Vincent A, Bruneton E, Beaudet-Savignat S, Guillet F, Minot B, Bouree F. Structural investigation of $La_{9.33}Si_6O_{26}$- and $La_9AESi_6O_{26+\delta}$-doped apatites-type lanthanum silicate (AE = Ba, Sr and Ca) by neutron powder diffraction. Journal of Solid State Chemistry 2006;179(8) 2602–2608.

[20] Fukuda K, Asaka T, Okino M, Berghout A, Béchade E, Masson O, Julien I, Thomas P. Anisotropy of oxide-ion conduction in apatite-type lanthanum silicate. Solid State Ionics 2012;217 40–45.

[21] Sumathi S, Buvaneswari G. Synthesis of apatite structure based $BiNaCa_3(PO_4)_3OH$ and its application for condensation reaction. Ceramics International 2012;38(5) 3547–3552.

[22] Kiyono H, Matsuda Y, Shimada T, Ando M, Oikawa I, Maekawa H, Nakayama S, Ohki S, Tansho M, Shimizu T, Florian P, Massiot D. Oxygen-17 nuclear magnetic resonance measurements on apatite-type lanthanum silicate ($La_{9.33}(SiO_4)_6O_2$). Solid State Ionics 2012;228 64–69.

[23] Harman G. Wire Bonding in Microelectronics 3 Edition by George Harman: Wire Bonding in Microelectronics. McGraw-Hill Professional, 2010. ISBN: 978-0-07-164265-1.

[24] Abram EJ, Kirk CA, Sinclair DC, West AR. Synthesis and characterisation of lanthanum germanate-based apatite phases. Solid State Ionics 2005;176(23-24) 1941–1947.

[25] Shaula AL, Kharton VV, Waerenborgh JC, Rojas DP, Marques FMB. Oxygen ionic and electronic transport in apatite ceramics. Journal of the European Ceramic Society 2005;25(12) 2583–2586.

[26] Shaula AL, Kharton VV, Marques FMB. Oxygen ionic and electronic transport in apatite-type $La_{10-x}(Si,Al)_6O_{26\pm\delta}$. Journal of Solid State Chemistry 2005;178(6) 2050–2061.

[27] Shaula AL, Kharton VV, Marques FMB. Ionic and electronic conductivities, stability and thermal expansion of $La_{10-x}(Si,Al)_6O_{26\pm\delta}$ solid electrolytes. Solid State Ionics 2006;177(19-25) 1725–1728.

[28] Helfferich FG. Ion Exchange. Dover Science Books. McGraw-Hill Series in Advanced Chemistry. Courier Corporation, 1962. ISBN: 978-0486687841

[29] Abram EJ, Sinclair DC, West AR. A novel enhancement of ionic conductivity in the cation-deficient apatite $La_{9.33}(SiO_4)_6O_2$. Journal of Material Chemistry 2001;11 1978–1979.

[30] Kharton VV, Marques FMB, Atkinson A. Transport properties of solid oxide electrolyte ceramics: a brief review. Solid State Ionics 2004;174(1-4) 135–149.

[31] Mössbauer RL. Kernresonanzfluoreszenz von Gammastrahlung in Ir191. Zeitschrift für Physik A 1958;151(2) 124–143.

[32] Greenwood NN. Mössbauer Spectroscopy. Springer Science & Business Media, 2012. ISBN: 978-9400956971

[33] Dickson PE, Berry FJ. Mössbauer Spectroscopy. Cambridge University Press, 2005. ISBN: 978-0521018104

[34] Miglierini M, Petridis D. Mössbauer Spectroscopy in Materials Science. Nato Science Partnership Subseries: 3 — Volume 66. Springer Science & Business Media, 2013. ISBN: 978-9401145480

[35] Yaremchenko AA, Shaula AL, Kharton VV, Waerenborgh JC, Rojas DP, Patrakeev MV, Marques FMB. Ionic and electronic conductivity of $La_{9.83-x}Pr_xSi_{4.5}Fe_{1.5}O_{26\pm\delta}$ apatites. Solid State Ionics 2004;171(1-2) 51–59.

[36] Yoshioka H, Nojiri Y, Tanase Sh. Ionic conductivity and fuel cell properties of apatite-type lanthanum silicates doped with Mg and containing excess oxide ions. Solid State Ionics 2008;179(30) 2165–2169.

[37] Yuan W, Gu Y, Li L. Synthesis and ionic conduction of cation-deficient apatite $La_{9.332x/3}M_xSi_6O_{26}$ doped with Mg, Ca, Sr*. Chinese Journal of Chemical Engineering 2008;16(3) 488–491.

[38] Irvine JTS, Corcoran DJD, Cull PA. Structure and ionic conduction in solids. Nordic Workshop on Materials for Electrochemical Conversion, Geilo Norway 2000; 37-41.

[39] Hosseini SM, Shvareva T, Navrotsky A. Energetics of lanthanum silicate apatite: Influence of interstitial oxygen and cation vacancy concentrations in $La_{9.33+x}(SiO_4)_6O_{2+3x/2}$ and $La_{10-x}Sr_x(SiO_4)_6O_{3-0.5x}$. Solid State Ionics 2013;233 62–66.

[40] Karato S. Deformation of Earth Materials: An Introduction to the Rheology of Solid Earth. Cambridge University Press, 2008. ISBN: 978-1139469562

[41] Kong LB, Huang YZ, Que WX, Zhang TS, Li S, Zhang J, Dong ZL, Tang DY. Transparent Ceramics. Topics in Mining, Metallurgy and Materials Engineering. Springer, 2015. ISBN: 978-3319189567

[42] Chung Yip-W. Introduction to Materials Science and Engineering. CRC Press, 2006. ISBN: 978-0849392634

[43] Kosevich AM. The Crystal Lattice: Phonons, Solitons, Dislocations, Superlattices. John Wiley & Sons, 2006. ISBN: 978-3527606931

[44] Tao Sh, Irvine JTS. Preparation and characterization of apatite-type lanthanum silicates by a sol-gel proces. Materials Research Bulletin 2001;36(7-8) 1245–1258.

[45] Li B, Liu W, Pan W. Synthesis and electrical properties of apatite-type $La_{10}Si_6O_{27}$. Journal of Power Sources 2010;195(8) 2196–2201.

[46] Masubuchi Y, Higuchi M, Takeda T, Kikkawa Sh. Preparation of apatite-type $La_{9.33}(SiO_4)_6O_2$ oxide ion conductor by alcoxide-hydrolysis. Journal of Alloys and Compounds 2006;408-412 641-644.

[47] Sansom JEH, Sermon PA, Slater PR. Synthesis and conductivities of the Ti doped apatite-type phases $(La/Ba)_{10-x}(Si/Ge)_{6-y}Ti_yO_{26+z}$. Solid State Ionics 2005;176(19-22) 1765–1768.

[48] Yuan W, Gu Y, Li L. Synthesis and conductivity of oxyapatite ionic conductor $La_{10-x}V_x(SiO_4)_6O_{3+x}$. Chinese Journal of Chemical Engineering 2010;18(2) 328–332.

[49] Mazza D, Tribaudino M, Delmastro A, Lebech B. Synthesis and neutron diffraction study of $La_5Si_2BO_{13}$, an analog of the apatite mineral. Journal of Solid State Chemistry 2000;155(2) 389–393.

[50] Arbib EH, Eloudi E, Chaminade JP, Darriet J. The crystal structure of the phosphate eulytite $Ba_3Bi(PO_4)_3$. Materials Research Bulletin 2000;35(5) 761–773.

[51] Naidu SA, Varadaraju UV, Raveau B. Eu^{3+} luminescence in $La_5Si_2BO_{13}$ with apatite related structure and magnetic studies in $Ln_5Si_2BO_{13}$ (Ln=Gd, Dy). Journal of Solid State Chemistry 2010;183(8) 1847–1852.

[52] Vieira MM, Oliveira JC, Shaula AL, Trindade B, Cavaleiro A. Structure and ionic conductivity of reactively sputtered apatite-type lanthanum silicate thin films. Surface and Coatings Technology 2014;247 14–19.

[53] Wang Shou-Q, Uda S. Phase relations around langasite ($La_3Ga_5SiO_{14}$) in the system La_2O_3-Ga_2O_3-SiO_2 in air. Journal of Crystal Growth 2003;250(3-4) 463–470.

[54] El Ouenzerfi R, Goutaudier C, Panczer G, Moine B, Cohen-Adad MT, Trabelsi-Ayedi M, Kbir-Ariguib N. Investigation of the CaO-La_2O_3-SiO_2-P_2O_5 quaternary diagram. Synthesis, existence domain, and characterization of apatitic phosphosilicates. Solid State Ionics 2003;156(1-2) 209–222.

[55] Wanmaker WL, ter Vrugt JW, Verlijsdonk JG. Luminescence of alkaline earth yttrium and lanthanum phosphate-silicates with apatite structure. Journal of Solid State Chemistry 1971;3(3) 452–457.

[56] Bonhomme C, Beaudet-Savignat S, Chartier Th, Maître A, Sauvet Anne-L, Soulestin B. Sintering kinetics and oxide ion conduction in Sr-doped apatite-type lanthanum silicates, $La_9Sr_1Si_6O_{26.5}$. Solid State Ionics 2009;180(36-39) 1593–1598,

[57] Mazza D, Ronchetti S. Study on the Al_2O_3-SiO_2-La_2O_3 ternary system at 1300°C. Materials Research Bulletin 1999;34(9) 1375–1382.

[58] Kuz'min VN, Belov KP. Doklady Akademii Nauk SSSR 1965;165, 88.

[59] McCarthy GJ, White WB, Roy R. Preparation of $Sm_4(SiO_4)_3$. Journal of Inorganic and Nuclear Chemistry 1967;29(1) 253–254.

[60] Felsche J. Naturwissenschaften 1969;56 325.

[61] Nakayama S, Sakamoto M. Preparation of apatite-type $La_{9.33}Ge_6O_{26}$ single-crystal from sintered ceramics by a seeding method and its oxide ionic conduction. Solid State Ionics 2013;253 47–52.

[62] Kendrick E, Slater PR. Synthesis of hexagonal lanthanum germanate apatites through site selective isovalent doping with yttrium. Materials Research Bulletin 2008;43(8-9) 2509–2513.

[63] Najib A, Sansom JE, Tolchard JR, Slater PR, Islam MS. Doping strategies to optimise the oxide ion conductivity in apatite-type ionic conductors. Dalton Transactions 2004;7(19) 3106–3109.

[64] Orera A, Headspith D, Apperley DC, Francesconi MG, Slater PR. Formation of apatite oxynitrides by the reaction between apatite-type oxide ion conductors, $La_{8+x}Sr_{2-x}$(Si/Ge)$_6O_{26+x/2}$, and ammonia. Journal of Solid State Chemistry 2009;182(12) 3294–3298.

[65] Emirdag-Eanes M, Pennington WT, Kolis JW. Synthesis, structural characterization, and magnetic properties of $NaRE_9(GeO_4)_6O_2$ (RE = Nd, Pr). Journal of Alloys and Compounds 2004;366(1-2) 76–80.

[66] Takahashi M, Uematsu K, Ye Zuo-G, Sato M. Single-crystal growth and structure determination of a new oxide apatite, $NaLa_9(GeO_4)_6O_2$. Journal of Solid State Chemistry 1998;139(2) 304–309.

[67] Chen P, Li RK. Two high terbium content apatites: $Tb_5Si_2BO_{13}$ and $Tb_{4.66}Si_3O_{13}$. Journal of Alloys and Compounds 2015;622 859–864.

[68] Xia MJ. Li RK. Structure and optical properties of a noncentrosymmetric borate $RbSr_4(BO_3)_3$. Journal of Solid State Chemistry 2013;197 366–369.

[69] Kazmierczak K, Höppe HA. Synthesis, crystal structure and optical spectra of europium borate fluoride $Eu_5(BO_3)_3F$. European Journal of Inorganic Chemistry 2010;18 2678–2681.

[70] Chen S, Hoffmann S, Carrillo-Cabrera W, Akselrud LG, Prots Y, Schwarz U, Zhao Jing-T, Kniep R. $Sr_{10}[(PO_4)_{5.5}(BO_4)_{0.5}](BO_2)$: Growth and crystal structure of a strontium phosphate orthoborate metaborate closely related to the apatite-type crystal structure. Journal of Solid State Chemistry 2010;183(3) 658–661.

[71] Yu Y, Wu QS, Li RK. Structure of two new borates $YCa_3(AlO)_3(BO_3)_4$ and $YCa_3(GaO)_3(BO_3)_4$. Journal of Solid State Chemistry 2006;179(2) 429–432.

[72] Kim HK, Ao Sio-I, Rieger BB. IAENG Transactions on Engineering Technologies: Special Edition of the World Congress on Engineering and Computer Science 2011. Springer Science & Business Media, 2012. ISBN: 978-9400747852

[73] Khomskii D. Transition Metal Compounds. Cambridge University Press, 2014. ISBN: 978-1107020177

[74] Lacroix C, Mendels P, Mila F. Introduction to Frustrated Magnetism: Materials, Experiments, Theory. Springer Series in Solid-State Sciences — Volume 164. Springer Science & Business Media, 2011. ISBN: 978-3642105890

[75] Werner F, Kubel F. Apatite-type $Pr_9K(SiO_4)_6O_2$ — A potential oxide ion conductor. Materials Letters 2005;59(28) 3660–3665.

[76] Morgan MG, Wang M, Mar A. Samarium orthosilicate oxyapatite, $Sm_5(SiO_4)_3O$. Acta Crystallographica Section E, 2010. DOI: 10.1107/S1600536802011868

[77] Sakakura T, Kamoshita M, Iguchi H, Wang J, Ishizawa N. Apatite-type $SrPr_4(SiO_4)_3O$. Crystallographica Section E, 2010. DOI: 10.1107/S1600536810033349

[78] Felsche J. Rare earth silicates with the apatite structure. Journal of Solid State Chemistry 1972;5(2) 266–275.

[79] Xiang J, Liu Zhan-G, Ouyang Jia-H, Zhou Y, Yan Fu-Y. Influence of doping with various cations on electrical conductivity of apatite-type neodymium silicates. Ceramics International 2013; 39(5) 4847–4851.

[80] Latshaw AM, Smith MD, zur Loye Hans-C. Crystal growth and structure of three new neodymium containing silicates: $Na_{0.50}Nd_{4.50}(SiO_4)_3O$, $Na_{0.63}Nd_{4.37}(SiO_4)_3O_{0.74}F_{0.26}$ and $Na_{4.74}Nd_{4.26}(O_{0.52}F_{0.48})[SiO_4]_4$. Solid State Sciences 2014;35 28–32.

[81] Wang Ch, Liu X, Fleet ME, Feng Sh , Xu R. High-pressure synthesis and single-crystal structure refinement of gadolinium holmium silicate hydroxyapatite $Gd_{4.33}Ho_{4.33}(SiO_4)_6(OH)_2$. Journal of Solid State Chemistry 2006;179(7) 2245–2250.

[82] Haile SM, Wuensch BJ, Siegrist T, Laudise RA. Hydrothermal synthesis of new alkali silicates I. Potassium neodymium phases. Journal of Crystal Growth 1993;131(3-4) 352–372.

[83] Haile SM, Wuensch BJ, Laudise RA. Hydrothermal synthesis of new alkali silicates II. Sodium neodymium and sodium yttrium phases. Journal of Crystal Growth 1993;131(3-4) 373–386.

[84] Boughzala K, Salem BE, Chrifa AB, Gaudin E, Bouzouita K. Synthesis and characterization of strontium-lanthanum apatites. Materials Research Bulletin 2007;42(7) 1221–1229.

[85] Małecka MA, Kępiński L. Synthesis and structure of nanocrystalline mixed Ce-Yb silicates. Materials Research Bulletin 2013;48(7) 2571–2577.

[86] Xiang J, Ouyang Jia-H, Liu Zhan-G. Microstructure and electrical conductivity of apatite-type $La_{10}Si_{6-x}WxO_{27+\delta}$ electrolytes. Journal of Power Sources 2015;284 49–55.

[87] Wang Sea-F, Hsu Yung-F, Lin Wan-J. Effects of Nb^{5+}, Mo^{6+}, and W^{6+} dopants on the germanate-based apatites as electrolyte for use in solid oxide fuel cells. International Journal of Hydrogen Energy 2013;38(27) 12015–12023.

[88] Institut kristallografii im. Shubnikova AV, Sheftal' NN. Growth of Crystals, Volume 6, Part 1. Consultants Bureau, 1965.

[89] Toropov NA, Bondar IA. In: Levin EM, Robbins CR, McMurdie HF (Eds.), Phase Diagrams for Ceramists, 1969.

[90] Toporov NA, Fedorov NF. Composition diagram of the Ca_2SiO_4-$Y_4(SiO_4)_3$ system. Zhurnal Neorganicheskoi Khimii 1965;10 666–668.

[91] Laczai N, Péter Á, Kovács L, Bencs L, Lőrincz E. Synthesis and characterization of $Y_2(SiO_4)O$:Ce with LiF and NaF additives. Solid State Sciences 2015;45 23–29.

[92] Mao H, Selleby M, Fabrichnaya O. Thermodynamic reassessment of the Y_2O_3-Al_2O_3-SiO_2 and its subsystems. Calphad 32 399-412. DOI: 10.1016/j.calphad.2008.03.003.

[93] Liddell K, Thompson DP. X-ray diffraction data for yttrium silicates. British Ceramic Transactions 1986;85 17–22.

[94] Warshaw I, Roy R. In: Progress in Science and Technology of Rare Earths, Vol. 1. New York: Pergamum Press, 1964, p. 203.

[95] Ito J, Johnson H. Synthesis and study of yttrialite. American Mineralogist 1968;53 1940–1952.

[96] Escudero A, Alba MD, Becerro AI. Polymorphism in the $Sc_2Si_2O_7$-$Y_2Si_2O_7$ system. Journal of Solid State Chemistry 2007;180 1436–1445.

[97] Wills RR, Cunningham JA, Wimmer JM, Stewart RW. Stability of the silicon yttrium oxynitrides. Journal of the American Ceramic Society 1976;59(5-6) 269–270.

[98] Cannas C, Musinu A, Piccaluga G, Deidda C, Serra F, Bazzoni M, Enzo S. Advances in the structure and microstructure determination of yttrium silicates using the Rietveld method. Journal of Solid State Chemistry 2005;178(5) 1526–1532.

[99] McColm I. Dictionary of Ceramic Science and Engineering. 2nd ed., Springer Science & Business Media, 2013. ISBN: 978-1475723212

[100] Suwa Y, Naka S, Noda T. Preparation and properties of yttrium magnesium silicate with apatite structure. Materials Research Bulletin 1968;3(2) 139–147.

[101] Lee FC, Marr J, Glasser FP. Compounds in the $Na_2O{\cdot}Y_2O_3{\cdot}SiO_2$ system. Ceramics International 1981;7(2) 43–47.

[102] Bondar IA, Galakhov FY. Izvestiya Akademii Nauk: Seriya Khimicheskaya 1964;7 1325–1326.

[103] Shannon RD. Revised effective ionic radii and systematic studies of interatomic distances in halides and chalcogenides. Acta Crystallographica Section A 1976;32(5) 751–767.

[104] Brown ID, Altermatt D. Bond-valence parameters obtained from a systematic analysis of the inorganic crystal structure database. Acta Crystallographica Section B 1985;41 244–247.

[105] Wills RR, Cunningham JA, Wimmer JM, Stewart RW. Stability of the silicon yttrium oxynitrides. Journal of the American Ceramic Society 1976;59(5-6) 269–270.

[106] Rakovan JF, Hughes JM. Strontium in the apatite structure: Strontian fluorapatite and belovite-(Ce). The Canadian Mineralogist 2000;38 839–845.

[107] Zuev MG, Karpov AM, Shkvarin AS. Synthesis and spectral characteristics of $Sr_2Y_8(SiO_4)_6O_2$: Eu polycrystals. Journal of Solid State Chemistry 2011;184(1) 52–58.

[108] Karpov AM, Zuev MG. Sol-gel synthesis and spectral characteristics of crystal phosphors $Sr_2Y_{8(1-x)}Eu_{8x}Si_6O_{26}$. Glass Physics and Chemistry 2012;38(4) 431–436.

[109] Holand W, Beall GH. Glass Ceramic Technology. 2nd ed., John Wiley & Sons, 2012. ISBN: 978-1118265925

[110] Ahlborg NL, Zhu D. Calcium-Magnesium-Aluminosilicate (CMAS) Reactions and Degradation Mechanisms of Advanced Environmental Barrier Coatings. National Aeronautics and Space Administration. NASA/TM-2013-218091

[111] Li G, Zhang Y, Geng D, Shang M, Peng C, Cheng Z, Lin J. Single-composition trichromatic white-emitting $Ca_4Y_6(SiO_4)_6O$: $Ce^{3+}/Mn^{2+}/Tb^{3+}$ phosphor: Luminescence and energy transfer. ACS Applied Materials & Interfaces 2012;4(1) 296–305.

[112] Wanmaker WL, Vrugt JW, Verlijsdonk JG. Synthesis of new compounds with apatite structure. Philips Research Reports 1971;26 373–381.

[113] van't Hoen C, Rheinberger V, Höland W, Apel E. Crystallization of oxyapatite in glass-ceramics. Journal of the European Ceramic Society 2007;27(2-3) 1579–1584.

[114] Gunawardane RP, Howie RA, Glasser FP. Structure of the oxyapatite $NaY_9(SiO_4)_6O_2$. Acta Crystallographica Section B 1982;38(5) 1564–1566.

[115] Redhammer GR, Roth G. Lithium and sodium yttrium orthosilicate oxyapatite, $LiY_9(SiO_4)_6O_2$ and $NaY_9(SiO_4)_6O_2$, at both 100 K and near room temperature. Acta Crystallographica Section C 2003;59(12) 120–124.

[116] Ptáček P, Bartoníčková E, Švec J, Opravil T, Šoukal F, Wasserbauer J, Másilko J. Preparation, kinetics of sinter-crystallization and properties of hexagonal strontium-yttrate-silicate apatite phase: $SrY_4[SiO_4]_3O$. Ceramics International 2015;41(1) 1779–1795.

[117] Heimann RB. Classic and Advanced Ceramics: From Fundamentals to Applications. Vydavatel John Wiley & Sons, 2010. ISBN: 978-3527630189

[118] Fang CM, de Wijs GA, de Groot RA, Metselaar R, Hintzen HT, de With G. O/N ordering in $Y_2Si_3O_3N_4$ with the melilite-type structure from first-principles calculations. Chemistry of Materials 2000;12(4) 1071–1075.

[119] Liddell K, Mandal H, Thompson DP. X-ray data for new Y-Si-Al-O-N glass ceramics. Journal of the European Ceramic Society 1997;17(6) 781–877.

[120] Ahmad S, Ludwig T, Herrmann M, Mahmoud MM, Lippmann W, Seifert HJ. Crystallisation studies of Si_3N_4-Al_2O_3-SiO_2-Y_2O_3 glass-ceramics under different heat-treatment conditions. Journal of the European Ceramic Society 2015;35(8) 2261–2268.

[121] Ching Wai-Y, Xu Yong-Ni, Ouyang L. Electronic structure and bonding in crystalline $Y_{10}[SiO_4]_6N_2$. Journal of the American Ceramic Society 2004;86(8) 1424–1426.

[122] Riedel R, Chen I-W. Ceramics Science and Technology, Structures. Ceramics Science and Technology — Volume 1. John Wiley & Sons, 2011. ISBN: 978-3527631933.

[123] Alper AM (Ed.), Kostorz G, Herman H. Phase Diagrams in Advanced Ceramics. Treatise on Materials Science and Technology. Elsevier, 1995. ISBN: 978-0080538723

[124] Jansen M (Ed.), Haubner R, Herrmann M, Lux B, Petzow G, Weissenbacher R, Wilhelm M. High Performance Non-Oxide Ceramics II. Structure and Bonding — Volume 102. Springer, 2003. ISBN: 978-3540456230

[125] Alper AM. Phase Diagrams: Materials Science and Technology, In: Alper AM (Ed.), Volume 6, Part 5. Academic Press, 1978. ISBN: 978-0120532032

[126] Rae AWJ, Thompson DP, Pikin NJ, Jack KH. pp. 347-360 In: Poper P (Ed.), Special Ceramics. British Ceramic Research Association, Stoke-on Trent, England, D.P. Thompson, p. 358, 1975.

[127] Wills RR, Holmquist S, Wimmer JM, Cunningham JA. Phase relationships in the system SiO_2-Y_2O_3-Si_3N_4. Journal of Materials Science 1976;11(7) 1305–309.

[128] Gauckler LJ, Hohnke H, Tien TY. The system Si_3N_4-SiO_2-Y_2O_3. Journal of the American Ceramic Society 1980;63(1-2) 35–37.

[129] Mitomo M, Izumi F, Haruichi S, Matsui Y. Phase relationships in the system Si_3N_4-SiO_2-La_2O_3. Journal of Materials Science 1982;17(8) 2359–2369.

[130] Buhl H. Advanced Aerospace Materials. Materials Research and Engineering. Springer Science & Business Media, 2012. ISBN: 978-3642501593

[131] Kizilyalli M, Gürbüz G. Lanthanum silicon oxynitrides, synthesis and X-ray diffraction and IR studies. Find out how to access preview-only content. NATO ASI Series 1990;185 45–67.

[132] Lange FF. Si_3N_4-Ce_2O_3-SiO_2 materials: Phase relations and strength. American Ceramic Society Bulletin 1980;59(2) 239–240.

[133] Thompson DP. Special Ceramics 1975;358.

[134] Mitomo N, Kuramoto N, Suzuki H. Journal of Materials Science Letters 1978;13 2523.

[135] Gaude J, Guyader J, Lang J. Comptes Rendus de l'Académie des Sciences, Serie C 1975;280 883–884.

[136] Sakai H, Soma T. Matsui M, Oda I. Oxidation and microstructure of sintered silicon nitride. In: Pask JA, Evans AG (Eds.). Ceramic Microstructures '86: Role of Interfaces, Materials Science Research — Volume 21. Springer Science & Business Media, 2013. ISBN: 978-1461319337

[137] International Aerospace Abstracts, Volume 33, Edition 4-6. American Institute of Aeronautics and Astronautics. Technical Information Service, Cambridge Scientific Abstracts, Inc., United States. National Aeronautics and Space Administration, Institute of the Aerospace Sciences. Technical Information Service, American Institute of Aeronautics and Astronautics, 1993.

[138] Physics Soviet, Crystallography, Volume 14, Pages 1-484. American Institute of Physics, American Crystallographic Association. American Institute of Physics, 1969.

[139] Benmoussa H, Mikou M, Lacout JL. Synthesis and physicochemical study of new rare-earth-containing vanadocalcic oxyapatites. Materials Research Bulletin 1999;34(9) 1429–1434.

[140] Wanmaker WL, ter Vrugt JW, Verlijsdonk JG. Luminescence of alkaline earth yttrium and lanthanum phosphate-silicates with apatite structure. Journal of Solid State Chemistry 1971;3(3) 452–457.

[141] Owada H, Yamashita K, Umegaki T, Kanazawa T, Nagai M. Humidity-sensitivity of yttrium substituted apatite ceramics. Solid State Ionics 1989;35(3-4) 401–404.

[142] Yamashita K, Owada H, Umegaki T, Kanazawa T, Katayama K. Protonic conduction in yttrium-substituted hydroxyapatite ceramics and their applicability to H_2-O_2 fuel cell. Solid State Ionics 1990;40-41(2) 918-921.

3

Phosphate Rocks

Petr Ptáček

Abstract

Apatite is the most abundant phosphate mineral which include more than 95% of al phosphorus in the Earth´s crust. The seventh chapter of this book provides brief description of sedimentary and igneous phosphate rocks and introduces basic ideas for characterization and classification of phosphate rocks. The chapter continues with description of biogenic apatites, description of phosphate rocks deposits and introduces other sources of phosphorus. Furthermore, geological role of apatite, cycle of phosphorus, weathering of apatite, fission track analysis and extraterrestrial apatites were described. The last section is dedicated to structure and properties important of non-apatitic phosphate minerals, such as atuanite, crandallite, lazulite, millisite, monazite, tobernite, xenotime etc.

Keywords: Apatite, Phosphate Rock, Biogenic Apatites, Fission Track, Extraterrestrial Epatite, Non-Apatitic Phosphate Minerals

Apatite [$Ca_5(PO_4,CO_3)_3(OH,F,Cl)$] is the most abundant phosphate mineral, which accounts for more than 95% of all phosphorus in the Earth's crust and is found as an accessory mineral in most rock types[1] on the Earth's surface, primarily because it is stable in a wide variety of geological conditions and over a range of different geological processes [1],[2],[3],[4],[5],[6],[7],[8]. According to the list of symbols for rock- and ore-forming minerals, the abbreviated symbol used for apatite is **Ap** [9].

However, exploitable deposits of apatite are mainly found in igneous rocks and also in sedimentary and metamorphic rocks. The former comprises the stratiform phosphorite deposits in shelf-type shale-carbonate sequences, which contain high phosphorus ores of microcrystal-

[1] A rock may be best defined as any mineral or aggregate of minerals that forms an essential part of the Earth [10]. Apatite is found in all classes of rock: igneous, metamorphic and sedimentary [2]. Also, much of phosphorus in coal is present in the form of apatite [7]. Phosphorus is the 10th most abundant element on Earth, with an average crustal abundance of 0.1% [8].

line CO_2-rich fluorapatite (francolite) and cryptocrystalline collophane. The igneous deposits comprise fluorapatite ores, which mostly accommodate carbonatites and other types of alkaline intrusions. The magmatic ores are generally of lower grade but give higher-quality beneficiation products with low contents of unwanted contaminants (Cd, Pb, As, U, Th, Mg and Al) [11], [12].

The beneficiation products of apatite ores as a commodity are traded as phosphate rock. It is the only significant global resource of phosphorus used dominantly in the manufacturing of nitrogen-phosphorus-potassium (NPK) fertilizers for food-crop nutrition and in the production of animal feed supplements. Only 10 – 15% of the world's production of phosphate rock has other applications (e.g. pharmaceuticals, ceramics, textiles and explosives) and represents an important alternative source of rare-earth elements (REE) [12],[13]. The REE contents in apatites are useful in paleoceanographic studies to identify the seawater masses and circulation patterns or to quantify the redox state of the ocean [14].

The composition of phosphate rocks varies from one deposit to another. Therefore, phosphate rocks from different sources may be expected to behave differently in beneficiation and acidulation processes. Phosphate rocks are primarily composed of the apatite group in association with a wide assortment of accessory minerals, mainly fluorides, carbonates, clays, quartz, silicates and metal oxides [13],[15],[16].

Si, Ca, Fe and Al are the most common companion elements in phosphate rocks, with the median abundances of 53.3 wt.%, 30.0 wt.%, 13.6 wt.% and 8.0 wt.%, respectively, compared

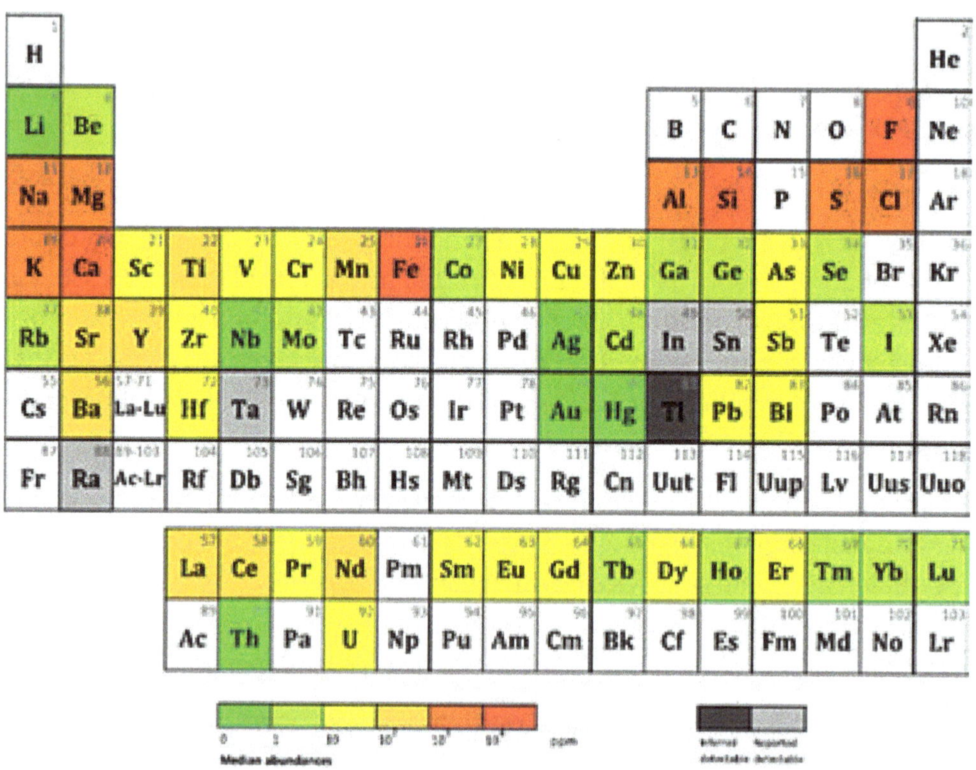

Fig. 1. The average distribution of trace elements in phosphate rock [17].

with P_2O_5. In some low-grade phosphate rock mines, the content of Fe and Al is even higher than that of P_2O_5, and it is usual that the P_2O_5 content in phosphate rock is less than the Si and Ca content. In addition, some elements that are very rare in the Earth's crust are found to be relatively abundant (**Fig. 1**) in phosphate rocks [17].

Characterization and classification of phosphate rocks

When rock contains phosphate components between 5 and 50% (by volume), then it is phosphatic, and the name of the main lithology is used as a suffix (phosphatic limestone, phosphatic claystone, etc.). In addition, the dominant textural form of the phosphate components in a phosphorite can be used in defining the rock name (e.g. peloidal[2]) (phosphorite, coprolitic phosphorite, etc.) [10],[18].

There are two main kinds of phosphate rocks deposits in the world [10],[20],[21],[22],[17],[23]:

1. **Sedimentary phosphate rocks**: marine phosphate deposit, metamorphic deposit, biogenic deposit (bird and bat guano accumulation) and phosphate deposit as the result of weathering. The sedimentary deposits contain the varieties of carbonate-fluorapatite that are collectively called as francolite (**Section**). The most common non-phosphatic accessory minerals associated with sedimentary phosphate rocks are quartz, clay and carbonates (calcite and dolomite). Phosphate rocks of high concentration of phosphates (10 – 15% of P_2O_5) are called phosphorites.

2. **Igneous phosphate rocks**: apatite is a common accessory mineral occurring in practically all types of igneous rocks (acid, basic or ultrabasic).

Depending on their origin (igneous or sedimentary), phosphate rocks have widely varying mineralogical, textural and chemical characteristics [23]. The locations of the major phosphate rocks deposit and producers are shown in **Fig. 2** [20],[24].

Sedimentary phosphate deposits are exploited to produce about 80% of the total world production of phosphate rocks. Igneous phosphate deposits are often associated with carbonatites[3] and/or alkalic (silica-deficient) intrusions. Igneous phosphate rock concentrates are produced from the deposits mainly exploited in Russia, the Republic of South Africa, Brazil, Finland and Zimbabwe. Igneous phosphate ores are often low in grade (less than 5% P_2O_5) but can be upgraded to high-grade products (from about 35% to over 40% P_2O_5) [22],[23].

The atom ratio of P:N = 1:15(16) in the oceans is not greatly different from that found in living organisms. The availability of soluble phosphate from weathering of apatite-containing rocks may initially has been the rate-determining factor in early live development. In most ecolog-

---

[2] Peloid is a comprehensive descriptive term for polygenetic grains composed of micro- and polycrystalline carbonate. The term was proposed in order to replace the widely used name "pellet," which for many authors had become a synonym for pelletal coprolites (fossilized faces). Peloids differ from ooids and oncoids by the absence of centrosymmetric or radial internal structures [19].

[3] Igneous rocks (intrusive or extrusive) that contain carbonates in the amount higher than 50%.

ical systems, the phosphate content is the limiting factor for growth. Nearly all igneous rocks contain some phosphate, even if it is only ~0.1% (0.2% P_2O_5 on average in lithosphere), with nearly all of it in the form of apatite. Sedimentary rocks generally contain rather less (~0.1% P_2O_5 on average). Sedimentary phosphorite is believed to originate from widely dispersed apatite mainly in igneous rocks [25].

Sedimentary phosphate rocks

Most marine sediments and rocks contain less than 0.3% of P_2O_5. However, periodically through geological time, phosphorites (with the content of P_2O_5 of 5% or greater) formed on the seafloor in response to specialized oceanic conditions and accumulated in sufficient concentrations to produce major deposits of regional extent[4] [26].

Marine phosphate formation and deposition represent the periods of low rates of sedimentation in combination with large supplies of nutrients. Phosphorus is then concentrated by various mechanisms, possibly bacterial (refer to discussion of **Fig. 7**), at either the sediment-water interface or within interstitial pore waters. This process leads to primary formation and growth of phosphate grains, which remain where they were formed or are transported as clastic particles within the environment of formation. During subsequent periods of time, some primary phosphate grains may be physically reworked into another sediment unit in response to either changing or different environmental processes [26].

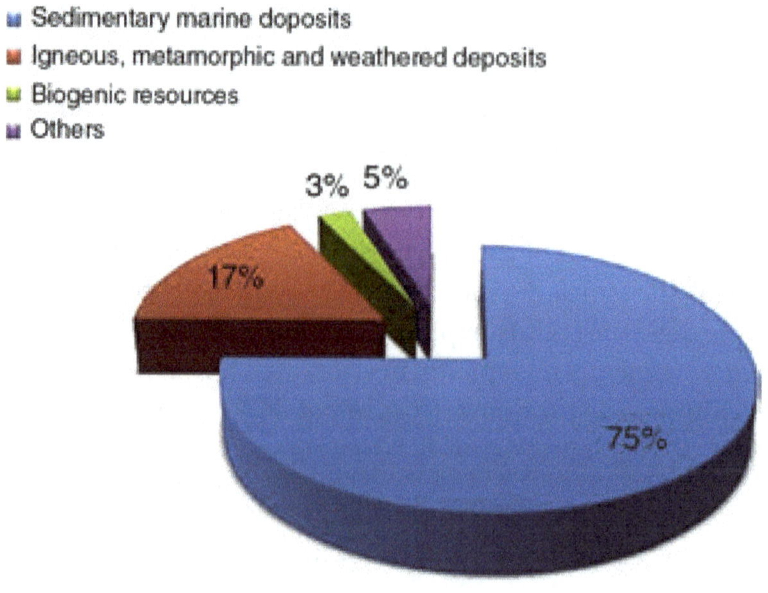

Fig. 2. The distribution of the world's phosphate resources [20].

[4] Most of the world's phosphate production comes from marine phosphorites [27].

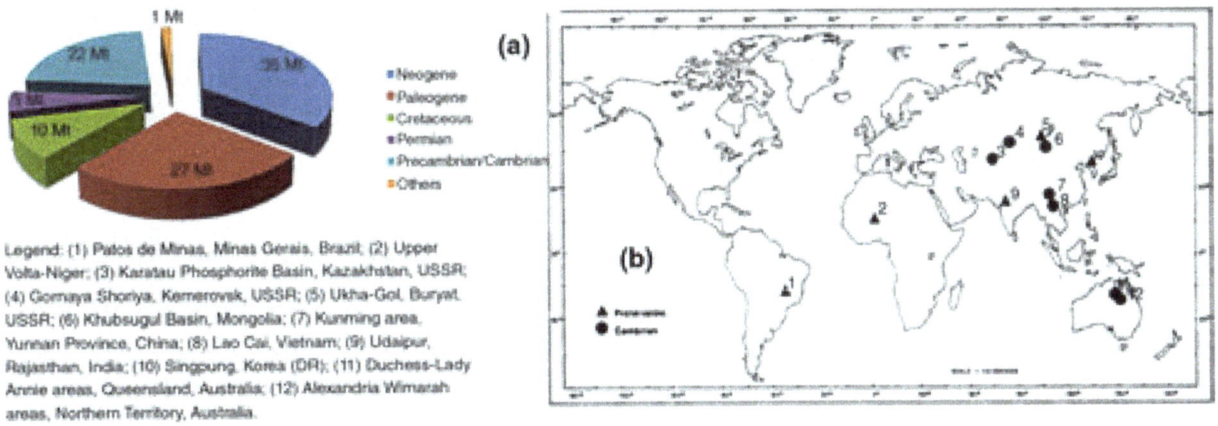

Fig. 3. Stratigraphic distribution of phosphorites based on 1982 production data (a) and distribution of major Proterozoic-Cambrian phosphorites (b) [28].

The stratigraphic distribution[5] [29] of phosphorites [28] is shown in **Fig. 3(a)**. The major discoveries in Proterozoic and Cambrian rocks were not made until late 1930s, initially in the USSR (Kazakhstan), Poland, Korea, China and northern Vietnam. Early discoveries were made in the course of regional geological mapping and exploration for metal deposits, but later deposits were found mainly using more direct exploration techniques. It is possible that the greatest global phosphogenic episode in geological history took place in Late Proterozoic and Cambrian. While pelletal phosphorites are common in most Cambrian deposits, the Proterozoic age phosphorites contain mudstone (**Fig. 3(b)**, microphosphorite) and stromatolitic phosphorite[6] [28],[30].

Some phosphates were formed during all major sea-level transgressions during 67 million years of Cenozoic history; however, some periods were more important than others with respect to producing large volumes of phosphorites and preserving them in the geologic column. During the Paleocene and Eocene, several major episodes of phosphogenesis occurred within the major episodes of phosphogenesis in the major east-west ocean, which included Tethys [31],[7] producing extensive amounts of phosphorites throughout the Middle East, Mediterranean and northern South American regions. By the Neogene, this circum-global ocean had been destroyed by the plate tectonic processes, and the north-south Pacific and

[5] Since the Earth is stratified, in a broad sense, all rocks and classes of rocks (sedimentary, igneous and metamorphic) fall within the scope of stratigraphy and stratigraphic classification. Rocks can be classified according to lithology, fossil content, magnetic polarity, electrical properties, seismic response, chemical or mineralogical composition, etc. Rocks can be also classified according to time of their origin or environment of genesis. Rock bodies can be classified into many different categories of stratigraphic classification, including lithostratigraphic units (1), biostratigraphic units (2), chronostratigraphic units (3), unconformity-bounded units (4) and magnetostratigraphic units (5). Please see work [29] for further details.

[6] Phosphate occurs as concentrated in stromatolite columns, laminar algal (stromatolitic) phosphorite, reworked fragments of stromatolites forming silicified conglomeratic or brecciated phosphorite, massive-bedded phosphorite with sandy and clayey laminate and disseminate pellets and nodules in dolomite [30].

[7] The Tethys Ocean divided the continental masses into northern and southern groups. It merged at both ends with the Panthalassa or proto-Pacific Ocean. The remnants of Tethys are now located within Alpine mountain belts from the Caribbean in the west to recent collision zone between Australia and Eurasia in the east and within still-growing Central Atlantic Ocean between Africa and North America [31].

Atlantic oceans dominated global circulation patterns. Upper Cenozoic phosphogenesis (**Table 1**) occurred along the north-south ocean-ways, which now contain modern continental margins. On the basis of the extent of known phosphate deposits, the Miocene was by far the most important episode of phosphate formation in Upper Cenozoic [26].

Geological period			**Age**	**Duration**	
Phanerozoic (541 *ma* to present)	**Quaternary**	**Holocene**	<3 *ma*	10 *ta*	
Cenozoic (66 *ma* to present)	258 *ma* to present	0.0117 *ma* to present			
		Pleistocene			
		2.58 – 0.0117 *ma*			
	Neogene	**Pliocene**	5 – 4 *ma*	1 *ma*	
	23.03 – 2.58 *ma*	5.333 – 2.58 *ma*			
		Miocene	19 – 13 *ma*	6 *ma*	
		23.03 – 5.333 *ma*			
		Paleogene	**Oligocene**	29 – 25 *ma*	4 *ma*
		66 – 23.03 *ma*	33.9 – 23.03 *ma*		

ma – million years ago, *ta* – thousand years

Table 1. Formation of phosphorites during Upper Cenozoic phosphogenesis [26].

Phosphorites and phosphatic sediments are known on the floor of the Pacific, Indian and Atlantic oceans. They occur in a number of inshore areas (the shelves and upper part of the continental slopes) and in pelagic zones, chiefly on seamounts. Most of the shelf phosphorites are localized in four very large oceanic phosphorite provinces [32],[26],[33]:

a. **East Atlantic**: Portugal, northwest Africa throughout South Africa and Agulhas Bank;

b. **West Atlantic**: North Carolina throughout Florida, Cuba, Venezuela and Argentina;

c. **East Pacific**: California throughout Baja California, Mexico and Peru throughout Chile;

d. **West Pacific**: Sakalin Island, Sea of Japan, Indonesia, Chathman Rise east of New Zealand and East Australian shelf.

Sedimentary rocks with the content of 18 – 20 wt.% of P_2O_5 are termed as phosphorites. The main phosphate mineral in phosphorites is carbonate-fluorapatite (CAF, francolite): $Ca_{10-a-b-c}Na_aMg_b(PO_4)_{6-x}(CO_3)_{x-y-z}(CO_3,F)_y(SO_4)_zF_2$, where $x = y + a + 2c$ and c denotes the number of Ca vacancies, present as grain or mud. Most phosphorites are of marine origin [34],[35],[36]. Phosphorites on the sea floor occur in two types of environments on [37]:

i. Continental margins in association with terrigenous;

ii. Submerged mountains in association with calcareous and volcanogenic rocks.

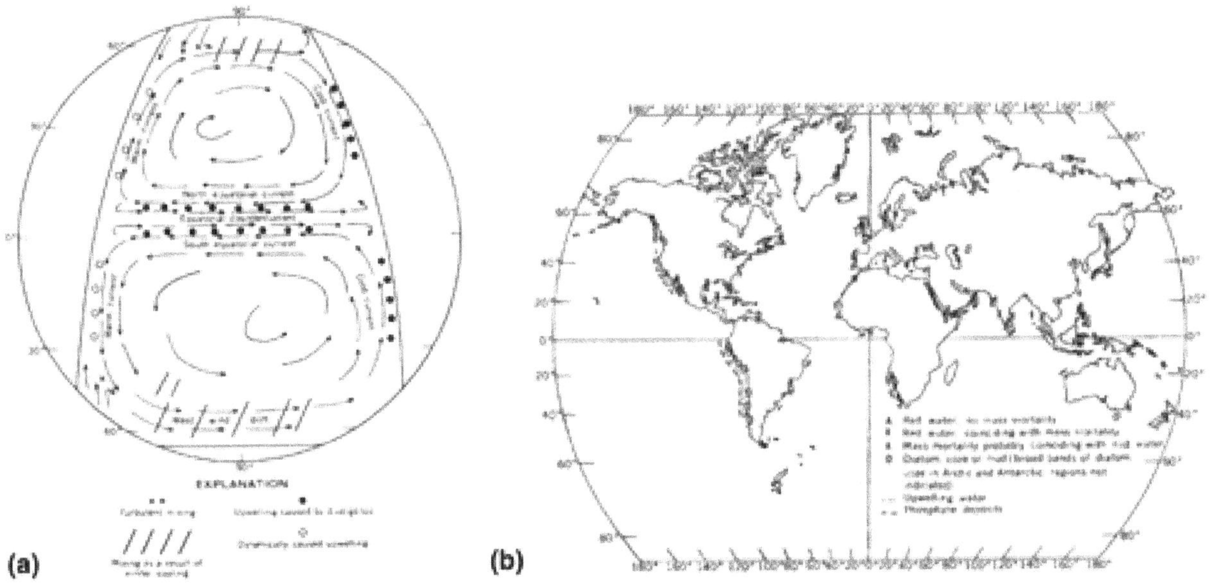

Fig. 4. The surface currents in an idealized ocean, showing the areas of ascending nutrient-rich water (a) and the distribution of upwelling water and related phenomena in modern oceans (b) [27].

Phosphorites consist mainly of phosphate cement enveloping small grains of phosphatic and non-phosphatic materials. Phosphate in the cavities of foraminifera is purer than that enveloping the grains [37].

The largest phosphorite-bearing regions are situated along the west coasts of Africa and America, at the east coast of the USA, off New Zealand and in the central part of the northern Pacific. The phosphatic matter of phosphorites consists of carbonate-fluorapatite and is intermixed with variable amounts of terrigenous, biogenic and diagenetic non-phosphatic components, which are the cause of a wide range of fluctuations in their chemical compositions. The age of sea-floor phosphorites varies from Cretaceous to Recent. Recent phosphorites are localized in the south west of Africa and at Peru-Chile shelves, which are the areas influenced by strong upwelling of nutrient-rich waters (**Fig. 4(a)**), resulting in high biological productivity, intensive biogenic sedimentation and diagenetic redistribution of geochemically active, mobile, organic-derived phosphorus in sediments. This phosphorus is accreted in the form of initially soft and friable nodules undergoing gradual lithification [27],[37].

Pronounced climatic, biological and geologic effects accompany upwelling, especially where it is produced by the divergence in coastal areas (**Fig. 4(b)**). The presence of cold waters along the coasts produces the coastal fogs and humid-air deserts, such as those of northern Chile and southwest Africa. The nutrient-rich waters that lie alongside these deserts are the lushest gardens of the sea, as the upwelling cold waters there support tremendous quantities of organisms. Most of large accumulations of guano (**Section**) are formed by the seafowl colonies feeding in these waters, and it is the extremely dry climate created by upwelling that makes the preservation of guano possible [27].

Sedimentary deposits usually contain varieties of carbonate-fluorapatite called francolite (described in **Section**). Francolite is defined as apatite that contains significant amount of

CO_2 with less than 1% of fluorine. Apatite associated with igneous source rocks may be of primary magmatic, hydrothermal or secondary origin. Primary apatite from igneous sources may be of fluorapatite, hydroxylapatite or chlorapatite varieties. Pure apatites from igneous deposits contain slightly over 42% of P_2O_5 [23].

Since a wide range of very different particles and processes of formation complicates the simple classification of phosphorites, there is not any unified phosphorite classification[8] [18],[36],[38]. The proposed classification schemes for phosphorites describe their constituent particles, such as pelletal phosphorite. Nonetheless, the descriptor pelletal indicates nothing more than rounded phosphate particles of any origin. A widely recognized distinction in phosphorites is based on the grain size and holds specifically among phosphorites where the phosphate particles are of sand- or coarse silt-size and those that are of clay- and fine silt-size [28]. The phosphorites can be classified as follows [37]:

1. **Non-conglomeratic** (also termed as **nodular**) phosphorites: consist of phosphatized limestones and gluconate-quartz sandstones. Two varieties of nodular phosphorites can be recognized:

   a. **Ferruginized** with glazed surface. The cement of ferruginized phosphorites is much richer in finely dispersed goethite than that of non-ferruginized phosphorites. Furthermore, ferruginized phosphorites do not contain the fragments of macrofauna.

   b. **Non-ferruginized** with rough surface. In non-ferruginized phosphorites, the cement is micrite-collophane and its color varies form yellow (collophane, described in **Section**) to gray (micrite[9] [19]). The chambers of foraminifera are filled with phosphate-carbonate cement, less often with glauconite or goethite (FeO(OH) [33]).

   Some nodules of phosphatized limestone are coated with a discontinuous layer of secondary phosphate with the thickness up to 1 cm.

2. **Conglomeratic phosphorites** consist of pebbles of phosphatized limestone (up to 50% of the rock) held together by the cement similar in composition to the phosphatized glauconite-quartz sandstones described above. In many samples of this type, two or three conglomerate layers are clearly visible, differing in size of pebbles and in content of glauconite. The bedding planes separating the layers with denser or less dense packing of grains are also distinguished in the cement. The surface of these planes is glazed and brown due to higher content of iron hydroxides and organic matter. Upon impact, the rock breaks along the planes. In addition, irregular microerosion surfaces are observed in

---

[8] The nature and origin of phosphorites have been a matter of much speculation since they were first discovered more than 150 years ago [38], and there is not any commonly accepted nomenclature [39].

[9] Micrite (the abbreviation for microcrystalline calcite) is characterized by crypto- to microcrystalline crystal texture. As the synonym of micrite, the names as lime mud, lime ooze, lime mudstone, calcimudstone and calcilutite are also used. The original definition sets a grain-size limit to < 4 μm, but current terminology distinguishes between minimicrite (<1 μm), micrite I (1 – 4 μm) and micrite II (4 – 30 μm). Furthermore, primary micrites (orthomicrites and nannoagorites), secondary micrites and pseudomicrites are recognized. Orthomicrites consist of subhedral polygonal calcite grains meeting at the interfaces. Nannoagorites are composed of calcareous pelagic biota. Secondary and pseudo micrites result from the diagenetic processes [19].

the conglomeratic phosphorites, which run across the grains of glauconite, shells and bedding planes [33].

The investigation of the microstructure of phosphorites by electron microscopy enables to recognize the following varieties [33]:

i. **Gel-like,** phosphatized diatomaceous oozes, analogous (except phosphate content) to the enclosing diatomaceous oozes.

ii. **Microgranular**, forming solid masses and globules from 1 to 3 mm in diameter. The rough surface of phosphate is caused by the fact that it consists of granules less than 0.1 mm in size.

iii. **Fibrous**, constituting inner parts of globules.

iv. **Ultramicrocrystalline phosphate**, forming a "jacket" on the surface of globules of amorphous phosphate. The size of crystal of apatite is 0.1 – 0.3 μm.

v. **Microcrystalline phosphate**, consisting of crystal of 1 – 3 μm in size. Euhedral crystals are often formed in the cavities within the carbonate grains.

vi. **Multiphase microgranular cement** consisting of carbonate, phosphate, quartz and layered silicates.

With regard to their texture and petrographic character, phosphorites can be classified according to the predominant size of the phosphorite component into four types [40]:

a. **Microgranular (oolitic microgranular) phosphorites**, conditionally including aphanite: 0.01 – 0.1 mm;

b. **Granular phosphorites**: 0.1 – 1 mm;

c. **Nodular phosphorites**: 1 – 5 mm;

d. **Shelly phosphorites**: 5 – 100 mm.

Although the types are named on the structural basis, the phosphate grains do not always have the dimensions given above. At the same time, the classified types fairly differ in many features, such as the association with various geological formations, the phosphate mineralogy and the stratigraphic sequence, thus being of lithologic character [40].

Most attempts to classify the phosphorite rocks adopt and modify the classification scheme for carbonates [36]. In 1962, Dunham [41] published the classification scheme for limestone. This scheme for carbonate rocks was modified for non-genetic classification of phosphorites (**Fig. 5**) [28].

<table>
<tr><td colspan="5">Depositional texture recognizable</td><td rowspan="6">Depositional texture not recognizable
Crystalline carbonate
(subdivide according to classification designed to bear on physical texture or diagenesis)</td></tr>
<tr><td colspan="4">Original components not bound together during deposition</td><td rowspan="4">Original components were bound together during deposition as shown by intergrown skeletal matter, lamination contrary to gravity, or sediment-floored cavities that are roofed over by organic matter and are to large to be interstices</td></tr>
<tr><td colspan="3">Contains mud (particles of clay and fine silt size)</td><td rowspan="3">Lacks mud and is grain supported</td></tr>
<tr><td colspan="2">Mud-supported</td><td rowspan="2">Grain-supported</td></tr>
<tr><td>Less than 10% grains</td><td>More than 10% grains</td></tr>
<tr><td>Mudstone</td><td>Wackestone</td><td>Packstone</td><td>Grainstone</td><td>Boundstone</td></tr>
<tr><td colspan="6">Classification of phosphorites</td></tr>
<tr><td colspan="2">"Non-pelletal phosphorites"</td><td colspan="2">"Pelletal phosphorites"</td><td rowspan="2">Boundstone phosphorite</td><td rowspan="2"></td></tr>
<tr><td>Mudstone phosphorite</td><td>Wackestone phosphorite</td><td>Packstone phosphorite</td><td>Grainstone phosphorite</td></tr>
</table>

Fig. 5. The classification scheme of carbonate rocks modified for phosphorites [28].

The macroscopic classification scheme for phosphate sediments suggested by Riggs [42] is shown in **Fig. 6(a)**. The ancient deposits are better characterized via the scheme in **Fig. 6(b)**, which was proposed by Kastern and Garrison [43]. In this model, three types of phosphates were recognized [39],[43]:

i. F-phosphates are friable, light-colored micronodules and peloids of carbonate-fluorapatite (CFAP); they were formed by the precipitation of CFAP in laminated diatom muds deposited within the oxygen-minimum zone.

ii. Phosphatic sands, termed as P-phosphates, consist of phosphatic peloids, coated grains and fish debris, often having an admixture of fine siliciclastic grains. These sands occur in thin layers and burrowed beds up to 2 m thick.

iii. Dark and dense phosphates, herein called D-phosphates, are the most abundant. They occur as nodules, gravels and hard grounds. These phosphates were formed through complicated cycles of CFAP precipitation during early diagenesis, erosion and exhumation and reburial and rephosphatization processes associated with changing energy conditions, which may reflect the effects of changes in the sea level.

CFAP cements in P- and D-phosphates are often replaced microbial structures, but our data do not reveal whether this microbial involvement was passive or active. F-phosphates are most common in deeper water, outer-shelf/upper-slope sites, whereas D- and P-phosphates tend to predominate at shallower shelf sites more subjected to episodic high-energy conditions, especially during the low stands of sea level. This concept reveals the paleoenvironmental and time relationships of various phosphate sediments [39],[43].

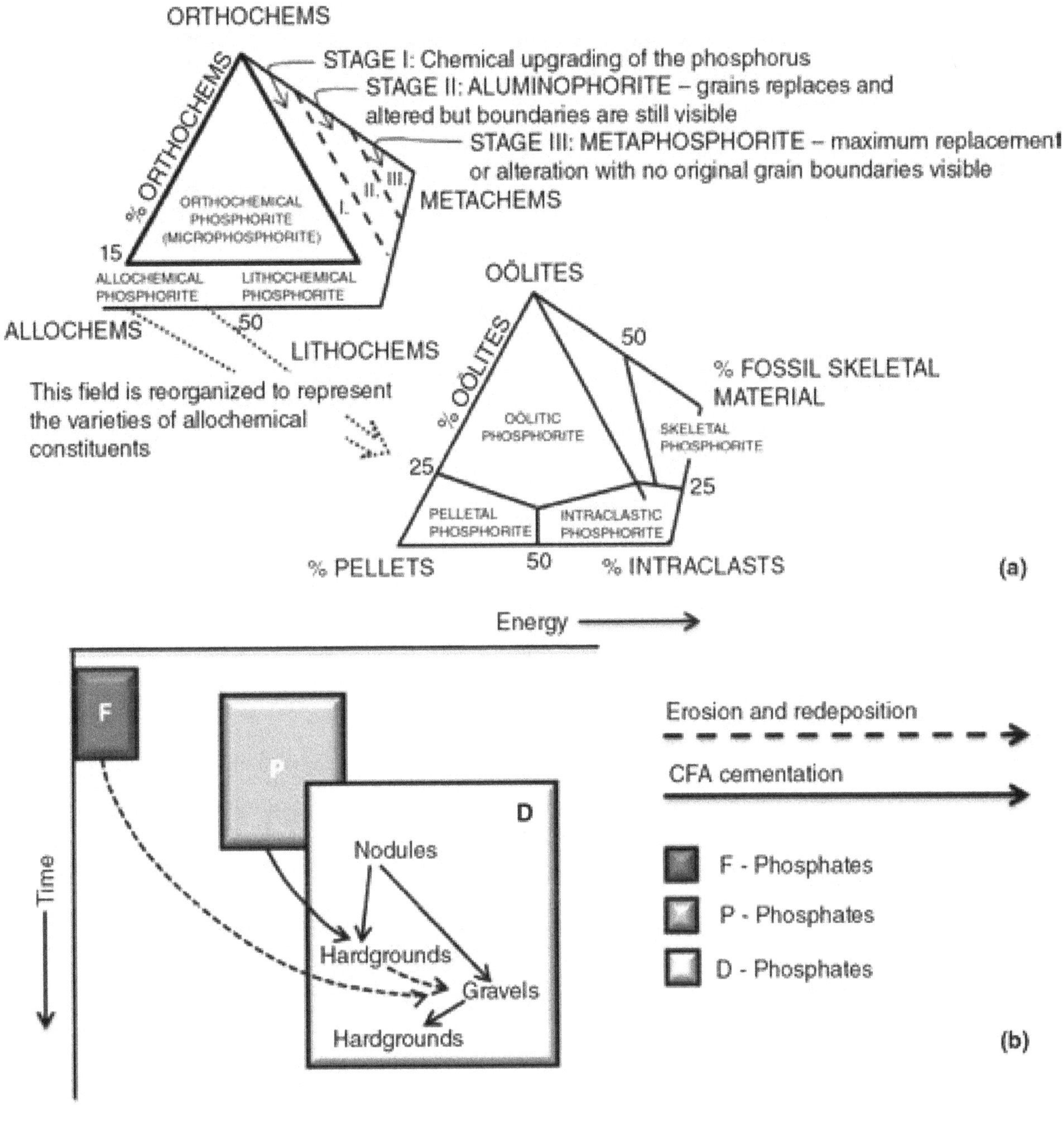

Fig. 6. The classification scheme of phosphorites proposed by RIGGS (a) and GARRISON and KASTNER [43].

Phosphorites can be formed in nature **authigenically** or **diagenically**. In authigenesis, phosphorite forms as a result of the reaction of soluble phosphate with calcium ions forming corresponding insoluble phosphate compound. The role of microbes in these processes may be one or more of the following [44]:

a. Making reactive phosphate available;

b. Making reactive calcium available;

c. Generating or maintaining the pH and redox conditions, which favor the precipitation of phosphate.

The models of authigenic phosphorite genesis (**Fig. 7**) assume the occurrence of mineralization of organic phosphorus in biologically productive waters, such as at ocean margins, that is, at shallow depths on continental slopes, shelf areas or plateaus [44].

Here, detrital accumulations may be mineralized at the sediment-water interface and in interstitial pore waters, liberating phosphate, some of which may then interact chemically with calcium in seawater to form phosphorite grains. These grains may be subsequently redistributed within the sediments units. The dissolution of fish debris (bones) is also considered an important source of phosphate in authigenic phosphorite genesis. The upwelling probably also plays an important role in many cases of authigenic formation of phosphorite. During non-upwelling period in winter, the phosphate-sequestering bacteria of oxidative genera *Pseudomonas* and *Acinetobacter* become dominant in the water column. Fermentative Vibrios and Enterobacteriaceae are dominant during upwelling in summer. It was suggested that *Pseudomonas* and *Acinetobacter*, which sequester phosphate as polyphosphate under aerobic conditions and hydrolyze polyphosphate under anaerobic conditions to obtain the energy of maintenance and to sequester volatile fatty acid from polyhydroxybutyrate formation,

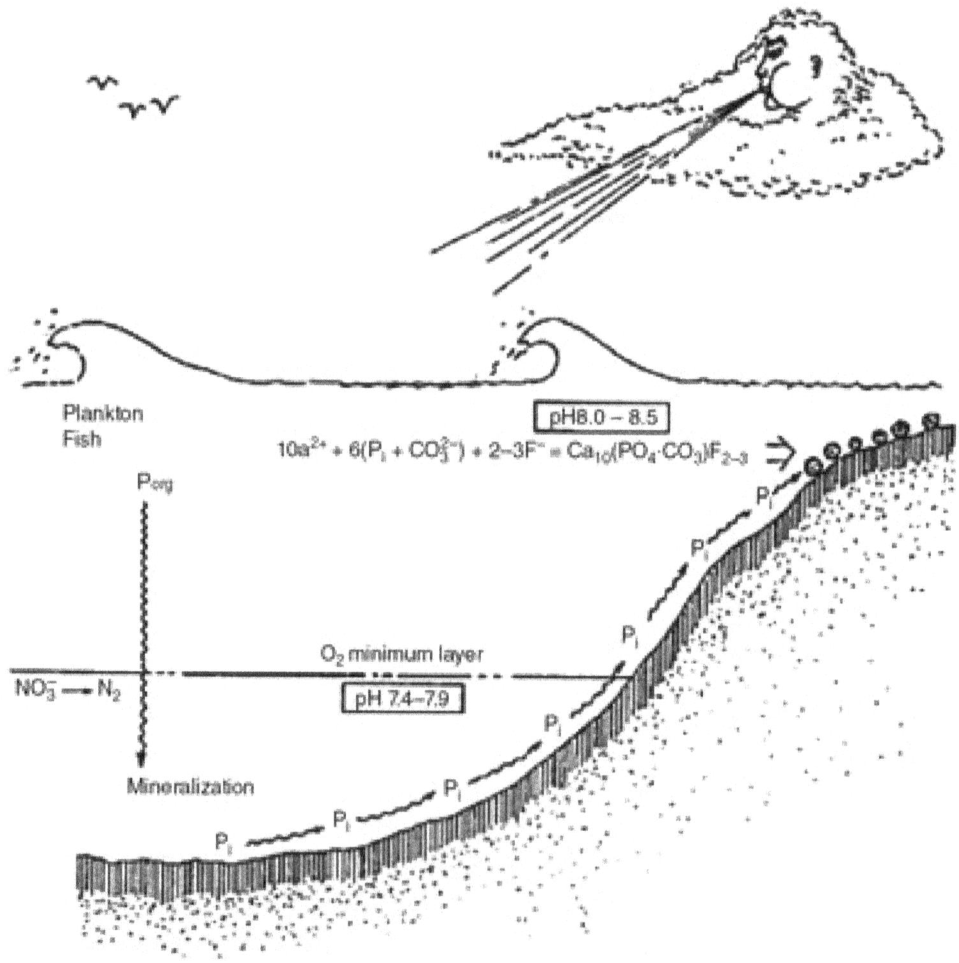

Fig. 7. The schematic presentation of formation of phosphorite in marine environment [44].

contribute to the phosphorite formation. Locally elevated, excreted orthophosphate becomes available for the precipitation as phosphorite by reacting with seawater calcium [44].

Authigenic phosphorite formation at some eastern continental margins, where upwelling, if occurred at all, was a weak and intermittent process that may have been formed more directly as a result of intracellular bacterial phosphate accumulation, which became transformed into carbonate-fluorapatite upon the death of cells accumulated in sediments in areas where the sedimentation rate was very low [44].

The model of diagenetic formation of phosphorite generally assumes the exchange of phosphate for carbonate in accretions that have the form of calcite and aragonite. The role of bacteria in this process is to mobilize phosphate by mineralizing detrital organic matter. The demonstration of this process in marine and freshwater environment under laboratory conditions leads to the hypothesis that the diagenesis of calcite to form apatite explains the origin of some deposits in the North Atlantic. The phosphorite deposits of Baja California and in the core of eastern Pacific Ocean seem to have formed as a result of partial diagenesis [44].

Igneous rocks

Apatites of igneous origin include hydrothermal veins and disseminated replacements, marginal differentiations near the boundaries of intrusions and pegmatites, but the largest deposits are intrusive masses or sheets associated with carbonatite, nepheline-syenite and other alkalic rocks [27]. Igneous rocks are classified on the basis of their [21],[45]:

1. **Color index**: means the volume percentage of dark-colored mineral and divides the rocks to **leucocratic** (color index varies from 0 to 30%), **mesocratic** (from 30 to 60%) and **melanocratic** (from 60 to 100%).
2. **Texture**: coarse-grained rocks, which crystallize at depth, are termed as **plutonic rocks** (as the rate of crystallization is slow, such rocks are holocrystalline[10]). Fine-grained rocks containing minerals embedded in glassy matrix[11] (often devitrified) are named as **volcanic** or **effusive rocks**. The crystallization of volcanic rocks takes place on the surface and is associated with rapid cooling and loss of volatile constituents of the lava. Their texture is therefore vesicular [46][12] (vesicle-rich) and hypocrystalline.[13]

   Rocks that crystallize partly at depth and partly near the surface are called **hypabyssal** [14] (**subvolcanic**). The term **porphyry** is also related to hypabyssal rocks, which are characterized by one or more than one minerals present as phenocrysts in fine-grained groundmass.

---

[10] The name for fully (100%) crystallized igneous rock [45].

[11] The matrix is defined as interstitial material between larger (skeletal) grains [46].

[12] The texture is characterized by many cavities (vesicles), which were formed by bubbles of volatile gasses during the decrease of pressure at extrusion of magma to the surface. Lava solidifies before bubbles of gases can escape to the atmosphere [46].

[13] The name for igneous rocks where the ratio of crystals to glassy phase is higher than 3:5. Rocks containing higher amount of glass are termed as hypohyaline or holohyaline [45].

[14] Denotes the intrusions of magma at shallow depths in the crust, often directly related to overlying volcanic edifices [45].

3. **Chemistry and mineralogy**: the rocks comprising more than 90 vol.% of ferromagnesian minerals, such as olivine, pyroxene, amphibole and biotite, are called **ultramafic (ultrabasic) rocks**. The rocks composed from essentially one or more ferromagnesian minerals are termed as **mafic** [15] (**basic**) **rocks**. In **mafelsic rocks**, the mafic and felsic minerals are present in approximately equal amounts. **Felsic rocks** [16] contain predominantly light-colored minerals, such as quartz, feldspar, feldspathoid and muscovite.

   An acidic rock contains > 60% SiO_2, whereas a basic rock is characterized by silica content ranging from 44 to 52% of SiO_2. Many of ultramafic rocks are ultrabasic with the content of silica < 44%, but such ultramafic rocks as pyroxenites and amphibolites are not ultrabasic, but they are rather basic [21].

Igneous rocks are formed by the solidification of silicate melt from high temperatures. Since the sequence of crystallization follows the liquidus-solidus phase relationships, the minerals of low content will normally crystallize the least, but diorite and granodiorite melts may have enough phosphorus present for the FAP phase field to intersect the liquidus and to allow early formation of fluorapatite. Later-crystallizing phases should form in the interstices between early-crystallizing phases of alkali-rich igneous rocks and should form an immiscible phosphate-rich liquid phase, which leads to large late-stage segregations of FAP, some of which are associated with magnetite [47].

Where the content of phosphorus is very low, phosphorus may remain in the fluid phase, and apatite will form during the time at which the rock re-reacts with this fluid. This reaction is termed as pneumatolitic, and formed crystals will be small and often euhedral (with crystal facets). They may be included inside preexisting mineral grains. This situation is often encountered in granites and other related siliceous igneous rocks. The concentration of apatite minerals in igneous rocks is rarely sufficient to yield the source for mining the deposits for the phosphorus content [47].

Biogenic apatites

The biogenic (endogenous) mineral deposits form in surface environments as the transformation of primary organic aggregates or as a result of biochemical processes. Since the organism produces many of the same substances that form inorganically in rocks, the biogenic minerals are not minerals in the conventional sense[1]. Biogenic minerals originate from living organisms or with their assistance (**Table 2**). These compounds are crystallized within living organisms as a result of cell activity and are surrounded by organic matter. Classical examples are the bones of vertebrates. The bones and teeth consist of fine fibers or platy crystals (**Section**) of a mineral closely related to carbonate-hydroxylapatite (**Section**). These crystals are suspended in organic collagen. The crystals of apatite, which often do not exceed 10 nm in length, comprise up to 70% of weight of dried bone. The proteins make up the remain-ing 30% [48].

---

[15] Term mafic is the abbreviation of names of elements magnesium and iron (Latin word ferrum) [45].

[16] Term felsic is the abbreviator of names of minerals **fel**dspar and **si**lica [45].

Composition		Plant or animal examples
Silica (opal, chalcedony, quartz)		Radiolaria, siliceous sponges, diatomic algae
Carbonate	Calcite	Archeocyatha, foraminifera, stromatoporoids, carbonate sponges, echinoderms, brachiopods, belemnites, ostracods, coccolithophora, cyanophycerae, purple algae, some mollusk shells, eggshells or birds and reptiles
	Calcite crystals	Eyes of trilobites and brittle star
	Aragonite	Corals, shells of mollusks and cephalopods
	Aragonite transforming to calcite	Corals, bryozoa, gastropods, pelecypods
Phosphate	Apatite	Bones, teeth, scales of vertebrates, brachiopods
	Barite, gypsum	Ear stones of animals
	Struvite	Kidney and gall stones
Oxalates	Whewellite, weddellite	Kidney and gall stones
Phosphate-bearing carbonates		Brachiopods
Magnetite		In brain tissue of birds and insects (carrier pigeons, bees, etc.), bacteria (*Magnetospirillum magnetotacticum*). Magnetite is used for navigation and orientation.
Fe-hydroxides		Shells of diatoms, pediculates of Protozoa

Table 2. Mineralogical composition of solid plant and animal tissues [48].

In addition to the occurrence within the bones and teeth of vertebrates, mineral-organic aggregates are also found in mollusk shells, solid tissues of foraminifera corals, trilobites and other arthropods, echinoderms, some algae, etc. Some other biogenic processes involve bacteria. Large deposits of native sulfur, manganese oxides and hydroxides and iron are attributed to bacterial activity. Bacterial activity is also involved in the weathering processes of sulfide oxidation and transformation of kaolinite into bauxites [48].

Biogenic apatite is one of the most promising authigenic phases in this respect, as it is present in most siliciclastic deposits and is strongly enriched in a large suite of trace elements. Biogenic (fish teeth and bones) and diagenetic apatites are essential repositories of sedimentary phosphorus. They occasionally form huge deposits, as in West Africa, which are actively mined to provide agricultural fertilizers. Some of these deposits, found in particular in the Late Precambrian of China, after chemical precipitates seem to be associated with the episodes of global glaciation. In low-temperature waters, phosphates form numerous complexes. The concentration of phosphorus in sea and river water is limited by very low solubility of apatite. Phosphate radicals often attach to the surface of iron oxyhydroxide colloids when they precipitate in estuaries [49].

Genetic classification of phosphate rocks

The genetic classification proposed by Egorov [50] includes three types of apatite ores according to the mineral assemblages [6],[51],[52],[53],[54],[55]:

i. **Silicate-apatite** (**ijolite**[17]): ijolite is a medium- to coarse-grained equigranular rock composed of nephelite and aegirite-augite, or other pyroxene, in nearly equal proportions. There are varieties richer in nephelite, approaching the composition of urtite.[18] Nephelite is an equant anhedrons, pyroxene forms euhedral crystals with zonal structure and margin of more sodic pyroxenes than that in the center, which is commonly titaniferous. Some varieties contain titaniferous melanite and iivaarite. Apatite is a prominent constituent and titanate is generally present in small amounts.

ii. **Silicate-magnetite-apatite** (**phoscorite**): phoscorites[19] are spatially and temporally associated with carbonatites, often forming multiphase phoscorite-carbonatite series [56]. The term "phoscorite" was originally used to describe the magnetite-olivine-apatite rock with a carbonate core by Russell et al [57],[58],[59].

iii. **Carbonate-apatite** (**carbonatite**): the crystalline products of low-volume and high-temperature carbonate melts that have been evolving from the upper mantle (**Fig. 8**) for at least the past 2 Ga. Many are associated with crustal complexes of alkali-rich silicate rocks from which they may have evolved by liquid immiscibility.

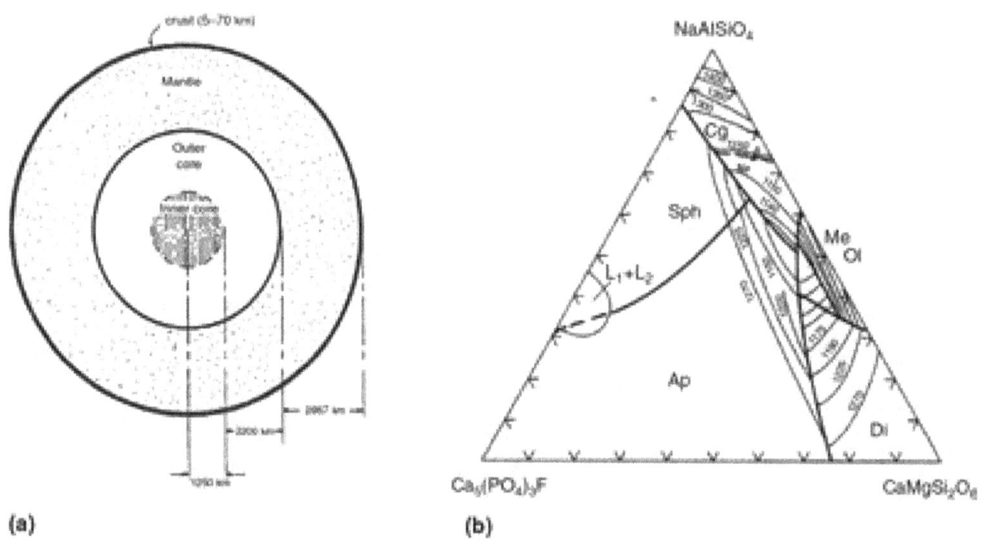

Fig. 8. The location of upper mantle on the cross-section of the Earth, core and mantle drawn to scale [44] (a). Ternary equilibrium phase diagram of the apatite bearing ijolite-urtite rock (b) approximated by the system $NaAlSiO_4$-$CaMgSi_2O_6$-$Ca_5(PO_4)_3F$ [6]: olivine (Ol), melilite (Me), silicophosphate (Sph), apatite (Ap) and immiscibility field (L_1+L_2).

---

[17] Ijolite, theralite and teschenite belong to phanerites (prelenic rocks characterized by equal or nearly equal amounts of femic components) [55].

[18] Urtite is medium-grained and light gray rock composed of about 70% nephelite, 25% aegirite or aegirite-augite [55].

[19] In the work [58], the name camaforite, i.e. Russian neologism [59], denotes the calcite-magnetite-forsterite assemblage, used as a synonym with phoscorite.

The endogenous apatite deposits can be classified[20] into the following types [60],[61]:

1. **Crystallized and concentrated in the late magmatic stage**, which can be further divided to (a) apatite-magnetite deposits (e.g. Kiruna deposit in Sweden) and (b) nepheline-apatite (e.g. Khibiny in Kola Peninsula in Russia).
2. **Skarns** (e.g. Ontario and Quebec in Canada).
3. **Carbonatites** (e.g. Sukulu in Uganda and Dorowa in Zimbabwe).
4. **Hypothermal veins with phlogopite** (e.g. Kaceres in Spain).
5. **Mesothermal type** (e.g. Toledo in Spain).
6. **Metamorphosed sedimentary phosphates** (e.g. Southern Peribaikalia, the Aldan massif in Russia (regional-metamorphosed subtype) and Karatau in Russia (contact-metamorphosis subtype)).
7. **Sedimentary phosphates**, which yield 85% of the world production, form in the sea by biochemical processes, in either:
   - **Geosynclinal sea**: e.g. in the Late Cretaceous and in Paleogene on the shelf of the Tethys geosyncline with the deposits in Morocco (Khouribga and Youssoufia), Algeria (Djebel Onk, El Kouit), Tunis (Gafsa), of the Permian geosyncline of the Rocky Mountains, with the deposits in the Phosphoria Formation in Idaho, Wyoming, Utah and Montana (USA), of the Caledonian geosyncline Karatau (Russia), large deposits are in the Upper Cretaceous in Kazakhstan, in the neighborhood of Aktyubinsk.
   - **Epicontinental sea**: e.g. in the Cenomanian of the south Russian digression and in the Jurassic of the Moscow region, at the margin of the African Shield in the Eocene complex near Hahotoe (Togo) and Taiba (Senegal), in the upper Cretaceous of Egypt [60].

Phosphate rock reserves

The Earth's crust contains about 0.27% of P_2O_5. About 200 minerals are known, which contain 1% or more P_2O_5. Minable concentrations of phosphate, containing from 5 to 35% of P_2O_5, are formed in all phases of the phosphate cycle (**Section**). The primary deposits include igneous apatites, sedimentary phosphorites and guano. The secondary deposits form from each of these as the result of weathering. Apatite is the principal primary mineral, but a number of others (**Section**) are common in the deposits formed during weathering of phosphate rocks and guano, e.g. brushite ($CaHPO_4 \cdot 2H_2O$), monetite ($CaHPO_4$), whitlockite (β-$Ca_3(PO_4)_2$), crandallite ($CaAl_3(PO_4)_2(OH)_5 \cdot H_2O$), wavellite ($Al_3(OH)_3(PO_4)_2 \cdot 5H_2O$), taranakite ($K_2Al_6(PO_4)_6(OH)_2 \cdot 18H_2O$), millisite ($(Na,K)CaAl_6(PO_4)_4(OH)_9 \cdot 3H_2O$), variscite ($AlPO_4 \cdot 2H_2O$) and strengite ($FePO_4 \cdot 2H_2O$) [27].

[20] The classification according to Dybkov and Karyakin.

Phosphate rock deposits

In general, phosphate rock reserves are non-metallic ores[21] that can be economically produced at the present time using existing technology. Phosphate rock resources include reserves and any other materials of interest that are not reserves. A reserve base is a portion of the resource from which future reserves may be developed. The classification applying to phosphate rocks include [23].

- **Resource** is defined as a concentration of naturally occurring phosphate material in such a form or amount for which the economic extraction of a product is currently or potentially feasible. The resources are divided into many categories depending on the amount of pertinent information available to define the amount of material potentially available and if it is economic, marginally economic or sub-economic to exploit these resources.
- **Reserve base** is the part of an identified resource that meets the minimum criteria related to current mining and production practices including grade, quality, thickness and depth.
- **Reserves** are the part of the reserve base that can be economically extracted or produced at the time of the determination. They may be termed as marginal, inferred or inferred marginal reserves. This does not signify that the extraction facilities are in place or functional.
- The grades of apatite deposits from the economic point of view are introduced in **Table 3**. The locations of the world's phosphate deposits are shown in **Fig. 9** and the average content of P_2O_5 is listed in **Table 4**.

The most known Miocene phosphate deposits (**Table 1**) are in North Carolina, Florida, Venezuela, California, Baja California and Peru. In several cases, these emerged deposits are only the up dip limit of a larger Miocene section that extends seaward beyond the coastal plain and constitute large portions of the upper sediment regime that built the modern continental shelves [26].

Grade	P_2O_5 [wt.%]	BLP [%]	Location
1 Economic	20	40.70	Florida and Moroccan sedimentary phosphorites, Kola and Palabora crystalline igneous apatites
2 Sub-economic	5 – 20	10.93 – 40.70	Western USA phosphoria, Russia nepheline-apatites
3 Non-economic	1 – 5	2.19 – 10.93	Low-grade ores, phosphatic limestones
4 Non-phosphatic	0.1 – 1.0	0.22 – 2.19	Widely distributed apatite in almost all igneous rocks

Table 3. Grades of apatite deposits [25].

[21] Ores of economic value can be classified as metallic or non-metallic according to the use of the mineral. Certain minerals may be mined and processed for more than one purpose. In one category, the mineral may be metal and non-metallic ore, e.g. bauxite used for the production of aluminum and ceramics, respectively [62].

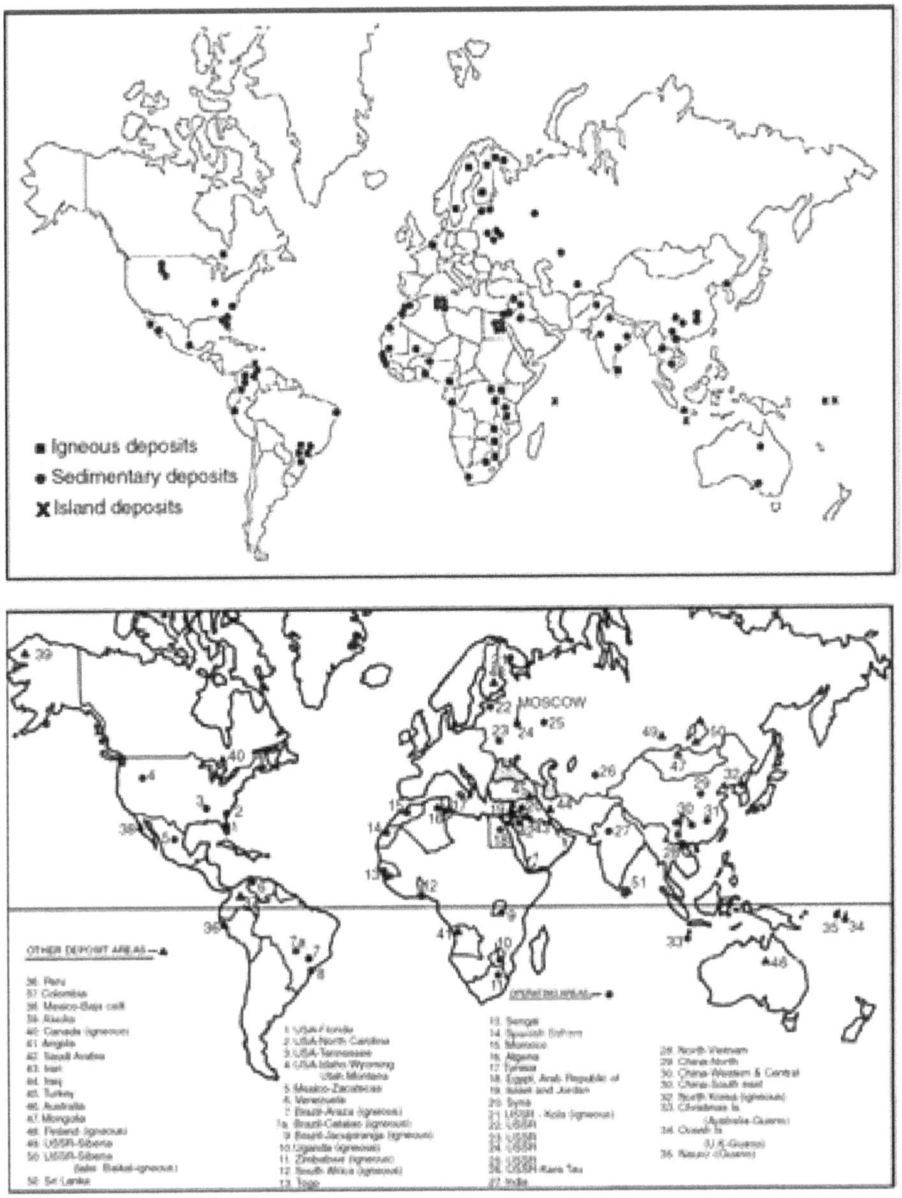

Fig. 9. The overview of the world's phosphate deposits (a) [20] and the location of the major phosphate rock-producing areas of some known deposits (b) [26],[63].

Locally, these Miocene sediments are exposed on the seafloor; however, generally, they are buried below thin covers of Plio-Pleistocene and Holocene surface sediments. Thus, there is high potential for discovering new phosphate deposits within the Miocene sediments on the world's continental shelves, because [26]:

a. Phosphate genesis is known to occur throughout the shelf in upper slope environments.

b. Thicker and more extensive sequences of Miocene sediments occur on the shelves than on adjacent coastal plains.

c. The shallow subsurface Neogene geology of most of the world's shelves is poorly known.

Source	%	Source	%	Source	%
Fluorapatite	42	Tunisia (sedimentary)	28	Basic slag	10 – 20
Kola (igneous)	40	West USA (phosphoria)	18 – 30	Bone meal	20
Nauru (phosphorite)	39	Queensland	16 – 30	Guano	12 – 15
Florida (sedimentary)	35	Venezuela	20	California (seabed)	30
Kazakhstan	23	China (Yunnan)	32 – 36	Australia (Queensland)	24
Morocco (sedimentary)	35	Kola (nepheline)	12 – 20	–	–

Table 4. Grades of apatite deposits [25].

Two important sediment relationships were developed concerning reworked phosphate in surficial sediments on the North Carolina continental shelf [26],[63]:

i. The distribution of phosphate in surface sediments closely reflects the distribution within underlying Miocene sediments units.

ii. The process of reworking significantly dilutes the concentration of phosphate in the surficial sediments.

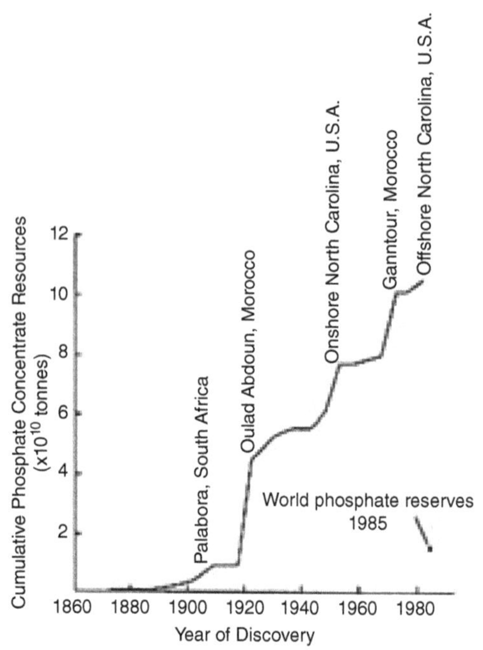

Fig. 10. The history of the discovery of the world's phosphate resources [22].

These relationships were also recognized on the shelves of northwest and southwest Africa and could represent important exploration tools for richer Tertiary[22] phosphorites occurring within the shallow subsurface on many continental shelves in the world [26].

[22] The term Tertiary (geologic period from 66 to 2.58 ma) is no longer recognized by the International Commission on Stratigraphy.

The deposits of apatite of igneous origin occur as intrusive masses or sheets, as hydrothermal veins or disseminated replacements, as marginal differentiations along or near the boundaries of intrusions or as pegmatites. Intrusive masses are the largest of these deposits. They are commonly associated with alkalic igneous rock complexes, many of which such as those in Africa, Brazil and Sweden are associated with the rift valley structures. Carbonatite, ijolite, nepheline-syenite and pyroxenite are common members of the rock assemblage. Many of these complexes have a ring like structure, with carbonatite as the central core [27].

Phosphate deposits have been discovered within the past 100 years at the rate far greater than the rate of consumption (**Fig. 10**). Since new phosphate deposits are expected to be discovered in the future, the oil exploration programs have probed most of the coastal sedimentary basins of the world during the past 20 – 30 years, and any large-scale discoveries of phosphate rock would probably have occurred in conjunction with these activities [22].

Other sources of phosphorus

Other important commercial sources of phosphorus (**Table 4**) include [25]:

i. **Guano** [25],[27],[64],[65],[66]: natural deposit (accumulation) formed from decaying bones and excreta from fish-eating birds.[23] Fresh seafowl droppings contain about 22% N and 4% P_2O_5. Bat guanos are the most abundant in the cave areas of temperate and tropical regions. Although many bat guano deposits were found and mined, most of them are measured in hundreds or thousands of tons, and only sporadic production is obtained from them now. Seafowl deposits are mainly confined to islands and coastal regions at low latitudes. The largest lie along the west coasts of lower California, South America and Africa and on islands near the equatorial currents.

   It was known that bird dung was utilized by the Carthaginians as early as 200 BC in order to improve crop yields. The content of P_2O_5 in guano can vary in dependence on the age of the deposit,[24] the local climate and the bird kind. Guano deposits are found in Chile, Peru, Mexico, Seychelles, Philippines, the Arabian Gulf and elsewhere, but they account for less than 2% of the world's phosphate production.

   Guano is used almost exclusively as fertilizers. The Nauru and Christmas Island phosphorite deposits may be guano in origin, but they are of very limited extent. It

---

[23] It is estimated that marine birds may take out as much as 5.10^{10} g of phosphorus from the ocean each year [67]. The Spanish name "guano" has an origin in Quechua word "huanu" (i.e. dung).

Guano is composed of bird droppings, and although birds existed as early as the mid-Mesozoic, their major development did not occur until the Cenozoic era, probably in the Eocene period. It follows that some seamount deposits are too old to have originated from bird droppings [64].

Most of large accumulations of guano are formed on the surface by seafowl, but smaller quantities are formed by bats and to a lesser extent by other cave-dwelling mammals and birds [27].

[24] Recent guano contains 10 – 12% of P_2O_5, but leached guano contains 20 – 32%. The mineralogy of guano is complex. Slightly decomposed deposits contain soluble ammonium and alkali oxalates, sulfates and nitrates and a variety of magnesium and ammonium-magnesium phosphates. Deeply decomposed guano consists chiefly of calcium phosphates (for example, monetite or whitlockite) [27].

is believed that rainwater can carry soluble phosphate from guano and trickle over rocks, where phosphate interacts to form phosphatic layers, e.g. phosphatized coral rock. Bird guano, mainly from Peru, achieved the greatest importance in about the middle of the 19th century, shortly before the phosphate rock industry began to establish itself.

ii. **Basic slag** [25]: is a minor source of phosphorus. This waste is the product from blast furnaces operating on iron ores with significant content of phosphorus. Basic slag contains tetracalcium phosphate ($Ca_3(PO_4)_2{\cdot}CaO$) and silicocarnotite ($Ca_3(PO_4)_2{\cdot}Ca_2SiO_4$), which are applied directly as fertilizers. Recorded world production is mainly from France, Germany and Luxembourg.

iii. **Meat and bone meal (MBM) or bone ash** [25],[68]: as an animal byproduct, MBM contains not only substantial amounts of phosphorus in soluble organic form but also calcium and microelements. Ground (bone meal) or calcined and ground (bone ash) bones were recognized as a source of phosphorus at an early date.

Other important commercial sources of phosphorus are casein and lecithin. Casein is obtained from bovine milk. Lecithin was extracted from soy bean oil [25].

Phosphorite weathering derivates

Weathering (**Section**) leads to the formation of enriched residual and replacing depos-its from phosphatic deposits not otherwise minable. The Tennessee "brown rock" phosphate deposits consist of nowadays residuum developed through the decomposition of phosphatic limestones of Ordovician age. The "river pebble" deposits prominent in early history of phosphate mining in Florida and South Carolina are mostly placers formed by alluvial concentration of phosphatic pebbles eroded from the phosphatic formations of adjacent terrain [27].

The Tennessee "white rock" and Florida "hard rock" deposits were formed by the redeposition of phosphate derived from the decomposition of apatite under more advanced weathering. The same decomposition-phosphatization process accounts for the formation of calcium aluminum phosphate and aluminum phosphate in the "leached zone" of the Bone Valley field and deeply weathered Cretaceous and Eocene deposits of west Africa [27].

Geological role of apatite

As was described above, the number of different elements can substitute into the structure of apatite,[25] and this mineral can contain a number of trace elements by the substitution in both anion and cation sites. This means that apatite can be used as an indicator of planetary halogen compositions. The quantitative ion microprobe measurements of apatite from lunar basalts showed that portions of the lunar mantle and/or crust are richer in volatile species than previously thought [4].

[25] The tools based on the isotope composition of apatite were already described in **Section**.

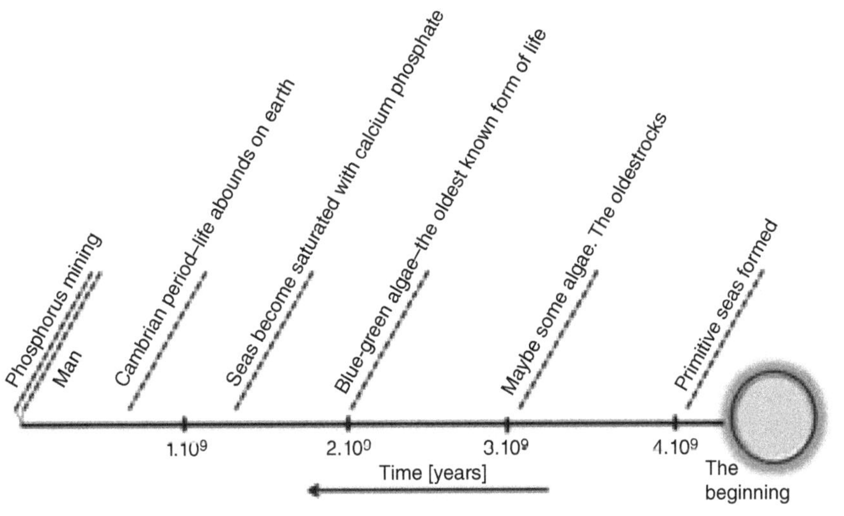

Fig. 12. The establishment of the Earth's cycle pattern of phosphorus [78].

precipitate formed when an island sea had invaded a land area, the new sediment became landlocked. The new landlocked sedimentary deposits are more easily leached than igneous rocks from which they are derived. When the seas recede sufficiently to expose the new sediments to the greater solvent action of fresh water, the sediments begin to weather and the cycle is complete. Best estimates of the cycle time of phosphorus in the oceans today are in the range of 50,000 years. This is a short period compared to $3{\cdot}10^9$ years, which were required for the saturation of oceans for the first time [78].

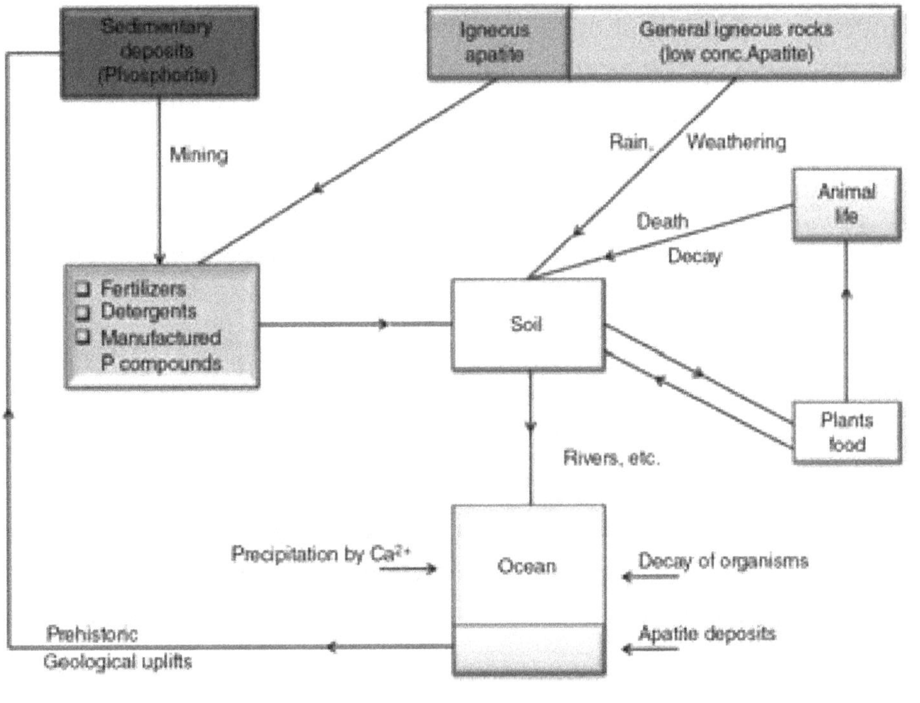

Fig. 13. Natural and artificial cycles of phosphorus [25].

Recently, man has only a slightly stronger influence on the total amount of the Earth's phosphorus than his prehistoric ancestors. If man made a significant alteration in the cycles of phosphorus, it had an impact on the cycles of fresh surface waters. The detergent phosphates have been blamed for degrading freshwater lakes and there is no doubt that several lakes have been overabundant with phosphates and sewage. Sewage treatment will alleviate most of the problems associated with point-source loading of lakes [78].

The overall natural and artificial cycles involving phosphorus are introduced in **Fig. 13**.

Weathering of apatite

Weathering and leaching processes from millions of years ago led to the transfer of phosphate to rivers and oceans where it was concentrated in shells, bones and marine organism that were deposited on the sea floor. Subsequent uplift and other geological movements led to these accumulations becoming dry land deposits [25],[78].

Generally, weathering of apatite occurs synergistically through biotic and abiotic processes and leads to the release of mineral phosphate. Inorganic phosphate cannot be assimilated by plants, but it can be converted to the bioavailable form orthophosphate ($HPO_4{}^{2-}$, $H_2PO_4{}^{-}$) by some species of phosphate-solubilizing fungi and bacteria. The main mechanism underlying the microbial phosphate solubilization is the secretion of organic acids that, by changing the soil pH and acting as chelators, may induce the dissolution of phosphorus from minerals and its release into the pore water of soils [79],[80]. The dissolution of apatite is described in **Section**.

Apatite represents an important source of inorganic P for natural ecosystems and may favor the establishment of microbial communities able to exploit it [79]. The microorganisms can cause the fixation or immobilization of phosphate, either by promoting the formation of inorganic precipitates or by the assimilation of phosphate into organic cell constituents on intracellular polyphosphate granules. Insoluble forms of inorganic phosphorus, e.g. calcium, aluminum and iron phosphates, may be solubilized through the microbial action. The mechanisms by which the microbes accomplish this solubilization vary [44]:

i. The first mechanism may be the production of inorganic or organic acids that attack the insoluble phosphates.

ii. The second mechanism may be the production of chelators such as gluconate and 2-ketogluconate, citrate, oxalate and lactate. All these chelators can complex the cation portion of insoluble phosphate salts and thus force their dissociation.

iii. The third mechanism of phosphate solubilization may be the reduction of iron in ferric phosphate, e.g. strengite ($Fe^{3+}PO_4{\cdot}2H_2O$ [81]), to ferrous iron by enzymes and metabolic products of nitrate reducers such as *Pseudomonas fluorescens* and *Alcaligenes* spp. in sediments.

iv. The fourth mechanism is the production of hydrogen sulfide (H_2S), which can react with iron phosphate and precipitate it as iron sulfide, thereby mobilizing phosphate, as in the reaction [44]:

$$2\ FePO_4 + 3\ H_2S \rightarrow 2\ FeS + 2\ H_3PO_4 + S^0 \quad (1)$$

The phosphate-solubilizing ability is a feature of many free-living and plant-symbiotic bacterial taxa, such as [79]:

- ***Pseudomonas*** [82]: the genus *Pseudomonas* sensu stricto comprises many species characterized by their metabolic diversity and by a wide range of niches that they can colonize.
- ***Rhizobium*** [83]: specific group of bacteria that have the capability of symbiotic nitrogen fixation.
- ***Burkholderia*** [84]: aerobic, non-spore-forming bacteria. *Burkholderia* is very versatile and occupies a wide range of ecological niches.

Microbial rock weathering is common in all climate zones and usually acts very slowly [80].

Fission track and apatite fission-track analysis

The fission-track (FT) dating is a radiometric dating method[26] based on the analysis of radiation damage trails (fission tracks) in uranium-bearing, non-conductive minerals and glasses. It is routinely applied to the minerals apatite, zircon and titanite. Fission tracks are produced continuously through geological time as a result of spontaneous fission track of ^{238}U atoms[27] that undergo spontaneous fission.[28] The atom splits into two parts that move rapidly in opposite directions, creating a long thin region of damage. The submicroscopic features with an initial width of approximately 10 nm and the length of up to 20 μm can be revealed by chemical etching.[29] Crucially, fission tracks are semi-stable features that can self-repair (shorten and eventually disappear) by the process known as annealing at a rate that is a function of both time and temperature. The extent of any track shortening (exposure to elevated temperatures) in a sample can be quantified by examining the distribution of fission-track lengths [6], [85],[86],[87],[88].

This unique sensitivity of the apatite fission-track system is now of considerable economic importance due to the coincidence between the temperature range over which annealing occurs and that over which liquid hydrocarbons are generated. Other applications include the determination of timing of emplacement and the thermal history of ore deposits. There is

---

[26] Radiometric dating techniques are, in general, complementary to one another, and each method produces an age with the special meaning, such as the last outgassing, the last melting accompanied by mixing with isotopically separate material and the last heating to remove the track. Fission-track dating is conceptually the simplest of several dating techniques that provide absolute measures of time from slow but statistically steady decay of radioactive nuclides [86].

[27] Most radiometric dating processes are based on the statistical regularity of the decay of one parent radionuclide into a daughter nuclide, for example, ^{40}K into ^{40}Ar, ^{87}Rb into ^{87} Sr or ^{238}U ,^{235}U and ^{232}Th into ^{206} Pb, ^{207}Pb and ^{208}Pb, respectively. The age of sample can be then determined by measuring relative abundance of parent and daughter in any pair. The major isotope of uranium (^{238}U) decays at a spontaneous rate of ~10^{-16} per year [86].

[28] Fission track can also be created artificially (induced track) by irradiating the mineral specimen with thermal neutrons in a nuclear [88].

[29] The mineral grain is ground and polished to expose a flat surface inside the crystal. It is then immersed in a chemical etchant that preferentially attacks the regions of damage, widening them and making them visible under optical microscope. The track appears in the apatite torch readily in 20 to 30 s when immersed in diluted nitric acid [88].

abundant literature on both fission-track dating and its use in evaluating the tectonic and thermal history of rocks [6],[89],[90],[91].

Apatite is the most frequently used material for fission-track dating [92]. Apatite fission-track (AFT) analysis serves as a thermochronological tool to investigate the low-temperature thermal history of rocks below ~120°C [93],[94]. The estimates of closure temperatures for fission-track retention in apatite are usually in the range from 75 to 120°C at cooling rates between 1 and 100°C/m.y. [6].

Thermochronology may be described as the quantitative study of the thermal histories of rocks using temperature-sensitive radiometric dating methods such as $^{40}Ar/^{39}Ar$ and K-Ar, fission track and (U-Th)/He. Among these different methods, apatite fission track and apatite (U-Th-Sm)/He (AHe) are now, perhaps, the most widely used thermochronometers, as they are the most sensitive to low temperatures (typically between 40 and 125°C for the durations of heating and cooling in the extent of 10^6 years). They are ideal for investigating the tectonic and climate-driven surficial interactions that take place within the top few (<5 km) kilometers of the Earth's crust. These processes govern the landscape evolution, influence the climate and generate the natural resources essential to the well-being of mankind [85],[95].

Extraterrestrial apatite

On Earth, magmatic volatiles (i.e. H_2O, F, Cl, C-species and S-species) play an important role in the physicochemical processes that control thermal stabilities of minerals and melts, in magma eruptive processes and in the transportation of economically important metals. On the Moon, magmatic volatiles in igneous systems are poorly understood, and the magmatic volatile inventory of lunar interior, aside from being very low, is not well constrained. Although the Moon is a volatile-depleted planetary body, there is evidence indicating that magmatic volatiles have played a role in igneous processes on the Moon. Specifically, magmatic volatiles were implicated as the propellants that drove fire-fountain eruptions, which produced the pyroclastic glass deposits encountered at the Apollo 15 and 17 sites [96]. That is supported by recent discoveries of water-rich apatite from lunar mare basalts [97],[98].

Apatite was found in a large number of samples of igneous lunar rocks, although it typically occurred in only trace amounts and is typically reported as coexisting with REE-merrillite [$(Mg,Fe)_2REE_2Ca_{16}P_{14}O_{56}$], and those two minerals make up the primary mineralogical budget for P on the Moon [96]. Merrillite, also known as the mineral whitlockite (or more precious and dehydrogenated whitlockite) [99], is one of the main phosphate minerals, along with apatite, which occur in lunar rocks, in Martian meteorites and in many other groups of meteorites. Significant structural differences between terrestrial whitlockite and lunar (and meteoritic) varieties require the use of "merrillite" name for the H-free extraterrestrial material, and the systematic enrichment of REE in lunar merrillite requires the use of "REE-merrillite". Lunar merrillite, ideally $(Mg,Fe^{2+},Mn^{2+})_2[Ca_{18-x}(Y,REE)_x]$ $(Na_{2-x})(P,Si)_{14}O_{56}$, contains high concentrations of Y + REE [100],[101].

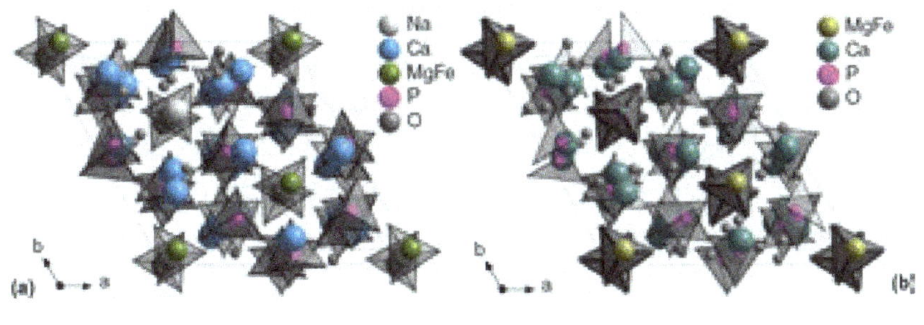

Fig. 14. The structure of lunar merrillite (a) [102] and terrestrial whitlockite (b).

Lunar merrillite (**Fig. 14(a)**, trigonal, the space group R3c with the cell parameters ***a*** = 10.2909 Å, ***c*** = 36.8746 Å, ***c***:***a*** = 3.5832 and ***V*** = 3381.93 Å^3) and terrestrial whitlockite (**Fig. 14(b)**, trigonal, the space group R3c with the cell parameters ***a*** = 10.3300 Å, ***c*** = 37.1030 Å, ***c***:***a*** = 3.5918 and ***V*** = 3428.79 Å^3) have largely similar atomic arrangements, but the phases differ due to the presence or absence of hydrogen. In whitlockite, H is an essential element and allows the charge balance. Hydrogen is incorporated into the whitlockite atomic arrangement by disordering one of the phosphate tetrahedra and forming the $PO_3(OH)$ group. Lunar merrillite is devoid of hydrogen; thus, no disordered tetrahedral groups exist. The charge balance for the substituents Y and REE (for Ca) is maintained by $Si^{4+} \leftrightarrow P^{5+}$ tetrahedral substitution and $\square \leftrightarrow Na^+$ substitution at Na site [99],[102].

A number of sources potentially contributed to the overall inventory of lunar water, including primary indigenous water acquired during lunar accretion, late addition of water through asteroidal and cometary impacts and solar wind implanting H into lunar soils. The average D/H ratios of apatite in norite (Apollo sample 78235) and in the granite clast (14303) are consistent with the estimates for the H isotopic composition of recent bulk-Earth and terrestrial mantle. By contrast, the average H isotopic composition of apatites in norite 77215 is lower. The content of water in norite parental melts provides strong evidence that the magmas involved in secondary crust production on the Moon were hydrated, in agreement with recent findings of water in lunar ferroan anorthosites. Water they contain, locked in the crystalline structure of apatite, is characterized by an H isotopic composition similar to that on Earth and in some carbonaceous chondrites [103].

Apatite preserves a record of halogen and water fugacities that existed during the waning stages of crystallization of planetary magmas, when they became saturated in phosphates. The thermodynamic formalism based on apatite-merrillite equilibria that makes it possible to compare the relative values of halogen and water fugacities in Martian, lunar and terrestrial basalts, accounting for possible differences in pressure, temperature and oxygen fugacities among the planets, was described by Douce and Roden [104].

They showed that planetary bodies have distinctive ratios among volatile fugacities at apatite saturation and that these fugacities are, in some cases, related in a consistent way to volatile fugacities in the mantle magma sources. Their analysis shows that the Martian mantle parental to basaltic SNC meteorites was dry and poor in both fluorine and chlorine compared to the terrestrial mantle. Limited data available from Mars show no secular variations in mantle halogen and water fugacities from ~4 Ga to ~180 Ma. Water and halogens found in recent

Martian surface rocks have thus resided in the planet's surficial systems since at least 4 Ga and may have been degassed from the planet's interior during the primordial crust-forming event. In comparison to the Earth and Mars, the Moon, and possibly the eucrite parent body too, appear to be strongly depleted not only in H_2O but also in Cl_2 relative to H_2O. The chlorine depletion is the strongest in mare basalts, perhaps reflecting an eruptive process characteristic with large-scale lunar magmatism [104].

Mars does not recycle crustal materials via the plate tectonics. For this reason, the magmatic water reservoir of the Martian mantle has not been affected by the surface processes, and the deuterium/hydrogen (D/H) ratio of this water should represent the original primordial Martian value. Following this logic, hydrous primary igneous minerals on the Martian surface should also carry this primordial D/H ratio, assuming no assimilation of Martian atmospheric water during the crystallization and no major hydrogen fractionation during the melt degassing. Hydrous primary igneous minerals, such as apatite and amphibole, are present in Martian meteorites here on Earth. Provided these minerals have not been affected by terrestrial weathering, Martian atmospheric water or shock processes after the crystallization, they should contain a good approximation of the primordial Martian D/H ratio. As Nakhla was seen to fall on the Egyptian desert in 1911, the terrestrial contamination is minimized in this meteorite. The nakhlites are also among the least shocked Martian meteorites. Therefore, apatite within Nakhla could contain primordial Martian hydrogen isotope ratios. The similar D/H ratios indicate that the Earth and Mars, and possibly the other terrestrial planets, accreted water from the same source [105].

Vesta, as the second most massive asteroid, has long been perceived as anhydrous. Recent studies suggesting the presence of hydrated minerals and past subsurface water have challenged this long-standing perception. The volatile components indicate the presence of apatite in eucrites. Eucritic apatite is fluorine rich with minor chlorine and hydroxyl (calculated by difference) [106],[107],[108].

The future of phosphate rocks

The biochemist and sci-fi author Izaak Asimov said [60]: "In the future coal will be probably substituted by nuclear energy, wood by plastics, meat by yeast and suspecting solitude by friendship, but there is no substitute for phosphorus."

The search for phosphate rock deposits became a global effort in the 20th century as the demand for phosphate rocks increased. The development of deposits further intensified in the 1950s and 1960s. The world production reached its peak in 1987 – 1988 and then again in 2008 with over 160 million metric tons (mmt) of the product. Phosphate rock mining has evolved over time, and worldwide, it relies on high volume and advanced technology using mainly open-pit mining methods and advanced transportation systems to move hundreds of millions of tons of overburden to produce hundreds of millions of tons of ore, which are beneficiated to produce approximately 160 mmt of phosphate rock concentrate per year. The concentrate of suitable grade and chemical quality is then used to produce phosphoric acid, the basis of many fertilizer and non-fertilizer products [23]. The world phosphate production rate since 1850 according to JASINSKI [109] and ABOUZEID et al [20] is shown in **Fig. 15**.

The estimates of the world's phosphate reserves and availability of exploitable deposits vary greatly and the assessments of how long it will take until these reserves are exhausted vary also considerably. Furthermore, it is commonly recognized that the high-quality reserves are being depleted expeditiously and that the prevailing management of phosphate, a finite non-renewable source, is not fully in accordance with the principles of sustainability. The depletion of current economically exploitable reserves is estimated to be completed in some 60 to 130 years. Using the median reserve estimates and under reasonable predictions, it appears that phosphate reserves would last for at least 100+ years [20].

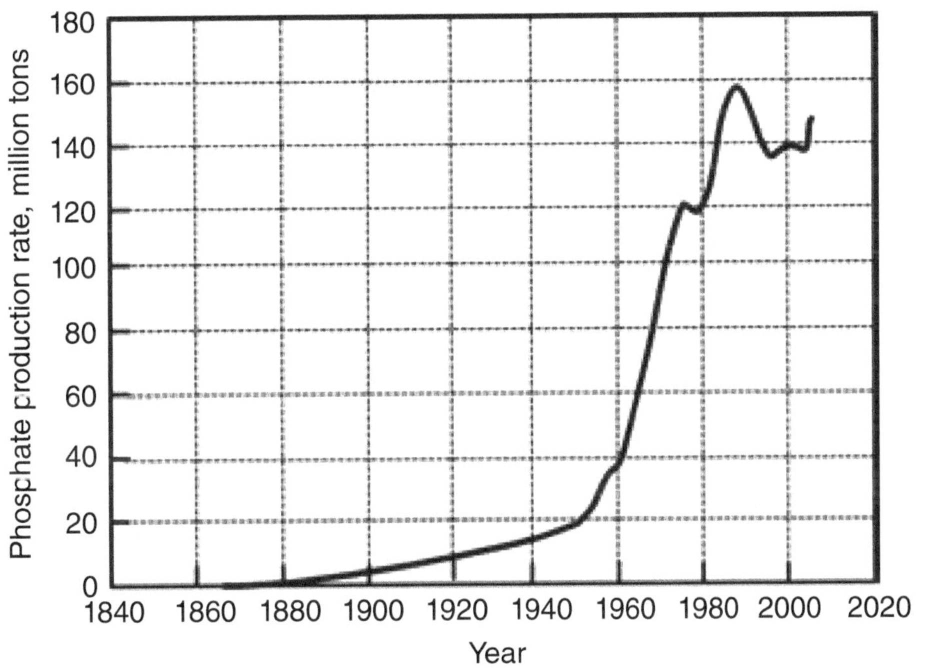

Fig. 15. World phosphate production rate [20].

Preliminary estimates of phosphate rock reserves range from 15,000 mmt to over 1,000,000 mmt, while the estimates of phosphate rock resources range from about 91,000 mmt to over 1,000,000 mmt. Using available literature, the reserves of various countries were assessed in the terms of reserves of concentrate. The IFDC[30] estimate of worldwide reserve is approximately 60,000 mmt of concentrate. Based on the data gathered, collated and analyzed for the IFDC report, there is no indication that a "peak phosphorus" event will occur in 20 – 25 years. Based on the data reviewed, and assuming current rates of production, phosphate rock concentrate reserves to produce fertilizers will be available for the next 300 – 400 years [23].

Phosphate rock prices will increase when the demand approaches the limits of supply. When the phosphate rock prices increase, some resources will become reserves, marginal mining

[30] International Fertilizer Development Center.

projects will become viable and the production will be stimulated. In the future, fuel and fuel-related transportation costs may become even more important components in the world phosphate rock production scenario. The political disruption is always an unknown factor, and it can profoundly influence the supply and demand for fertilizer raw materials on a worldwide basis [22].

Non-apatitic phosphate minerals

Apart from those in the supergroup of apatite minerals, the well-known phosphate minerals include [25]:

1. **Autunite** [110],[111]: is orthorhombic mineral (space group I4/MMM) of the composition of ($Ca[(UO_2)(PO_4)]_2(H_2O)_{11}$, **Fig. 16**), which crystallizes in the space group Pnma with the cell parameters: $\boldsymbol{a}$ = 14.0135 Å, $\boldsymbol{b}$ = 20.7121 Å, $\boldsymbol{c}$ = 6.9959 Å, $\boldsymbol{V}$ = 2030.55 Å^3 and $\boldsymbol{Z}$ = 4. It belongs to the most abundant and widely distributed uranyl phosphate minerals. The structure contains the well-known autunite-type sheet with the composition $[(UO_2)(PO_4)]^-$ resulting from the sharing of equatorial vertices of uranyl square bipyramids with phosphate tetrahedra. Calcium atom in the interlayer is coordinated by seven H_2O groups and two longer distances to uranyl apical O atoms. Two symmetrically independent H_2O groups are held in the structure only by hydrogen bonding. The bond-length-constrained refinement provides a crystal-chemically reasonable description of the hydrogen bonding.

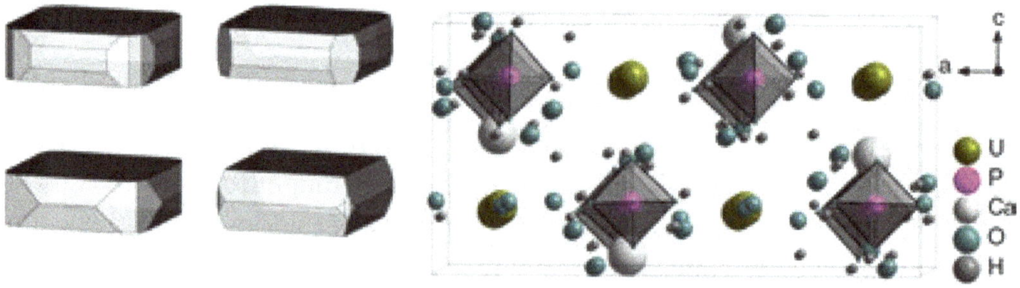

Fig. 16. The examples of forms and the structure of mineral atunite [110] viewed along the b-axis.

The mineral was named by Henry J. Brooke and William H. Miller in 1854 after the typical locality at Saint Symphorien, Autun District, Saône-et-Loire, France. Autunite dehydrates rapidly in air (except for high relative humidity) to tetragonal meta-autunite (P4/NMM, $\boldsymbol{a}$ = 6.96 Å, $\boldsymbol{c}$ = 8.40 Å and c:$\boldsymbol{a}$ = 1.21) [112]. The loss of O12 and O13 from autunite results in the formula $Ca[(UO_2)(PO_4)]_2(H_2O)_7$ (**Fig. 17**) [110].

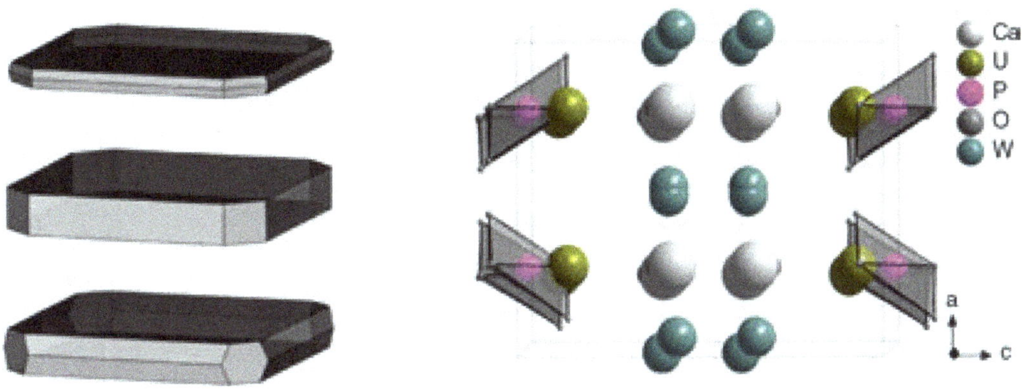

Fig. 17. The examples of forms and the structure of mineral meta-atunite [112] viewed along the b-axis.

2. **Crandallite** [113]: $CaAl_3(OH)_6[PO_3(O_{1/2}(OH)_{1/2}]_2$, has hexagonal structure (a = 7.005 Å and c = 16.192 Å), which is analogous with alunite. The structure of mineral (**Fig. 18**) consists of corner-sharing Al octahedra, which are linked into trigonal and hexagonal rings to form the sheets perpendicular to the c-axis. Ca ions, surrounded by 12 oxygen and hydroxyl ions, lie in large cavities between the sheets. Each phosphate tetrahedron shares three corners with three Al octahedra from a trigonal ring in the sheet. The unshared corner is turned away from the trigonal hole towards the adjacent sheet to which it is hydrogen bonded. The mineral "deltaite" was found to be identical to crandallite, within the accuracy of the structural results.

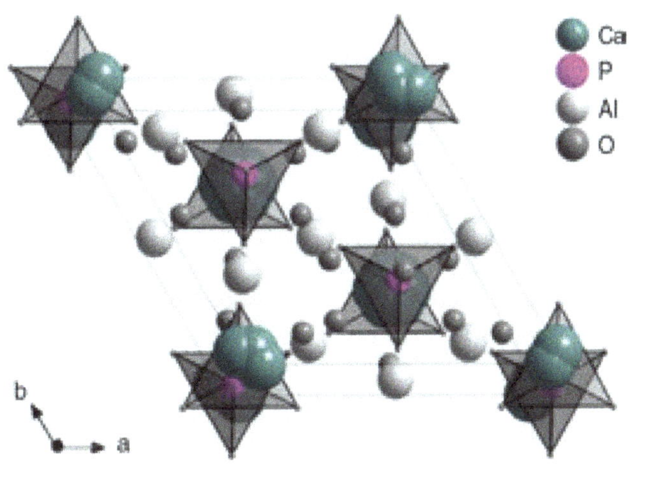

Fig. 18. The structure of mineral crandallite [113] viewed along the c-axis.

3. **Lazulite** [114],[115]: monoclinic mineral of the composition of $MgAl_2(PO_4)_2(OH)_2$, which crystallizes in the $P2_1/c$ space group and has the cell parameters: $\boldsymbol{a}$ = 7.16 Å, $\boldsymbol{b}$ = 7.26 Å and c = 7.24 Å and β = 120.67°. The mineral is the magnesium analogue of scorzalite ($Fe^{2+}Al_2(PO_4)_2(OH)_2$). Named in 1795 by Marten H. Klaproth from the Arabic word meaning "heaven," in allusion to its color.

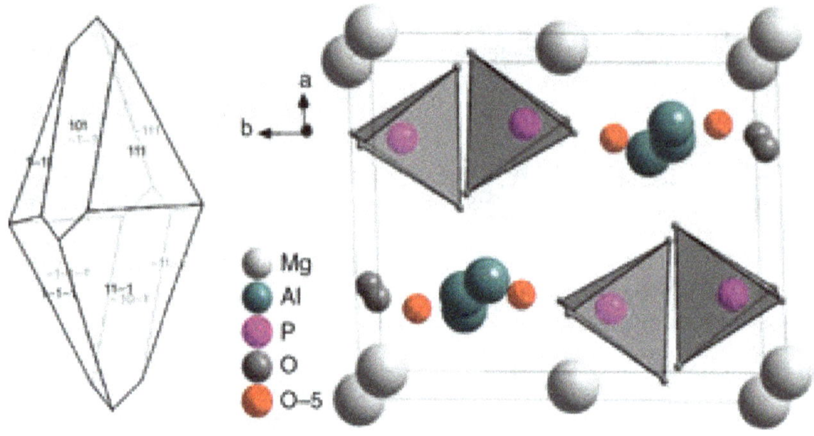

Fig. 19. The form and the structure of lazulite [114] viewed along the c-axis.

4. **Millisite** [116]: tetragonal mineral of the composition of $NaCaAl_6(PO_4)_4(OH)_9{\cdot}3H_2O$ (the space group $P4_12_12$, $\boldsymbol{a}$ = 7.00 Å and $\boldsymbol{c}$ = 18.99Å, $\boldsymbol{a{:}c}$ = 1:2.713 and **Z** = 2), which was named in 1942 by Esper Signius Larsen III in the honor of Mr. F.T. Millis of Lehi, Utah, who had collected the first specimens.

5. **Monazite** [117],[118]: is natural light rare-earth element phosphate that generally contains large amounts of uranium and thorium. The monazite-type compounds (AXO_4, **Fig. 21**) form an extended family that is described in this review in the terms of field of stability versus composition. They crystallize in a monoclinic lattice with the space group $P2_1/N$ (**Z** = 4).

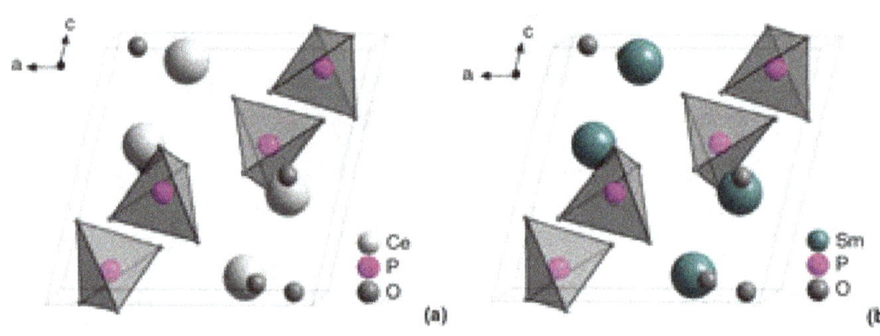

Fig. 20. The structure of monazite-(Ce): $\boldsymbol{a}$ = 6.7902 Å, $\boldsymbol{b}$ = 7.0203 Å, $\boldsymbol{c}$ = 6.4674 Å and $\boldsymbol{\beta}$ = 103.38° (a) and monazite-(Sm) (b): $\boldsymbol{a}$ = 6.6818 Å, $\boldsymbol{b}$ = 6.8877 Å, $\boldsymbol{c}$ = 6.3653 Å and $\boldsymbol{\beta}$ = 103.386° (b) [118] viewed along the b-axis.

Monazite (**Fig. 20**) and xenotime dimorphs[31] [119] are the most ubiquitous rare-earth (REE) minerals. Monazite incorporates preferentially larger, light rare-earth elements (LREEs, here, La-Gd), whereas xenotime tends to incorporate smaller, heavy rare-earth elements (HREEs, here, Tb-Lu, and Y).

[31] One chemical compound is capable of crystallizing in two different systems, e.g. $CaCO_3$ can occur as calcite or aragonite [119].

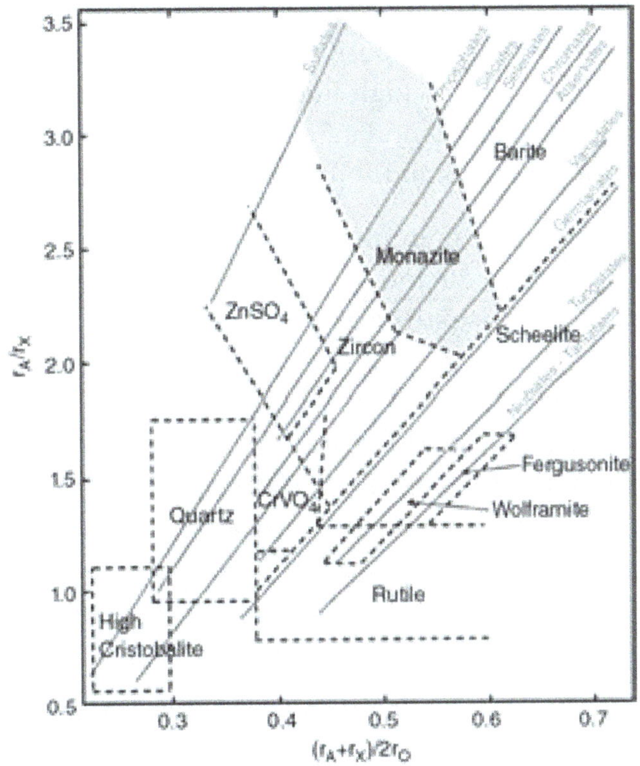

Fig. 21. The classification diagram of AXO_4-type compounds. The monazite stability domain is colored by gray [117].

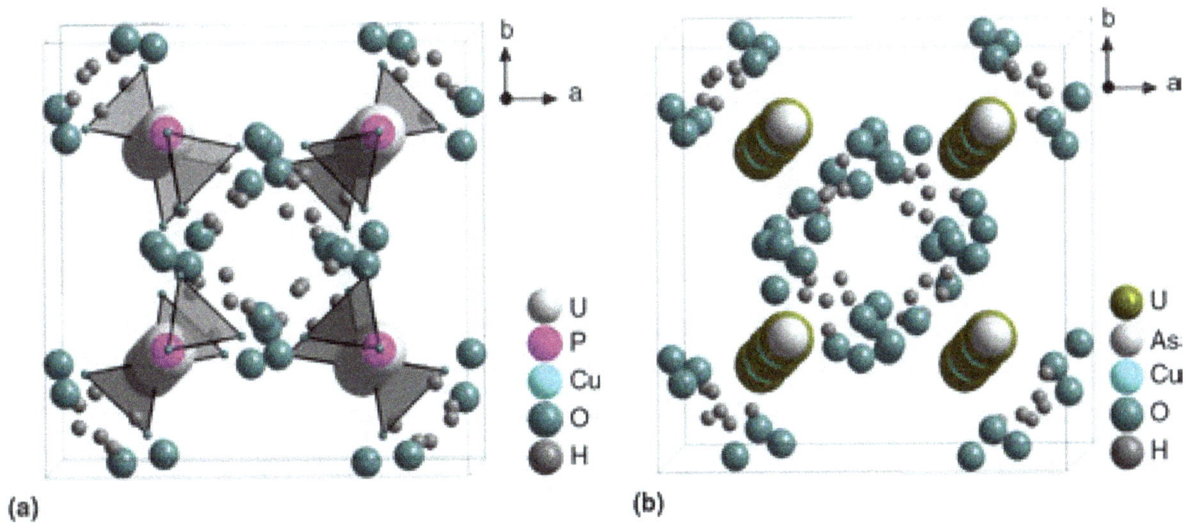

Fig. 22. The structure of tobernite and isostructural zeunerite [120] viewed along the c-axis.

6. **Tobernite** [120],[121]: tetragonal torbernite (P4/NNC, a = 7.0267 Å, c = 20.8070 Å, c:a = 2.9611, $Cu[(UO_2)(PO_4)]_2(H_2O)_{12}$, **Fig. 22(a)**) and zeunerite ($Cu[(UO_2)(AsO_4)]_2(H_2O)_{12}$, **Fig. 22(b)**), as well as metatorbernite ($Cu[(UO_2)(PO_4)]_2(H_2O)_8$) and metazeunerite ($Cu[(UO_2)(AsO_4)]_2(H_2O)_8$), belong to the autunite and meta-autunite groups, respectively, which

make up together approximately 40 mineral species of hydrated uranyl phosphates and arsenates. The structures, compositions and stabilities of minerals of the autunite and meta-autunite groups are of high interest because of their environmental significance. They are widespread and abundant and exert an impact on the mobility of uranium in phosphate-bearing systems and soils contaminated by actinides.

These minerals contain the autunite-type sheet, of the composition of $[(UO_2)(PO_4)]^-$, which involves the sharing of equatorial vertices of uranyl square bipyramids with phosphate tetrahedra. In each of these structures, Cu^{2+} cations are located between the sheets in Jahn-Teller[32] [122] distorted (4 + 2) octahedra, with short bonds to four H_2O groups in a square-planar arrangement and two longer distances to oxygen atoms of uranyl ions. A symmetrically independent H_2O group is held in each structure only by H-bonding, and in torbernite (and in zeunerite), it forms the square-planar sets of interstitial H_2O groups both above and below the planes of Cu^{2+} cations. In metatorbernite (and in metazeunerite), the square-planar sets of interstitial H_2O groups are either above or below the planes of Cu^{2+} cations. The bond-length-constrained refinement provides the crystal-chemically reasonable descriptions of H-bonding in those four structures [120].

7. **Turquoise** [123],[124]: $CuAl_6(PO_4)_4(OH)_8{\cdot}4H_2O$ is a copper analogue of triclinic mineral fausite $ZnAl_6(PO_4)_4(OH)_8{\cdot}4H_2O$ (the space group P1 with the cell parameters a = 7.410 Å, b = 7.633 Å, c = 9.904 Å, α = 68.4°, β = 69.65° and γ = 65.05°). The structure (**Fig. 23**) consists of distorted MO_6 polyhedra (M= Zn, Cu), AlO_6 octahedra and PO_4 tetrahedra. By the edge- and corner-sharing of these polyhedra, a fairly dense three-dimensional framework is formed, which is further strengthened by a system of hydrogen bonds. The metal atoms in the unique MO_6 (M = Zn or Cu) polyhedron show a distorted [2 + 2 + 2] coordination, the distortion being more pronounced in turquoise. About 10% of the M site is vacant in both minerals. In turquoise, a previously undetected structural site with very low occupancy of (possibly) Cu is present at the position (1/2,0,1/2).

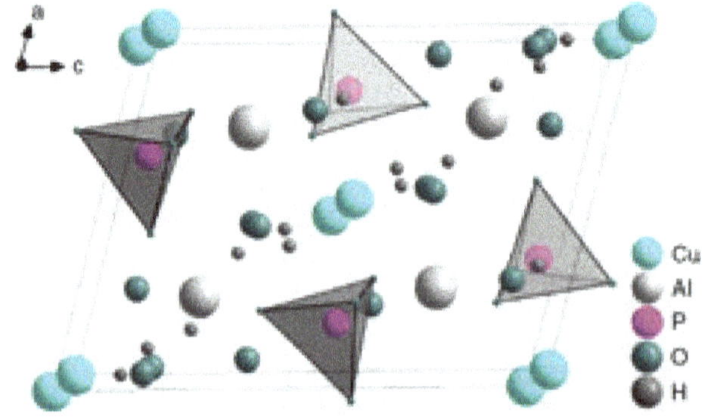

Fig. 23. The structure of turquoise viewed along the b-axis [124].

[32] According to the Jahn-Teller theorem, any nonlinear molecule in a degenerate electronic state will be unstable and will undergo some kind of distortion that will lower its symmetry so as to remove the degeneracy of the electronic state and also to attain lower energy. The Jahn-Teller effect is termed as static when there is permanent distortion in the structure of molecule [122].

8. **Vivianite** [125],[126],[127]: is monoclinic mineral of the composition of $Fe_3(PO_4)_3 \cdot 8H_2O$, which crystallizes in the space group of C1 $_2$ /M with the cell parameters $\boldsymbol{a}$ = 10.08 Å, $\boldsymbol{b}$ = 13.43 Å, $\boldsymbol{c}$ = 4.70 Å and $\boldsymbol{\beta}$ = 104.50° (**Fig. 24**). The mineral was named by ABRAHAM GOTTLOB WERNER in 1817 after JOHN HENRY VIVIAN. Vivianite belongs to the simplest group of minerals with the composition given by general formula: $A_3(XO_4)_2 \cdot 8H_2O$, where A = Mg, Zn, Ni, Co or Fe and X is P or As [128].

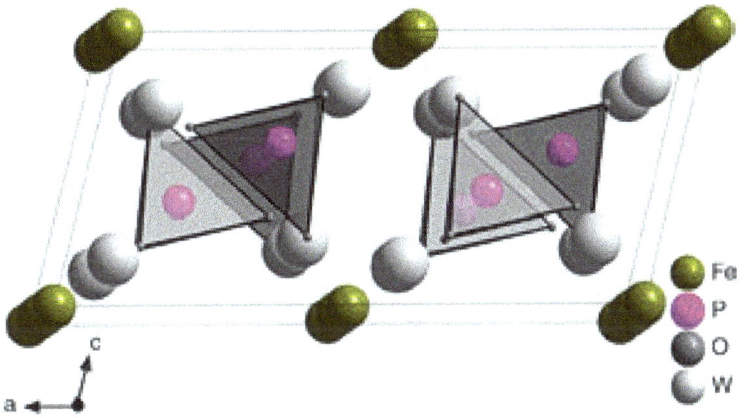

Fig. 24. The structure of vivanite viewed along the b-axis [125].

9. **Wavellite** [129]: the mineral of the composition of $Al_3(PO_4)_2(OH)_3 \cdot 4.5\text{-}5H_2O$ (the space group *Pcmn*, $\boldsymbol{a}$ = 9.62 Å, $\boldsymbol{b}$ = 17.36 Å and $\boldsymbol{c}$ = 6.99 Å). The two aluminum atoms in the structure are octahedrally coordinated (**Fig. 25**): one is bonded to two O atoms, two –OH groups and two H_2O molecules and the other to three O, two (–OH) and one H_2O. Phosphorus is in tetrahedral coordination with oxygen. Al octahedra, linked through (OH) corners, form chains parallel to the c-axis, and P tetrahedra are attached to this chain by sharing O atoms of subsequent octahedra. An extra H_2O molecule occupies the large cavity between the chains, and as indicated by a high temperature factor, it has a statistical distribution within this cavity.

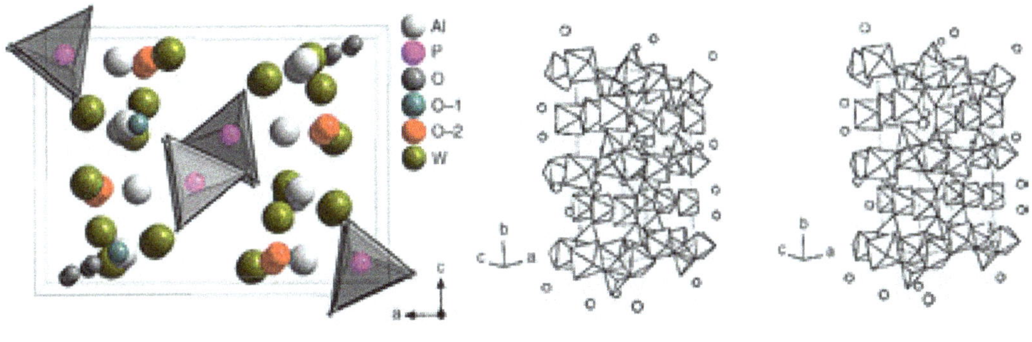

Fig. 25. The structure of wavellite viewed along the b-axis (a) and stereoscopic view of the wavellite structure (b) [129].

10. **Xenotime** [118]: monazite (**Fig. 26**) is isostructural with zircon ($I4_1/AMD$). Monazite atomic arrangement as well as that of xenotime is based on [001] chains of intervening phosphate tetrahedra and RE polyhedra, with REO, polyhedron in xenotime that accommodates heavy lanthanides (Tb-Lu in the synthetic phases) and REO polyhedron in monazite that preferentially incorporates larger light rare-earth elements (La-Gd). As the structure "transforms" from xenotime to monazite, the crystallographic properties are comparable along the [001] chains, with the structural adjustments to the different sizes of REE atoms occurring principally in (001).

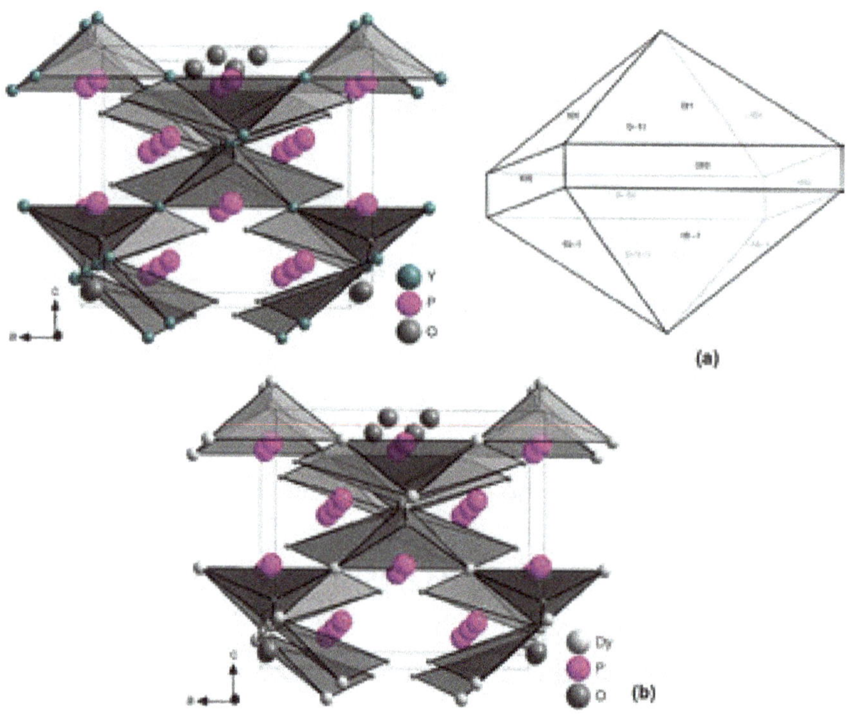

Fig. 26. The structure of xenotime-(Y): a = 6.8947 Å, b = 6.8947 Å and c = 6.0276 Å (a) and xenotime-(Dy) (b) a = 6.9052 Å, b = 6.9053 Å and c = 6.0384 Å (b) [118] viewed along the b-axis.

Isostructural arsenate analogues of many phosphate minerals are known, and in some cases, vanadates too. Some orthophosphates capable of forming complete ranges of solid solutions with the corresponding orthoarsenates are [25]:

- **Variscite group**: $MXO_4{\cdot}2H_2O$, where M = Fe, Al and X = P or As;
- **Fairfieldite group**: $Ca_2M(XO_4)_2{\cdot}2H_2O$, where M = Mn, Fe, Mg, Ni, Zn, Co and X = P, As;
- **Vivianite group**: $M_3(XO_4)_2{\cdot}8H_2O$, where M = Fe, Mn, Mg, Zn, Co, Ni and X = P, As;
- **Monazite group**: MXO_4, where M = Ce, La, Nd, Th, Bi and X = P, As;
- **Rhabdophane group**: $MXO_4{\cdot}H_2O$, where M = Ce, La, Nd, Th and X = P, As;
- **Xenotime group**: MXO_4, where M = Y, Ce, Bi and X = P, V;

- **Autunite group**: $M[(UO_2)_2(XO_4)]\cdot nH_2O$, where M = Ca, Ce, Ba, K, NH_4^+, Sr, Pb, Mg, Na, Zn and X = P, As, V;
- **Crandallite group**: $MM'_3(XO_4)_2(OH)_6\cdot H_2O$, where M = Ca, Sr, Ba, M'= Al, Fe and X = P, As.

Phosphate minerals, like silicate minerals, are found with a great variety of cations. Unlike the latter group that contains numerous types of condensed silicate anions, almost all phosphate minerals are orthophosphates that contain PO_4^{3-} anion. Non-phosphorus anions, such as O^{2-}, OH^-, F^-, Cl^-, SO_4^{2-}, SiO_4^{4-} and AsO_4^{3-}, may also be present in these stoichiometric (or as occluded) materials [25].

Author details

Petr Ptáček

Brno University of Technology, Czech Republic

References

[1] Bishop AC, Woolley AR, Woolley WRH. Cambridge Guide to Minerals, Rocks and Fossils. 2nd ed., Cambridge: Cambridge University Press, 1999. ISBN: 0-521-77881-6

[2] Clarke FW. The Data of Geochemistry. Bulletin (Geological Survey (U.S.)), U.S. Government Printing Office, 1920.

[3] McClellan GH, Lehr JR. Crystal chemical investigation of natural apatites. American Mineralogist 1969;54 1374–1391.

[4] Ding T, Ma D, Lu J, Zhang R. Apatite in granitoids related to polymetallic mineral deposits in southeastern Hunan Province, Shi-Hang Zone, China: Implications for petrogenesis and metallogenesis. Ore Geology Reviews 2015;69 104–117.

[5] Watson EB. Apatite and phosphorus in mantle source regions: an experimental study of apatite/melt equilibria at pressures to 25 kbar. Earth and Planetary Science Letters 1980;51(2) 322–335.

[6] Deer WA. Rock-Forming Minerals: Non-silicates. Volume 5B, 2nd ed., Geological Society of London, 1998. ISBN: 978-1897799901

[7] Raask E. Mineral Impurities in Coal Combustion: Behavior, Problems, and Remedial Measures. Taylor & Francis, 1985. ISBN: 978-3540138174

[8] Jacobson M, Charlson RJ, Rodhe H, Orians GH. Earth System Science: From Biogeochemical Cycles to Global Changes. International Geophysics — Volume 72. Academic Press, 2000. ISBN: 978-0080530642

[9] Spear FS. Metamorphic Phase Equilibria and Pressure-Temperature-Time Paths. Mineralogical Society of America short course notes. 2nd ed., Mineralogical Society of America, 1993. ISBN: 978-0939950348

[10] Kemp JF. A Handbook of Rocks for Use Without the Microscope. BiblioBazaar, 2009. ISBN: 978-1110466528

[11] Drouet Ch. A comprehensive guide to experimental and predicted thermodynamic properties of phosphate apatite minerals in view of applicative purposes. Journal of Chemical Thermodynamics 2015;81 143–159.

[12] Ihlen PM, Schiellerup H, Gautneb H, Skår Ø. Characterization of apatite resources in Norway and their REE potential — A review. Ore Geology Reviews 2014;58 126–147.

[13] Santana RC, Farnese ACC, Fortes MCB, Ataíde CH, Barrozo MAS. Influence of particle size and reagent dosage on the performance of apatite flotation. Separation and Purification Technology 2008;64(1) 8–15.

[14] Pucéat E, Reynard B, Lécuyer C. Can crystallinity be used to determine the degree of chemical alteration of biogenic apatites? Chemical Geology 2004;205(1-2) 83–97.

[15] Zafar ZI, Anwar MM, Pritchard DW. Optimization of thermal beneficiation of a low grade dolomitic phosphate rock. International Journal of Mineral Processing 1995;43(1-2) 123–131.

[16] dos Santos MA, Santana RC, Capponi F, Ataíde CH, Barrozo MAS. Effect of ionic species on the performance of apatite flotation. Separation and Purification Technology 2010;76(1) 15–20.

[17] Chen M, Graedel TE. The potential for mining trace elements from phosphate rock. Journal of Cleaner Production 2015;91 337–346.

[18] Al-Bassam KS, Aba-Hussain AA, Mohamed IQ, Al-Rawi Yehya T. Petrographic classification of phosphate components of East Mediterranean phosphorite deposits. Iraqi Bulletin of Geology and Mining 2010;6(1) 59–79.

[19] Flügel E. Microfacies of Carbonate Rocks: Analysis, Interpretation and Application. Springer Science & Business Media, 2013. ISBN: 978-3662087268

[20] Abouzeid A-ZM. Physical and thermal treatment of phosphate ores — An overview. International Journal of Mineral Processing 2008;85(4) 59–84.

[21] Gupta AK. Igneous Rocks. Allied Publishers, 1998. ISBN: 978-8170237846

[22] Manual Fertilizer, UN Industrial Development Organization, International Fertilizer Development Center. Springer Science & Business Media, 1998. ISBN: 978-0792350323

[23] World Phosphate Rock Reserves and Resources. International Center for Soil Fertility and Agricultural Development, 2010. ISBN: 978-0880901673

[24] Emich GD. Phosphate rock. Industrial Minerals and Rocks 1984;2 1017–1047.

[25] Derek EC. Chemistry, Biochemistry and Technology. 6th ed., CRC Press, 2013. ISBN: 978-1439840887

[26] Teleki PG, Dobson MR, Moore JR, von Stackelberg U. Marine Minerals: Advances in Research and Resource Assessment. Nato Science Series C: Volume 194. Springer Science & Business Media, 2012. ISBN: 978-9400938038

[27] McKelvey VE. Phosphate Deposits. Geological Survey Bulletin 1252-D. A Summary of Salient Features of the Geology of Phosphate Deposits, Their Origin, and Distribution. Washington: United States Government Printing Office, 1967.

[28] Cook PJ, Shergold JH. Phosphate Deposits of the World: Volume 1: Proterozoic and Cambrian Phosphorites. Cambridge Earth Science Series. Cambridge University Press, 2005. ISBN: 978-0521619219

[29] Salvador A. International Stratigraphic Guide: A Guide to Stratigraphic Classification, Terminology, and Procedure. Publication (International Union of Geological Sciences). 30th ed., Geological Society of America, 1994. ISBN: 978-0813774015

[30] Walter MR. Stromatolites. Developments in Sedimentology. Elsevier, 1976. ISBN: 978-0080869322

[31] Nairn AEM, Ricou Luc-E, Vrielynck B, Dercourt J. The Tethys Ocean. Ocean Basins and Margins — Volume 8. Springer Science & Business Media, 2013. ISBN: 978-1489915580

[32] Notholt AJG, Sheldon RP, Davidson DF. Phosphate Deposits of the World: Volume 2, Phosphate Rock Resources. Cambridge Earth Science Series. Revised edition, Cambridge University Press, 2005. ISBN: 978-0521673334

[33] Gualtieri A, Venturelli P. In situ study of the goethite-hematite phase transformation by real time, synchrotron powder diffraction, sample at T = 25 C. American Mineralogist 1999;84 895–904.

[34] Nathan Y. Phosphate minerals. In: Nriagu JO, Moore PB (Eds.), Chapter. The Mineralogy and Geochemistry of Phosphorites. Springer Science & Business Media, 2012. ISBN: 978-3642617362

[35] Rao VP. Phosphorites from the Error Seamount, northern Arabian Sea. Marine Geology 1986;71(1-2) 177–186.

[36] Glenn CR, Föllmi KB, Riggs SR, Baturin GN, Grimm KA, Trape J. Phosphorus and phosphorites: Sedimentology and environments of formation. Eclogae Geologicae Helvetiae 1994;87(3) 747–788.

[37] Baturin GN, Bezrukov PL. Phosphorites on the sea floor and their origin. Marine Geology 1979;31(3-4) 317–332.

[38] Cook PJ, Shergold JH, Burnett WC, Riggs SR. Phosphorite research: A historical overview. Geological Society, London, Special Publications 1990;52 1–22.

[39] Trappe J. Phanerozoic Phosphorite Depositional Systems: A Dynamic Model for a Sedimentary Resource System. Lecture Notes in Earth Sciences — Volume 76, ISSN 0930-0317. Springer, 1998. ISBN: 978-3540696049

[40] Bogdanov NA. Non-metallic Mineral Ores. Proceedings of the 27th International Geological Congress — Volume 15. VSP, 1984. ISBN: 978-9067640244

[41] Dunham RJ. Classification of carbonate rocks according to depositional texture, in: Ham WE. Classification of carbonate rocks. American Association of Petroleum Geologists Memoir 1962;1 108–121.

[42] Riggs SR. A petrographic classification of sedimentary phosphorites. In: Cook PJ, Shergold JH. (Eds.), Proterozoic–Cambrian Phosphorites. IGCP, 156, 1st Field Workshop and Seminar. Canberra: National Library of Australia, 1979.

[43] Suess E, von Huene R. Proceedings of the Ocean Drilling Program, Scientific Results, Volume 112. Phosphatic Sediments and Rocks Recovered from the Peru Margin during ODP LEG 1121. The Program, 1986.

[44] Ehrlich HL, Newman DK. Geomicrobiology. 5th ed., CRC Press, 2008. ISBN: 978-0849379079

[45] Tyrrell GW. The Principles of Petrology: An Introduction to the Science of Rocks. Springer Science & Business Media, 2012. ISBN: 978-9401160261

[46] Gill R. Igneous Rocks and Processes: A Practical Guide. John Wiley & Sons, 2011. ISBN: 978-1444362435

[47] Brown PW, Constanz B. Hydroxyapatite and Related Materials. CRC Press, 1994. ISBN: 978-0849347504

[48] Wenk Hans-R, Bulakh A. Minerals: Their Constitution and Origin. Cambridge University Press, 2004. ISBN: 978-0521529587

[49] Laenen B, Hertogen J, Vandenberghe N. The variation of the trace-element content of fossil biogenic apatite through eustatic sea-level cycles. Palaeogeography, Palaeoclimatology, Palaeoecology 1997;132(1-4) 325–342.

[50] Egorov LS. Ijolite-Carbonatite Plutonism (in Russian). Nedra, Leningrad, 1991.

[51] Isakova AT, Panina LI, Rokosova EY. Carbonatite melts and genesis of apatite mineralization in the Guli pluton (northern East Siberia). Russian Geology and Geophysics 2015;56(3) 466–475.

[52] Dawson JB, Steele IM, Smith JV, Rivers ML. Minor and trace element chemistry of carbonates, apatites and magnetites in some African carbonatites. Mineralogical Magazine 1996;60 415–425.

[53] Aspden JA. The mineralogy of primary inclusions in apatite crystals extracted from Alnö ijolite. Lithos 1980;13(3) 263–268.

[54] Cordeiro PFO, Brod JA, Dantas EL, Barbosa ESR. Mineral chemistry, isotope geochemistry and petrogenesis of niobium-rich rocks from the Catalão I carbonatite-phoscorite complex, Central Brazil. Lithos 2010;118(3-4) 223–237.

[55] Iddings JP. Igneous Rocks: Composition, Texture and Classification, Description and Occurrence, Volume 1. Рипол Классик, 1920. ISBN: 978-5883586766

[56] Wall F, Zaitsev AN. Phoscorites and Carbonatites from Mantle to Mine: The Key Example of the Kola Alkaline Province. Mineralogical Society of Great Britain & Ireland, 2004. ISBN: 978-0903056229

[57] Russell HD, Hiemstra SA, Groeneveld D. New aspects of the Lower Katanga rocks of the Kafue Gorge. Transactions of the Geological Society of South Africa 1955;58 197–208.

[58] Le Maitre RW, Streckeisen A, Zanettin B, Le Bas MJ, Bonin B, Bateman P. Igneous Rocks: A Classification and Glossary of Terms: Recommendations of the International Union of Geological Sciences Subcommission on the Systematics of Igneous Rocks. 2nd ed., Cambridge University Press, 2002. ISBN: 978-1139439398

[59] Reviews Book: Igneous Rocks: A Classification and Glossary of Terms (Recommendations of the IUGS Subcommission on the Systematics of Igneous Rocks). 2nd ed., LeMaitre RW (ed.). Cambridge University Press, New York, N.Y., 2002. ISBN: 0-521-66215-X. The Canadian Mineralogist 2002;40 1737-1742.

[60] Kucera M. Industrial Minerals and Rocks: Developments in Economic Geology. Elsevier, 2013. ISBN: 978-0444597502

[61] Tatarinov PM. Nonmetallic mineral deposits course (in Russian). Moscow: Nedra, 1969.

[62] Wills BA. Mineral Processing Technology: An Introduction to the Practical Aspects of Ore Treatment and Mineral Recovery (in SI/Metric Units). International Series on Materials Science and Technology. 4th ed., Elsevier, 2013. ISBN: 978-483286679

[63] Iijima A, Abed AM. Siliceous, Phosphatic and Clauconitic Sediments of the Tertiary and Mesozoic. Proceedings of the 29th International Geological Congress: Kyoto, Japan, 24 August-3 September 1992 — Part 3. VSP, 1994. ISBN: 978-9067641753

[64] Cronan DS. Marine Minerals in Exclusive Economic Zones. Topics in the Earth Sciences — Volume 5. Springer Science & Business Media, 1992. ISBN: 978-0412292705

[65] Cushman GT. Guano and the Opening of the Pacific World: A Global Ecological History. Studies in Environment and History. Cambridge University Press, 2013. ISBN: 978-1107004139

[66] Hollett D. More Precious Than Gold: The Story of the Peruvian Guano Trade. Associated University Press, 2008. ISBN: 978-0838641316

[67] Chang Jen-S, Kelly A, Crowley JM. Handbook of Electrostatic Processes. CRC Press, 1995. ISBN: 978-1420066166

[68] Nogalska A, Zalewska M. The effect of meat and bone meal on phosphorus concentrations in soil and crop plants. Plant Soil Environment 2013;59(12) 575–580.

[69] O'Reilly SY, Griffin WL. Apatite in the mantle: implications for metasomatic processes and high heat production in Phanerozoic mantle. Lithos 2000;53 217–232.

[70] Douce AEP, Roden MF, Chaumba J, Fleisher C, Yogodzinski G. Compositional variability of terrestrial mantle apatites, thermodynamic modeling of apatite volatile contents, and the halogen and water budgets of planetary mantles. Chemical Geology 2011;288 14–31.

[71] Sha LK, Chappell BW. Apatite chemical composition, determined by electron microprobe and laser-ablation inductively coupled plasma mass spectrometry, as a probe into granite petrogenesis. Geochimica et Cosmochimica Acta 1999;63(22) 3861–3881.

[72] Joosu L, Lepland A, Kirsimäe K, Romashkin AE, Roberts NMW, Martin AP, Črne AE. The REE-composition and petrography of apatite in 2 Ga Zaonega Formation, Russia: The environmental setting for phosphogenesis. Chemical Geology 2015;395 88–107.

[73] Follmi KB. The phosphorus cycle, phosphogenesis and marine phosphate-rich deposits. Earth-Science Reviews 1996;40(1-2) 55–124.

[74] Glenn CR, Föllmi KB, Riggs Sr, Baturin GN, Grimm KA, Trappe J. Phosphorus and phosphorites: sedimentology and environments of formation. Eclogae Geologicae Helvetiae 1994;87(3) 747–788.

[75] Jarvis I, Burnett WC, Nathan Y, Almbaydin FSM, Attia AKM, Castro LN, Flicoteaux R, Hilmy ME, Husain V, Qutawnah AA, Serjani A, Zanin YN. Phosphorite geochemistry — State-of-the-art and environmental concerns. Eclogae Geologicae Helvetiae 1994;87(3) 643–700.

[76] Alexander EB. Soils in Natural Landscapes. CRC Press, 2013. ISBN: 978-1466594364

[77] Jørgensen SE. Global Ecology. Academic Press, 2010. ISBN: 978-0444536273

[78] CIBA Foundation Symposium. Phosphorus in the Environment: Its Chemistry and Biochemistry. Novartis Foundation Symposia — Volume 919. John Wiley & Sons, 2009. ISBN: 978-0470718070

[79] Ragot S, Zeyer J, Zehnder L, Reusser E, Brandl H, Lazzaro A. Bacterial community structures of an alpine apatite deposit. Geoderma 2013;202-203 30-37.

[80] Puente ME, Rodriguez-Jaramillo MC, Li CY, Bashan Y. Image analysis for quantification of bacterial rock weathering. Journal of Microbiological Methods 2006;64(2) 275–286.

[81] Nies A. Strengit, ein neues Mineral. Neues Jahrbuch für Mineralogie, Geologie und Palaontologie 1877; 8-16.

[82] Cornelis P. *Pseudomonas*: Genomics and Molecular Biology. Horizon Scientific Press, 2008. ISBN: 978-1904455196

[83] Pongslip N. Phenotypic and Genotypic Diversity of Rhizobia. Bentham Science Publishers, 2012. ISBN: 978-1608054619

[84] Vanlaere E. *Burkholderia cepacia* complex. DCL Print & Sign, 2008. ISBN: 978-9087560379

[85] Lisker F, Ventura B, Glasmacher UA. Apatite thermochronology in modern geology. Geological Society, London, Special Publications 2009;324 1–23. DOI: 10.1144/SP324.1.

[86] Fleischer RL, Price PB, Walker RM. Nuclear Tracks in Solids: Principles and Applications. University of California Press, 1975. ISBN: 978-0520026650

[87] Fleisher RL, Price PB. Techniques for geological dating of minerals by chemical etching of fission fragment tracks. Geochimica et Cosmochimica Acta 1964;28(11) 1705–1714.

[88] Galbraith RF. Statistics for Fission Track Analysis. CRC Press, 2005. ISBN: 978-1420034929

[89] Li W, Wang L, Lang M, Trautmann C, Ewing RC. Thermal annealing mechanisms of latent fission tracks: Apatite vs. zircon. Earth and Planetary Science Letters 2011;302(1-2) 227–235.

[90] Li W, Lang M, Gleadow AJW, Zdorovets MV, Ewing RC. Thermal annealing of unetched fission tracks in apatite. Earth and Planetary Science Letters 2012;321-322 121-127.

[91] Chen H, Hu J, Wu G, Shi W, Geng Y, Qu H. Apatite fission-track thermochronological constraints on the pattern of late Mesozoic-Cenozoic uplift and exhumation of the Qinling Orogen, central China. Journal of Asian Earth Sciences 2014. In press. DOI: 10.1016/j.jseaes.2014.10.004.

[92] Hejl E. Evidence for unetchable gaps in apatite fission tracks. Chemical Geology 1995;122(1-4) 259–269.

[93] Peternell M, Kohlmann F, Wilson CJL, Seiler C, Gleadow AJW. A new approach to crystallographic orientation measurement for apatite fission track analysis: Effects of crystal morphology and implications for automation. Chemical Geology 2009;265(3-4) 527–539.

[94] Enkelmann E, Ehlers TA, Buck G, Schatz Ann-K. Advantages and challenges of automated apatite fission track counting. Chemical Geology 2012;322-323 278-289.

[95] van den Haute P, De Corte F. Advances in Fission-Track Geochronology. Solid Earth Sciences Library — Volume 10. Springer Science & Business Media, 2013. ISBN: 978-9401591331

[96] McCubbin FM, Jolliff BL, Nekvasil H, Carpenter PK, Zeigler RA, Steele A, Elardo SM, Lindsley DH. Fluorine and chlorine abundances in lunar apatite: Implications for heterogeneous distributions of magmatic volatiles in the lunar interior. Geochimica et Cosmochimica Acta 2011;75(17) 5073–5093.

[97] Boyce JW, Tomlinson SM, McCubbin FM, Greenwood JP, Treiman AH. The lunar apatite paradox. Science 25;344(6182) 400-402.

[98] Crotts APS. Lunar outgassing, transient phenomena, and the return to the Moon. Part 1: Existing data. The Astrophysical Journal 2008;687 692–705.

[99] Calvo C, Gopal R. The crystal structure of whitlockite from the Palermo quarry. American Mineralogist 1975;60 120–133.

[100] McConnell D. Apatite: Its Crystal Chemistry, Mineralogy, Utilization, and Geologic and Biologic Occurrences: Applied Mineralogy (Volume 5). Springer Science & Business Media, 2012. ISBN: 978-3709183144

[101] Jolliff BL, Hughes JM, Freeman JJ. Zeigler RA. Crystal chemistry of lunar merrillite and comparison to other meteoritic and planetary suites of whitlockite and merrillite. American Mineralogist 2006;91 1583–1595. DOI: 10.2138/am.2006.2185 1583.

[102] Hughes JM, Jolliff BL, Gunter ME. The atomic arrangement of merrillite from the Fra Mauro Formation, Apollo 14 lunar mission: The first structure of merrillite from the Moon. American Mineralogist 2006;91 1547–1552.

[103] Barnes JJ, Tartèse R, Anand M, McCubbin FM, Franchi IA, Starkey NA, Russell SS. The origin of water in the primitive Moon as revealed by the lunar highlands samples. Earth and Planetary Science Letters 2014;390 244–252.

[104] Douce AEP, Roden M. Apatite as a probe of halogen and water fugacities in the terrestrial planets. Geochimica et Cosmochimica Acta 2006;70(12) 3173–3196.

[105] Hallis LJ, Taylor GJ, Nagashima K, Huss GR. Magmatic water in the Martian meteorite Nakhla. Earth and Planetary Science Letters 2012;359-360 84-92.

[106] Sarafian AR, Roden MF, Patiño-Douce AE. The volatile content of Vesta: Clues from apatite in eucrites. Meteoritics & Planetary Science 2013,48(11) 2135–2154.

[107] Scully JEC, Russell CT, Yin A, Jaumann R, Carey E, Castillo-Rogez J, McSween HY, Raymond CA, Reddy V, Le Corre L. Geomorphological evidence for transient water flow on Vesta. Earth and Planetary Science Letters 2015;411 151–163.

[108] Treiman AH, Lanzirotti A, Xirouchakis D. Ancient water on asteroid 4 Vesta: evidence from a quartz veinlet in the Serra de Magé eucrite meteorite. Earth and Planetary Science Letters 2004;219(3-4) 189–199.

[109] Jasinski SM. Phosphate rock. U.S. Geological Survey Mineral Commodity Summaries 2007; 120-121.

[110] Locock AJ, Burns PC. The crystal structure of synthetic autunite, $Ca[(UO_2)(PO_4)]_2(H_2O)_{11}$. American Mineralogist 2003;88 240–244.

[111] Phillips W, Brooke HJ, Miller WH. Autunite. In: An Elementary Introduction to Mineralogy. London: Longman, Brown, Green, and Longman, 1852; 519-519.

[112] Makarov ES, Ivanov VI. The crystal structure of meta-autunite, $Ca(UO_2)_2(PO_4)_2{\cdot}6H_2O$. Doklady Akademii Nauk SSSR 1960;132 601–603.

[113] Blount AM. The crystal structure of crandallite. American Mineralogist 1974;59 41–47.

[114] Lindberg ML, Christ CL. Crystal structures of the isostructural minerals lazulite, scorzalite and barbosalite. Acta Crystallographica 1959;12 695–697.

[115] Giuseppetti G, Tadini C. Lazulite, $(Mg,Fe)Al_2(OH)_2(PO_4)_2$: structure refinement and hydrogen bonding. Neues Jahrbuch fur Mineralogie, Monatshefte 1983; 410-416.

[116] Larsen ES, Shannon EV. The minerals of the phosphate nodules from near Fairfield, Utah. American Mineralogist 1930;15 307–337.

[117] Clavier N, Podor R, Dacheux N. Crystal chemistry of the monazite structure. Journal of the European Ceramic Society 2011;31(6) 941–976.

[118] Ni Y, Hughes JM, Mariano AN. Crystal chemistry of the monazite and xenotime structures. American Mineralogist 1995;80 21–26.

[119] RutleyF.MineralogyMurby's"ScienceandArtDepartment"seriesoftextbooks.Thomas Murby & Company, 1802.

[120] Locock AJ, Burns PC. Crystal structures and synthesis of the copper-dominant members of the autunite and meta-autunite groups: Torbernite, zeunerite, metatorbernite and metazeunerite. The Canadian Mineralogist 2003;41 489–502.

[121] Hennig C, Reck G, Reich T, Robberg A, Kraus W, Sieler J. EXAFS and XRD investigations of zeunerite and meta-zeunerite. Zeitschrift fur Kristallographie 2003;218 37–45.

[122] Sathyanarayana DN. Electronic Absorption Spectroscopy and Related Techniques. Universities Press, 2001. ISBN: 978-8173713712

[123] Cid-Dresdner H. Determination and refinement of the crystal structure of turquoise, $CuAl_6(PO_4)_4(OH)_8{\cdot}4H_2O$. Zeitschrift fur Kristallographie 1965;121 87–113.

[124] Kolitsch U, Giester G. The crystal structure of faustite and its copper analogue turquoise. Mineralogical Magazine 2000;64 905–913.

[125] Werner AG. Vivianit, in Letztes Mineral-System, bey Craz und Gerlach, und bey Carl Gerold (Freiberg und Wien) 1817; 41-42.

[126] FejdiP, PoullenJF, GasperinM. Affinement de la structure de la vivianite $Fe_3(PO_4)_2{\cdot}8H_2O$. Bulletin de Mineralogie 1980;103 135–138.

[127] Bartl H. Water of crystallization and its hydrogen-bonded crosslinking in vivianite $Fe_3(PO_4)_2{\cdot}8H_2O$; a neutron diffraction investigation. Zeitschrift fur Analytische Chemie 1989;333 401–403.

[128] Mori H, Ito T. The structure of vivianite and symplesite. Acta Crystallographica 1950;3 1–6.

[129] Araki T, Zoltai T. The crystal structure of wavellite. Zeitschrift fur Kristallographie 1968;127 21–33.

4

Identification, Characterization and Properties of Apatites

Petr Ptáček

Abstract

The parameters of unit cell, structure, refractive index, solubility data, PO_4/CO_3 ratio, surface area, etc., are important parameters for characterization of phosphate rocks. Third chapter of this book introduces methods for identification, characterization and properties of apatites in four main sections. The first part describes techniques used for identification and investigation of properties of phosphate minerals, including X-ray diffraction analysis, powder neutron diffraction, X-ray fluorescence as well as spectroscopic and microscopic methods. Some of these techniques are then demonstrated on the fluorapatite specimen in the second part. The third part of this chapter deals with thermodynamic properties of apatite-type compounds and introduces some of thermodynamic predictive methods. The fourth part is dedicated to dissolution of apatite, where the reaction between solids and liquids according to different dissolution models is described. Chapter ends with methods for the evaluation of reactivity of phosphate rocks.

Keywords: Apatite, Carbonate to Phosphate Ratio, Dissolution of Apatite, Reactivity of Apatite, Citrate Solubility

Techniques used for identification of phosphate minerals

There are several parameters which are used in characterization of phosphate rocks, including the following [1],[2]:

a. Unit-cell parameters

b. Refractive index

c. Solubility data based on chemical extraction methods, e.g. neutral ammonium citrate solubility (NAC, **Section**)

d. PO_4/CO_3 ratio as a measure of carbonate substitution in phosphate minerals

e. Surface area and pore size distribution indicating the potential reactivity

The most important techniques used for the identification and characterization of phosphate minerals include methods for identification of phase composition, chemical composition, structure, surface properties, etc.

X-ray diffraction analysis

The diffraction of a beam of X-rays by a crystalline material is the process of beam scattering[1] by electrons associated with atoms in the crystal and of the interference of these scattered X-rays because of the periodic arrangement of atoms in the crystal and its symmetry.[2] X-ray diffraction analysis (XRD) is used for the determination of mineralogical composition and quantitative X-ray diffraction analysis (Rietveld method) for the refinements of structure of apatite from measured data using specialized software [3],[4],[5],[6].

Following the discovery of X-rays by RÖENTGEN[3] [8],[10],[11] in 1895 and the proof that X-rays have the wave properties and diffract from a periodic atomic array by VON LAUE [12] and his students in 1912–1913, the analytical application of X-ray diffraction has developed slowly over the next 20 years. Most of the earliness efforts were aimed at the solution of crystal structures of common phases. DEBYE and SHEERER (1916) and HULL (1917) suggested that powder diffraction patterns could be used for the identification of quantification of crystalline compounds. However, because most of the early developments were directed toward solving single-crystal structures, it was really the middle 1930s when the powder diffraction method began to attract the follower with the publication of the procedure of HANAWALT and RINN and the database of patterns by HANAWALT, RINN and FREVEL (1938). With the conversion of data sets into the first set of the Powder Diffraction File in 1941, the phase identification applications expanded, and the modern counter diffractometer was developed by PARRISH, HAMACHER and LOWITZSCH [5].

The phase identification was one of the first applications to grow to useful level. Other major applications of diffraction analysis include following phase changes under nonambient conditions and atmospheres. The first diffraction experiments were actually done on single crystals. The method is primarily directed toward determining the crystal periodicity and symmetry and solving the arrangement of atoms in the material because this information is

[1] Scattering is the process where the beam of radiation or particles is deviated from its initial trajectory by the inhomogeneity in the medium which it transverses [7].

[2] Other kinds of radiation commonly used for diffraction analysis are neutrons (**Section**) and electrons

[3] Wilhelm Conrad Röentgen (1845–1923) was the rector of the University of Würzburg [8]. The first X-ray photography was published in 1896 [9].

difficult to obtain from powder experiment. Powder diffraction is one of the most important material characterization techniques in the material research and industry [5].

Single-crystal diffraction studies are not limited to crystal structure analysis. The diffraction topography is a large field that has provided much information on the perfection of crystals used in industry as integral parts of devices. The examples include crystals used in sensing and control devices, substrates for electronic components, tools and dies, turbine blades and many other applications [5].

Powder neutron diffraction

Neutron powder diffraction (PND) or elastic neutron scattering enables to determine nuclear and magnetic structure of solids. Most of the information on the nature of ordered magnetic phases or magnetic structures comes from neutron diffraction experiments. Neutrons have no electric charge and interact with the nuclei rather than with the charge distribution of atoms in matter. They have the wavelength in the range of interatomic distances. They have magnetic moment and interact with the magnetic moment of atoms in matter. The mass of neutrons is similar to that of atomic nuclei; hence, they have energy and momentum similar to those of atoms in solid and fluid materials [13],[14]. The first neutron diffraction experiments were performed in 1945 by EO Wollan in the graphite reactor at Oak Ridge National Laboratory, USA [15].

Neutron scattering (NS) results from the interactions with atomic nuclei, i.e. overscattering lengths (distances) of the order of 10^{-15} m (1 fm).[4] Although scattering amplitude decreases greatly with the scattering vector (it is inverse to the scattering length), there are insignificant variations of scattering amplitude in the same range of scattering vector for neutrons. Consequently, powder diffraction with neutrons can resolve very fine structural and textural details of complex atomic structures. Moreover, the weak interaction of neutrons with matter results in very low attenuation offering a unique advantage for nondestructive, in situ work and bulk analysis (for polycrystalline materials, no crushing is required to obtain the patterns) [14].

Neutron diffraction was used to determine atomic arrangement in material. Inelastic neutron scattering measures the vibrations of atoms and small-angle neutron scattering[5] (SANS) is used to study larger structures such as polymers and colloids. The technique of surface reflection (reflectometry) was used to study layered materials. The technique of SANS provides the information about the size, shape and domain orientations; conformational changes and/or flexibility; and molecular associations is solution. For the structural studies, the elastic scattering effects, where there is no energy exchange between the radiation and atoms, are exploited [7],[14],[16].

Neutron powder diffraction is a method often used for the structure refinement of apatite or apatite type compounds from measured data using specialized software [17],[18],[19],[20],

---

[4] The interaction with electrons during X-ray analysis takes place over the distances of 10^{-10} m (1 Å) [14].

[5] Small angle scattering (SAS) of X-rays is abbreviated as SAXS [7].

[21],[22],[23],[24] and the effect of substitutions in the apatite structure [25]. Neutron diffraction data enable to explain the oxygen over-stoichiometry in the structure of $La_{9.67}(SiO_4)_6O_{2.5}$ apatite [17]. Since neutrons make possible the accurate determination of the thermal factors and provide the visualization of the diffusion paths in ionic conductors, powder neutron diffraction is also used for the characterization of solid oxide fuel cell materials [26]. This method is also used to investigate apatite in hard tissues where it provides the evidence about the deficiency of hydroxyl ion in bone apatites [27] and reconfirms that the inorganic portion is basically a hydroxylapatite-like material [27],[28].

X-ray fluorescence analysis and total-reflection X-ray fluorescence analysis

X-ray fluorescent analysis (XRF) is a method for the determination of sample composition [29], [30]. The origin of characteristic X-ray spectra can be described as follows. When sufficient energy is introduced into the atom, the electrons may be knocked out of one the inner shells. The atom is then in an excited (ionized) state and returns to the ground state within 10^{-8} s. The place of the missing electron is filled by an electron from a neighboring other shell, the place of which, in turn, is filled by an electron from more outer shell. The atom then returns to the ground state in steps. In every step, i.e. in every electron jump, the electron from a higher energy level goes into a lower energy level emitting excess energy in the form of an X-ray quantum. The energy of emitted radiation is characteristic for the atomic number of emitting element as well as for particular electron transition taking place within the electron shell of the atom. By measuring the energy or the wavelength of emitted radiation, the particle element can be identified unambiguously [31].

The energy that is necessary for the atom to get to excited state can be introduced either by the collision with a high-energy electron (sample is bombarded by electrons which are accelerated by high-voltage) or by the absorption of an energy-rich photon, i.e. the X-ray quantum (sample is irradiated by X-ray or gamma rays). In modern X-ray fluorescence analysis, the sample is irradiated by polychromatic radiation from an X-ray tube. In analogy to the optical case, this technique is referred to as fluorescence, which is responsible for the name X-ray fluorescence analysis as the technique of spectrochemical analysis with X-rays [31].

There are two types of instruments (**Fig. 1**) used for X-ray fluorescence spectrometry [32],[33]:

1. **Wavelength-dispersive XRF (WDXRF) or total reflection XRF (TRXRF):** the method is also often abbreviated as XRF. X-rays impinge on the sample (**Fig. 1**(**a**)) and generate fluorescent X-rays. These are then diffracted on a crystal. A goniometer selects the geometry between the crystal and detector that controls the detection of X-ray from the element of interest. Different crystals have different sensitivities. Many of commercial WDXRF instruments have two detectors and up to six crystals to optimize the conditions for each element.

2. **Energy-dispersive XRF (EDXRF):** the method is also abbreviated as EDX. The EDXRF instruments use much less energetic X-ray tube. Emitted X-ray radiation from the sample impinges directly on a detector, typically Si(Li), which generates pulses on an incident beam. These pulses are sorted and counted by a multichannel analyzer (**Fig. 1**(**b**)).

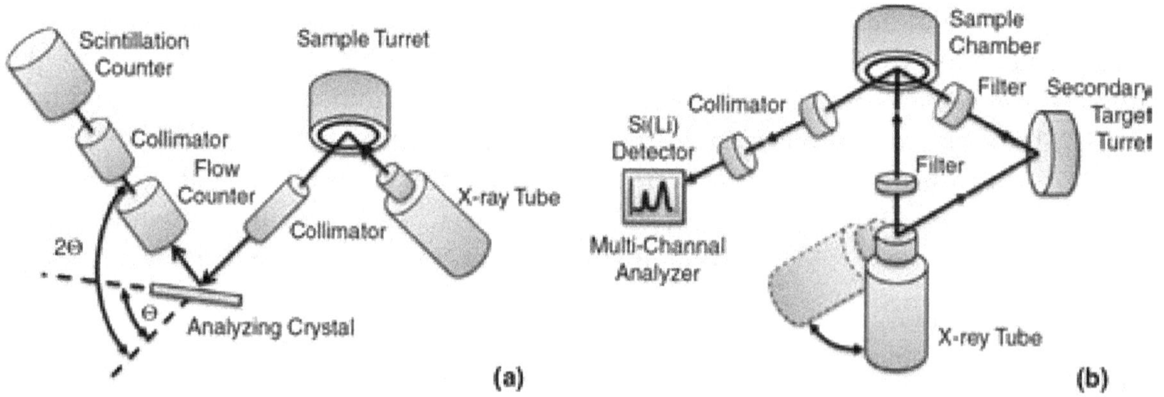

Fig. 1. Schematic representation of X-ray fluorescence analyzer: (a) wavelength-dispersive (XRF) and (b) energy dispersive (XRF).

Simultaneous determination for all elements, the atomic number of which is greater than Mg is possible.

The resolution and sensitivity of EDXRF is typically an order of magnitude worse than that for WDXRF.

Synchrotron radiation X-ray fluorescence (SRXRF) microprobe, a promising technique, is a nondestructive and qualitative to semiquantitative analysis of minerals and single fluid inclusions [34]. Synchrotron radiation (SR) is a powerful advanced light source (synchrotron radiation source, SRS) compared to conventional X-ray tube radiation and has many unique properties, such as high intensity, natural collimations, well-defined polarization, wide spectral range and energy tenability [35]. SRXRF is a widely applied technique for microscopic analysis of chemical elements. High-resolution requirements can be achieved using microbeam synchrotron radiation X-ray fluorescence (μ-SRXRF). Synchrotron radiation X-ray fluorescence can also provide the information about the oxidation state and coordination environment of metals using techniques known as X-ray absorption of near-edge structure (XANEX) or by micro-XANEX spectroscopy [37]. The unique tool for studying, the local structure around selected elements is X-ray absorption fine structure (XAFS) [38].

X-ray fluorescence is usually used to investigate the composition of apatite rocks and minerals for the purpose of their characterization [36],[39],[40], estimation of naturally occurring radionuclides in fertilizers [41] and analysis of phosphate ore at various stage of processing [42], e.g. flotation [43],[44].

Inductively coupled plasma spectrometry

Prior to inductively coupled plasma[6] (ICP), the flame, direct current-arc and controlled-waveform spark were used for the atomization (i.e. decomposition of sample to individual

[6] Plasma is defined as an electrically neutral gas which consists of positive ions and free electrons. Plasma have sufficiently high energy to atomize, ionize, and excite virtually all elements in the periodic table, which are intentionally introduced into it for the purpose of elemental chemical analysis [45].

atoms) and excitation of sample in elemental analysis. ICP denotes the technique that uses atmospheric pressure argon inductively coupled plasma[7] (ICP) for the atomization and excitation of sample. This plasma is a highly energetic media consisting of inert ionized gas with equivalent temperatures from 7000 to 10,000 K. Inductively coupled plasma are formed by coupling energy produced by RF generator (typically 700–1500 W) to the plasma support gas with an electromagnetic fields [45]. The cross-section of typical ICP torch is shown in **Fig. 2.**

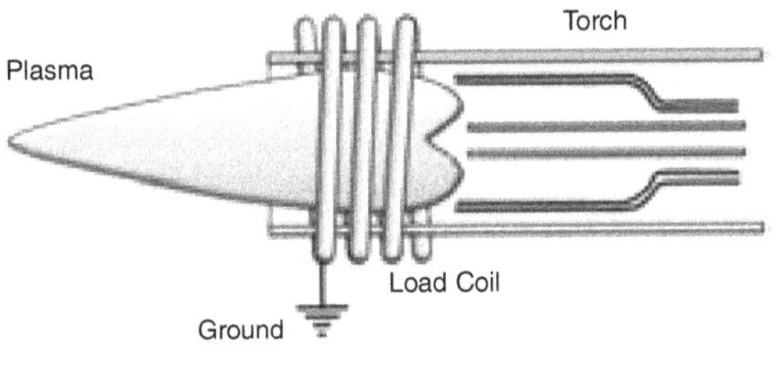

Fig. 2. Plasma with torch assembly and load coil [45].

The treatment of sample before the quantification includes vaporization, atomization, excitation and ionization.[8] The introduction of analyzed sample into inductively coupled plasma was applied in analytical techniques including [29],[45],[46],[47],[48]:

i. **Inductively coupled plasma atomic emission spectroscopy** (ICP-AES): the method connects atomic emission spectrometry with ICP. AES is based on spontaneous emission of free atoms or ions when the excitation is performed by thermal or electric energy. The method can identify and determine the concentration of up to 40 elements simultaneously with the detection limit of parts per billion (ppb). Serious limitations of this technique are the spectral interferences. Despite rapid growth of ICP-MS, ICP-AES still plays a dominant role in elemental analysis of geological, environmental, biological and other materials.

ii. **Inductively coupled plasma atomic fluorescence spectroscopy** (ICP-AFS): ASF is an analytical technique for the determination of elements in small quantities. It is based on the emission of free atoms when the excitation is performed by radiation energy.

iii. **Inductively coupled plasma-mass spectroscopy** (ICP-MS): is analytical technique for the determination of elemental composition of virtually any material. A sample, usually in the form of an aqueous solution, is converted to an aerosol by a neutralization process and transported to the plasma by an argon gas stream. In the plasma, the elements of analyte are atomized, followed immediately by ionization. The composition of ion population in the plasma is proportional to the concentration of

[7] Although there are many types of plasma, such as direct current, microwave induced, etc., the ICP is considered the most useful technique for analytical spectroscopy [45].

[8] Each element has characteristics first and second ionization potential, which depends on specific electronic structure of given element. Higher ionization potential means that more externally applied energy is required for ionization (thermal radiation, collision with other ion or electron, or exposure to high-energy photons) [45].

analyte species in original sample solution. Ions produced by ICP are representatively sampled and extracted from the plasma; next they are separated and measured by a quadrupole or time-of-flight mass spectrometer

The method known as laser ablation—inductively coupled plasma—mass spectrometry (LA-ICP-MS) is a coupling technique of laser ablation with ICP-MS technique [49],[50],[51],[52],[53]. Multicollector inductively coupled plasma mass spectrometry (MC-ICP-MS) was applied as the benchmark method for isotopic analysis [54] and for the determination of heavy rare earth elements in apatites [55],[56],[57].

Thermal ionization mass spectroscopy

Thermal ionization mass spectroscopy (TIMS) is highly specialized technique of mass spectroscopy used for very precise determination of isotope ratios and, as such, is widely used for the determination of stable isotope ratios in isotope geology and for the analysis of nuclear materials [58]. In this method, the solid sample is thermally ionized in solid-source mass spectrometer. Ions are accelerated into the mass analyzer and then transported to the detector [59]. Multicollector thermal ionization mass spectrometry (MC-TIMS) uses laminated magnetic sector filet for high speed peak jumping and low hysteresis for the mass of ion beams [60],[61].

The basic variants of the method of thermal ionization mass spectroscopy are as follows:

1. Isotope dilution–thermal ionization mass spectrometry method (ID-TIMS) [62],[63] is used for accurate determination of element concentration and is generally considered to be the definitive method to other techniques [58]. The method was applied for direct measurement of uranium, thorium, lead, etc., concentrations in the determination of single grain fission-track ages.
2. Chemical abrasion–thermal ionization mass spectrometry method (CA-TIMS)

The ability for precise determination of isotope ratio in apatite predetermines this method for geochronological investigations, i.e. studying of chronologic records in accessory minerals of igneous rocks [61],[63],[64],[65].

Secondary ion mass spectrometry

Secondary ion mass spectroscopy (SIMS) is a method[9] for the characterization of solid surface elemental composition and isotope distribution. The technique can be applied to all elements and allows quantitative analysis of solid surfaces, including monolayers. Energetic ion bombardment of a solid surface (primary ions, e.g. Ar^+, Cs^+, O^{2+}, …) causes that atoms of the sample are shifted from their original states as positive and negative ions which are termed as secondary ions. These ions are then analyzed by mass spectrometer (e.g. quadrupole mass spectrometer) to determine the composition of the surface of sample [66],[67]. The applica-

[9] The method was originally developed in the 1950s and 1960s by Herzog et al and Honig et al to analyze metals and oxides [69], but the basis of SIMS can be traced back to the beginning of the twentieth century with the first experimental evidence of secondary ions given by JJ Thomson in 1910 [68].

tions of SIMS can be broadly subdivided into static and dynamic SIMS. Static SIMS (SSIMS) is used to investigate the composition of the outermost monolayer on any solid. Dynamic SIMS (DSIMS) examines the concentration profile as the function of depth [68],[69],[70].

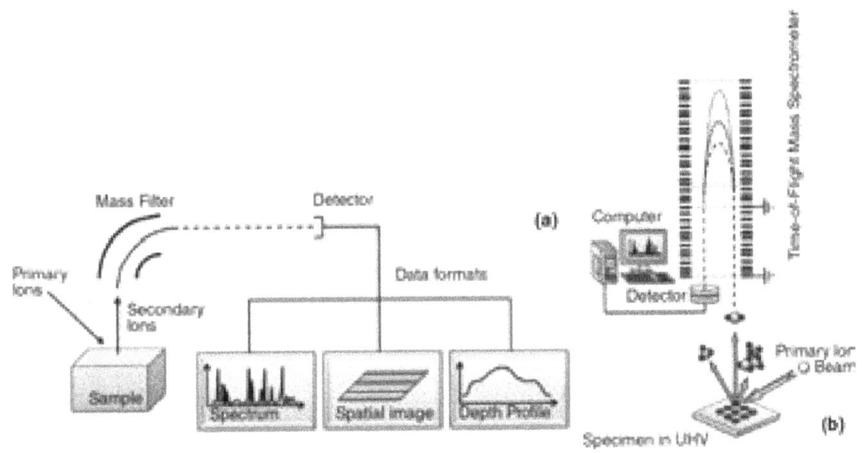

Fig. 3. Main components of SIMS [68] and schematic representation of ion microprobe imaging time-of-flight secondary ion mass spectrometry with reflectron-based mass analyser [69].

The mass filter of mass spectrometer of SIMS instruments defines the type of instruments [68], [70]:

1. Magnetic sector SIMS (M-SIMS) instruments (**Fig. 3(b)**, similar to those used in original mass spectrometer)
2. Quadrupole SIMS (Q-SIMS) instruments (**Fig. 4(a)**, first appeared in the 1970)
3. Time-of-flight SIMS (ToF-SIMS) instruments (**Fig. 4(b)**, first appeared in the 1980)

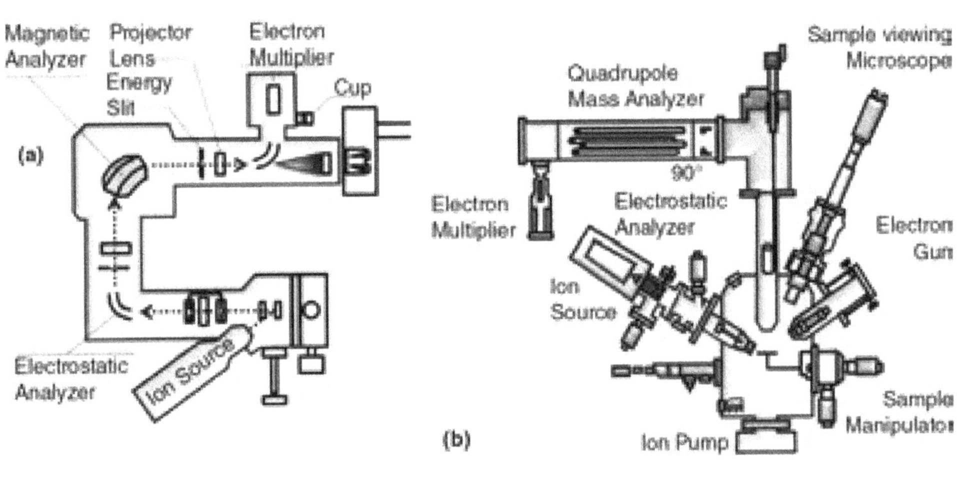

Fig. 4. Schematic representation of magnetic sector and quadrupole SIMS instruments [70].

Secondary ion emission begins when the primary ions energy exceeds some threshold level. This energy ranges from 30 to 80 eV and is much higher than the energies for sputtering of neutral particles. The most important features of the method are very low sensitivity limit for majority of elements, the possibility to determine the concentration profile, the possibility of isotope analysis and the identification of all the elements and isotopes starting from hydrogen [66].

The SIMS method is often used for the measurement of deuterium/hydrogen (D/H) ratio and hydroxyl (OH) content in anhydrous minerals and melt inclusions in Martian meteorites [71] and lunar materials (apatite grains, glass beds, melt inclusions and agglutinates in soils) [72]. The method is widely used to investigate biological apatites and collagen apatite composites [73] and the analysis of micrometer-sized samples like, e.g. interplanetary dust, presolar grains and small inclusions in meteorites, has become more and more important in cosmochemistry [74]. The method is also utilized for U-Th-Pb dating of apatites as common accessory in igneous rocks (**Chapter**), based on the radioactive decay of U and Th [75],[76],[77].

Laser secondary neutral mass spectrometry

Laser secondary neutral mass spectroscopy (laser-SNMS) can be further divided to nonresonant laser-SNMS (NR-laser-SNMS) and resonant laser-SNMS (R-laser-SNMS). In NR-laser-SNMS (**Fig. 5**(**b**)), an intense laser beam is used to nonselective ionization of all atoms and molecules within the volume intersected by the laser beam. Sufficient laser power density, which is necessary to saturate the ionization process, is typically achieved in a small volume. It limits the sensitivity of the method and leads to the problems with quantification due to the differences between effective ionization volumes of different elements. Laser-SNMS method has significantly improved ionization efficiency over SIMS (**a**) [78],[79].

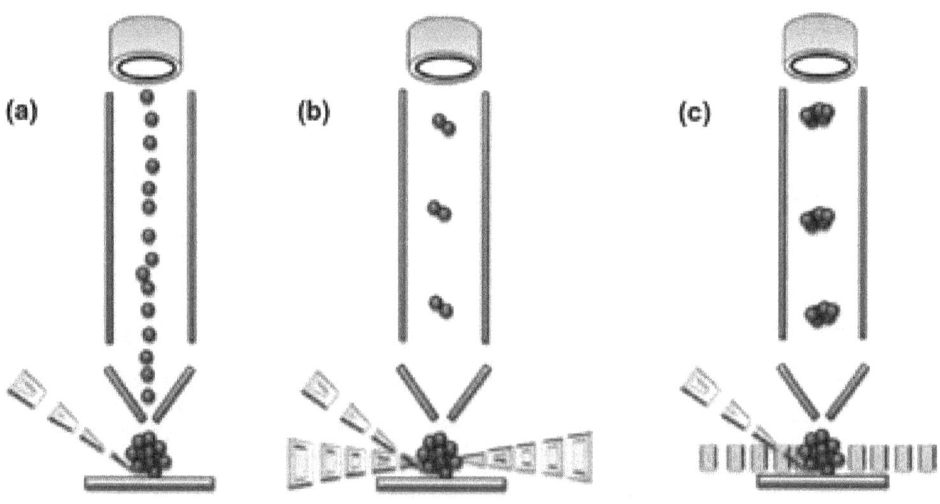

Fig. 5. Comparison of ToF-SIMS and laser-SNMS: (a) direct analysis of secondary ions (ToF-SIMS), (b) nonresonant laser-post-ionization of secondary neutrals (NR-laser-SNMS) and (c) resonant post-ionization of secondary neutrals (R-laser-SNMS) [78].

Resonant laser-SNMS uses a resonance laser ionization process, which selectively and efficiently ionizes atoms and molecules over a relatively large volume (**Fig. 5**(**c**)). The method has unit ionization efficiency for over 80% of elements in periodic table, i.e. the overall efficiency is greater than that for NR-laser-SNMS, the quantification is also simpler and extremely high selectivity prevents almost all isobaric and molecular interferences [78],[79].

The method of NR-laser-SNMS was used by DAMBACH et al [80] to investigate different states of biomineralization in vitro. The results indicate that in the vicinity of single osteoblasts, extracellular enrichment of potassium typically occurs during initial stages of mineralization. Potassium may interact with matrix macromolecules and prevent an uncontrolled apatite deposition. However, apatite biomineral formation is correlated with a potassium release. In conclusion, potassium seems to be involved in the process of extracellular matrix biomineralization.

Electron paramagnetic resonance

The concept of electron paramagnetic resonance (EPR) spectroscopy is very similar to more familiar nuclear magnetic resonance (NMR). Both methods deal with the interaction between electromagnetic radiation and magnetic moments. In the case of EPR, the magnetic moments arise from electron rather than nuclei. The term EPR was introduced as a designation taking into account contributions from electron orbital as well as spin angular momentum. The term electron spin resonance (ESR) was also widely used because in most cases the absorption is linked primarily to the electron-spin angular momentum [82],[83]. EPR spectrum is a diagram in which the absorption of microwave frequency radiation is plotted against the magnetic field intensity [83].

The technique of electron paramagnetic resonance spectroscopy may be regarded as the consequence of the STERN–GERLACH experiment. They showed (in 1920) that an electron magnetic moment in an atom can take on only discrete orientation in a magnetic field, despite the sphericity of the atom. Subsequently, UHLENBECK and GOUDSMIT liked the electron magnetic moment with the concept of electron spin angular momentum. In hydrogen atom, there is additional angular momentum arising from the proton nucleus. BREIT and RABI described the resultant energy levels of hydrogen atom in a magnetic field. RABI et al [81] studied the transition between levels induced by an oscillating magnetic field, and this experiment was the first observation of magnetic resonance. The first observation of electron paramagnetic resonance peak was made in 1945 by ZAVOISKY, who detected the radiofrequency absorption line from $CuCl_2 \cdot 2H_2O$ sample using the radiofrequency (RF) source at 133 MHz [82].

The major components of EPR spectrometer are shown in **Fig. 6**. The microwave bridge supplies the microwaves at controlled frequency and power, which are transmitted to the sample cavity via the waveguide. The sample cavity is placed perpendicular to applied magnetic field, which can be varied in controlled way. In addition to this main magnetic field, a controlled but smaller oscillating magnetic field is superimposed on the cavity via the Zeeman modulation frequency. The ideal way to perform the experiment would be to apply a fixed magnetic field and vary the microwave frequency. However, microwave generators are only tunable over very limited ranges. Thus, the microwave frequency is fixed and applied magnetic field is varied. The magnetic field is applied until it reaches the value at which the sample will absorb some of the microwave energy, i.e. and EPR transition occurs [84],[85].

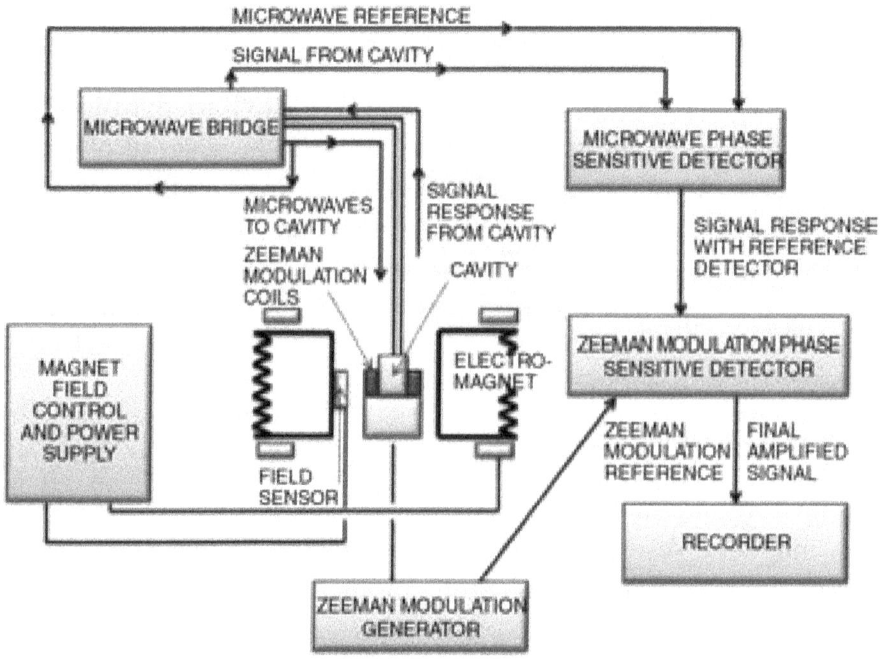

Fig. 6. The block diagram for typical continuous wave EPR spectrometer [84].

Electron paramagnetic resonance (EPR) spectrum of X-irradiated sodium and carbonate containing synthetic apatites has been studied by Moens et al [86]. Observed spectra were decomposed in terms of five theoretical curves representing O^- radical, two CO^{3-} radicals (surface and bulk) and two CO^{2-} radicals (surface and bulk). These species were also described in A-type and B-type carbonate-apatites [87],[88], tooth enamel [89],[90],[91],[92], [93] and bone [94],[95], apatites, renal stones [96], etc.

Nuclear magnetic resonance

Solid-state nuclear magnetic resonance (NMR) is a technique for accurate measurement of nuclear magnetic moments where the resonance frequency depends on its chemical environment [97],[98],[99]. The method can provide useful information on the number of molecules in the asymmetric unit and on the site symmetry of the molecule in the lattice to assist in the refinement of powder X-ray diffraction (**Section**) data. The method can distinguish between different polymorphs. Alternatively, solid-state NMR can be used for direct and accurate measurement of internuclear distances. For amorphous and disordered solids, such as inorganic glasses and organic polymers, solid-state NMR provides structural information that cannot be obtained by any other technique [100],[101]. NMR is also the diagnostic method used in veterinary science and medicine particularly in clinical research of human brain by magnetic resonance imaging (MRI) [102].

The solution-state NMR method was developed for the investigation of structure of soluble proteins [103]. Solution and solid-state NMR are both excellent methods for the determination of chemical composition [100].

The structural information of apatites is usually investigated from ^{1}H, ^{19}F and ^{31}P NMR spectra of apatites [104]. The ^{31}P solid-state NMR spectroscopy is a useful tool to investigate structural information about apatites on bone organic and inorganic mineral components, as well as to investigate the crystallinity and compositional changes in carbonated apatites [105]. Intact bone is a demanding tissue for structural studies. Serious experimental problems arise from the morphological diversity of bone and from the co-existence, interrelationship and great complexity of its organic and inorganic components. Furthermore, one has to perform noninvasive analysis because bone samples are very sensitive to physical effects and chemical treatment. Solid-state ^{31}P NMR gives us a unique opportunity to look specifically at the minerals of whole bone without any chemical pretreatment, thus avoiding the intervention into the bone structure [106].

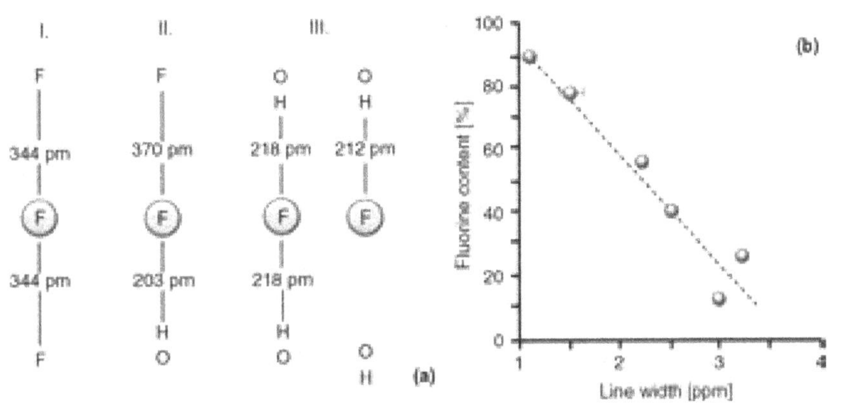

Fig. 7. Nuclear separation along parallel chains (the crystallographic c axis) in various apatites (a): FFF group (I), FFH group (II) and HFH group (III). Correlation between observed 19F line width and fluorine content of fluorinated hydroxylapatite (b) [104].

The ^{19}F NMR spectrum of fluorinated calcium hydroxylapatite ($Ca_{10}(PO_4)_6F_{2x}(OH)_{2-2x}$, where x is the fraction of OH^- replaced by F^-) indicates the correlation between ^{19}F chemical shift tensor parameters and the content of fluorine in apatite. The presence of OH^- groups induces perturbations of fluorine environments, involving the displacements of both fluorine and hydroxyl groups from their normal positions. This leads to a distortion of the electronic environment with regard to the investigated fluorine nucleus and gives reasons for observed change in the ^{19}F chemical shift tensor of fluoridated hydroxylapatite with different fluorine content. Furthermore, the presence of OH-group destroys the fluoride long-range structure and that results in an isotropic chemical shift distribution. This leads to observed increase in the ^{19}F line width in the case of low fluorine content [104],[107].

Scanning electron microscopy, structure and elemental analysis

The scanning microscope (SEM) permits the observation and characterization of heterogeneous organic and inorganic materials on a nanometer (nm) to micrometer (μm) scale. In SEM, the area to be examined or the volume to be analyzed is irradiated with finely focused electron beam, which may be swept in a raster across the surface of the specimen to form an imager or may be static to obtain the analysis at the position. The type of signals produced from the interaction of the electron beam (primary electron, PE) with the sample (**Fig. 8(a)**) includes secondary electrons (SE, with energy ≤50 eV), backscattered electrons (BSE, $E > 50$ eV), Auger electrons (AE), X-ray characteristics (X) and other photons of various energies such as continuum X-rays and heat. Low-loss electrons (LLE) show the energy losses of a few hundreds of eV. These signals are obtained from specific emission volumes within the sample and can be used to examine many characteristics of the sample such as surface topography, crystallography, composition, etc. [108],[109],[110].

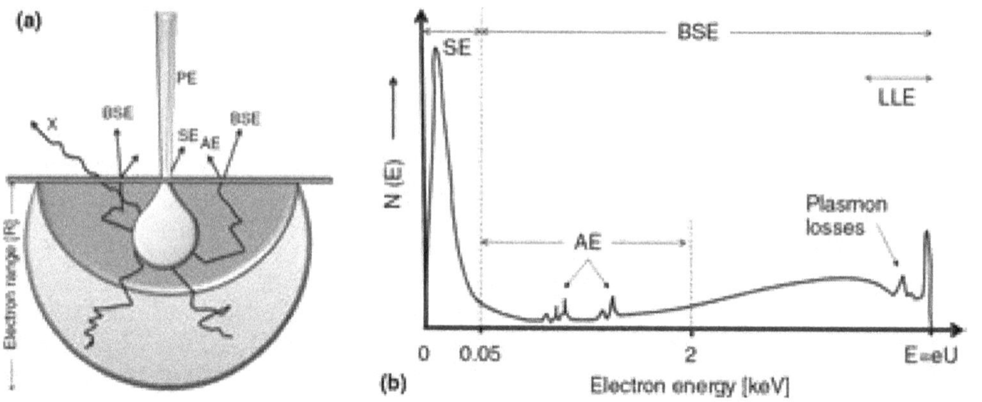

Fig. 8. Electron–specimen interaction (a) and schematic energy spectrum (a) [109].

Secondary and Auger electrons are highly susceptible to elastic and inelastic scattering and can leave the specimen only from a very thin surface layer of the thickness of a few nanometers. The most probable energy of BSE falls into the broad part of the spectrum in **Fig. 8(b)**, but they also show more or less pronounced elastic peak followed by plasnom losses, which depend on the primary energy, the take-off angle and the tilt of the specimen. Continuously slowing-down approximation assumes that the mean electron energy decreases smoothly with decreasing path length of the electron trajectories inside the specimen. The maximum information depth of BSE is of the order of half the electron range. Characteristic X-rays will only be excited in the volume in which the electron energy exceeds the ionization energy of the inner shell involved. Inelastic scattering in semiconductors results in the generation of electron-hole pairs. The recombination can take place without radiation but may result in the emission of light quanta (cathodoluminescence, CL) [111].

The method known as electron backscattering diffraction (EBDS) enables to determine the crystal structure and grain orientation of crystals on the surface of specimen. To collect maximum intensity in the diffraction pattern, the surface of specimen is stipple tilted at an angle of typically 70° from the horizontal (**Fig. 22(a)**). The intensity of backscatter **Kikuchi patterns** (please see the pattern of fluorapatite in **Fig. 23**) is rather low, as is the contrast of the signal, so extremely sensitive cameras and contrast enhancement facilities are required. This pattern allows to identify the phases and shows the misorientation across the grain boundaries [108].

Scanning electron microscope can be also used to determine compositional information using characteristic X-ray. The development of instruments for obtaining localized chemical analysis of solid samples, i.e. electron probe microanalyzer (EMPA), occurred at the same time as the development of SEM.

Scanning electron microscopy (SEM) is used for grain interactions and spot analysis [98],[112], electron microprobe microanalysis (EPMA) for the distribution of elements in the matrix, investigation of the effects of impurities on the properties of apatites and investigation of reaction interface [113],[114],[115],[116].

Fourier transform infrared and Raman spectroscopy

Infrared (IR) spectroscopy is one of the most important analytical techniques that can be used for the investigation of any sample in any state. Liquids, solutions, pastes, powders, films, fibers, gases and surfaces can be examined with judicious choice of sampling technique. Infrared spectrometers have been commercially available since the 1940s [117].

Fourier transform infrared (FT-IR or FTIR) spectroscopy is divided into three regions according to the increasing wavelength [118]:

1. Near-IR (NIR) spectroscopy, abbreviated as FT-NIR
2. Mid-IR (MIR) spectroscopy, abbreviated as FT-MIR
3. Far-IR (FAR) spectroscopy, abbreviated as FT-FAR

The spectral ranges of near-, mid- and far-infrared spectroscopy are shown in **Fig. 9(a)**.

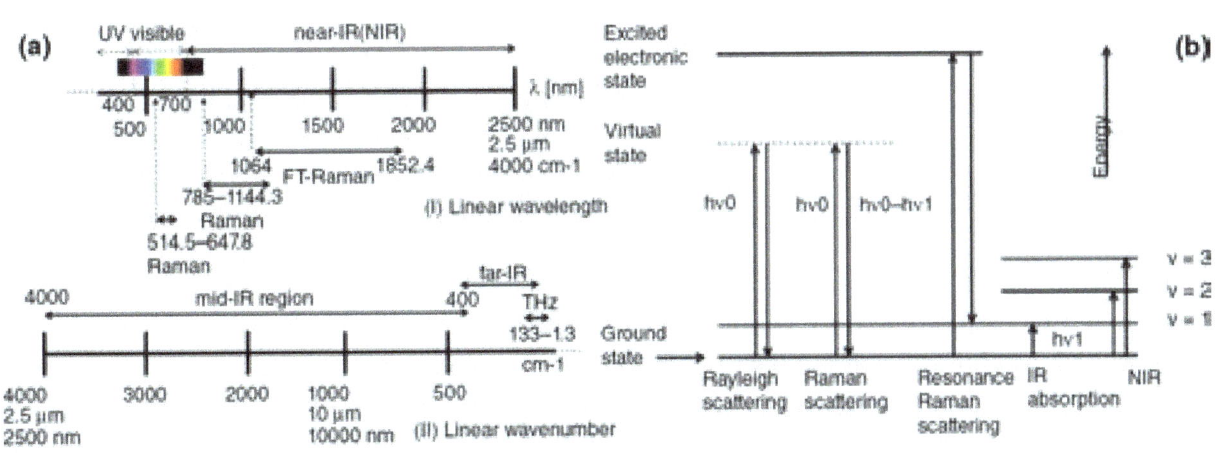

Fig. 9. Schematic illustration of relationships between the ranges of (a) vibrational spectroscopy and electromagnetic spectrum [118] and (b) spectroscopic transitions underlying several types of vibrational spectroscopy. v_0 indicates the laser frequency, while v is the vibrational quantum number. The virtual state is a short-lived distortion of the electron distribution by the electric field of the incident light [119].

The background for Raman spectroscopy was given by the discovery of Raman scattering by Krishna and Raman in 1928. Until approximately 1986 when Fourier transform (FT)–Raman was introduced, physical and structural investigations dominated in literature over relatively few reports of Raman spectroscopy applied in chemical analysis [119],[120].

When monochromatic light with the energy $h\nu_0$ encounters the matter (gas, solid or liquid), there is a small probability that it will be scattered at the same frequency (**Fig. 9(b)**). If the object, e.g. molecule is much smaller than the wavelength of the light, the scattering is Rayleigh scattering. The "*virtual state*" is not necessarily a true quantum state of the molecule but can be considered a very short-lived distortion of the electron cloud caused by oscillating electric field of the light. Since blue light is more efficiently scattered than red one, Rayleigh scattering is responsible for the blue color of sky. The electron cloud of the molecule is also perturbed by molecular vibrations, and it is possible for the optical and vibration oscillations to interact, leading to Raman scattering. Raman scattering is shown in (**Fig. 9(b)**) in which the scattered photon is lower in energy by an amount equal to the vibration transition. Raman spectrum consist of scattered intensity plotted versus energy and each peak corresponds to given Raman shift from the incident light energy $h\nu_0$ [119].

Just like Rayleigh scattering, Raman scattering depends on the polarizability of scattering molecules. IR band, on the other hand, arises from a change in the dipole moment. In many cases, the transitions that are allowed in Raman are forbidden in IR, so these techniques are often complementary (please compare **Fig. 18(a)** and (**b**)). In polarizable molecules, incident light can excite the vibrational modes, leading to scattered light diminished in energy by the amount of vibrational transition energies (same as in fluorescence). Scattered light under these conditions reveals the satellite lines below the Rayleigh scattering peak at the incident frequency–Stokes lines (Stokes part of spectrum). If there is enough energy, it is also possible to see anti-Stokes lines. Since anti-Stokes lines are usually weaker than Stokes lines, only the Stokes part of spectrum is usually measured [121].

The method combining Raman spectrometer with microscopic tools, typically an optical microscope, is known as micro-Raman spectroscopy (μRS) or also Raman microscopy. The μRS is nondestructive and noncontact method for the characterization of organic and inorganic materials [122].

Infrared [97],[98],[123],[112],[124],[125],[126], Raman [97],[98] and micro-Raman spectroscopies [125] were often used to identify and investigate the structure and extent of substitution and to optimize the synthesis conditions of minerals from the supergroup of apatite. Since carbonate ions exhibit clear vibrational signature in infrared spectrum, infrared spectroscopy is widely used to investigate the structure and to evaluate the carbonate/phosphate ratio ($r_{c/p}$) and the amount of carbonate ions in carbonate-apatites [127]:

$$CO_3^{2-}\ [\text{wt.\%}] = 28.62\ r_{c/p} + 0.0843 \qquad (1)$$

The example of infrared and Raman spectrum of fluorapatite is described in **Section**.

Infrared spectra of phosphate minerals in the pyromorphite series are described by ADLER [128]. In the pyromorphite series, the equilibrium internuclear X-O distance in XO_4^{3-} ion (PO_4^{3-},

AsO_4^{3-} and VO_4^{3-}) is primarily a function of the ionic radium of X atom. Since Pb, in this case, is always the dominant externally coordinated cation, for various members, there is no significant change in the interaction between the molecular vibration and the external environment. Bradger's equation [128],[129],

$$k_0 = 1.86 \cdot 10^5 / \left(R - d_{ij}\right)^3 \tag{2}$$

although specifically applicable to internuclear distances in diatomic molecules, reflects generally the inverse relationship between the force constant k_0 and the internuclear distance R. Symbol d_{ij} denotes the constant the values of which depend on the nature of bonded atoms. The molecular vibration frequency v is dependent on the restoring forces, measured in terms of k_0, between participating atoms as well as on the masses of these atoms. The relationship may be expressed approximately by the equation:

$$v = \frac{1}{2}\pi c \sqrt{\frac{k}{u}} \tag{3}$$

where the vibration frequency v is a function of the force constant k and the reduced mass u of vibrating atoms, all other terms being invariant.

The spectral frequency differences between pyromorphite, mimetite and vanadinite are explicable and to a considerable degree predictable in terms of these parameters. On complete substitution of As or V for P the effect of reduced force constants is reinforced by increases in mass, thereby shifting v_3 and v_1 to lower frequencies. Because of opposing mass and force-constant effects and perhaps also because of dissimilarities in orbital configurations, the relative positions of absorption bands are less predictable for mimetite and vanadinite than for pyromorphite and mimetite. The theoretical frequency trends are depicted in **Fig. 10** [128].

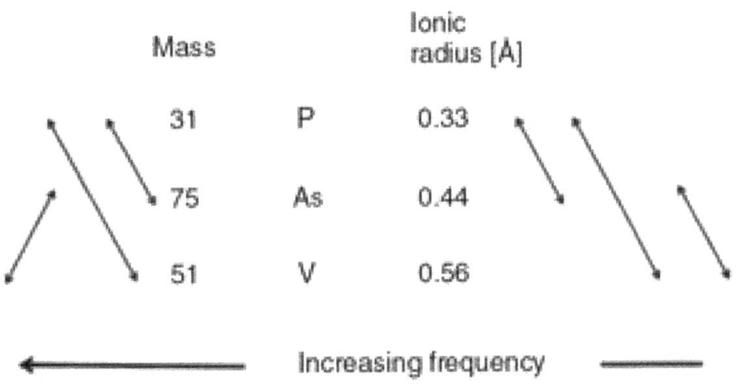

Fig. 10. Theoretical effect of change in mass and ionic radius on infrared vibration frequency of tetrahedral XO_4^{3-} ions, where $X = P^{5+}$, As^{5+} of V^{5+} [128].

Thermal analysis

Thermal analysis (TA) refers to a group of techniques[10] in which the property of a sample is monitored against time or temperature while the temperature of the sample, in a specified atmosphere[11] is programmed. These methods study the relationship between sample property and its temperature as the sample is heated or cooled in a controlled manner. The individual techniques are divided according to the measured property[12] [131],[132], [130] as is introduced in **Table** 1.

Property	Technique and abbreviation		Notes
Heat	Scanning calorimetry	–	–
Temperature	Thermometry	–	Also can be described as heating or cooling curve
Temperature difference	Differential thermal DMA analysis	DTA	A technique where the temperature difference between the sample and reference material is measured
Heat flow rate difference	Differential scanning calorimetry	DSC	A technique where the difference between heat flow rates into the sample and reference material is measured
Mass	Thermogravimetry or thermogravimetric analysis	TG or TGA	The abbreviation TG has been used, but should be avoided, so that it was not confused with T_g (temperature of glass transition)
Dimensional and mechanical properties	Dynamic mechanical analysis	DMA	Moduli (storage/loss) are determined
	Thermomechanical analysis	TMA	Deformations are measured
	Thermodilatometry	TD	Dimensions are measured
Electrical properties	Dielectric thermal analysis	DEA	Dielectric constant/dielectric loss are measured
	Thermally stimulated current	TSC	Current is measured
Magnetic properties	Thermomagnetometry	TM	Often combined with TGA
Gas flow	Evolved gas analysis	EGA	The composition and/or the amount of gas/vapor is determined
	Emanation thermal analysis	ETA	Trapped radioactive gas within the sample is released and measured
Pressure	Thermomanometry	–	Evolution of gas is detected by pressure change

[10] The definition of terms in thermal analysis was developed by ICTAC (Confederation for Thermal Analysis and Calorimetry).

[11] Gaseous environment of the sample, which may be controlled by the instrumentation or generated by the sample [130].

[12] Resulting dependence, i.e. any graph of any combination of property vs. time or temperature derived from a thermal analysis technique, should be termed as thermal curve, which is a simplified form of more correct term thermoanalytical curve. The first mathematical derivation of any curve with respect to temperature or time leads to the derivative thermoanalytical curve [130]. Since the name thermogram has medical usage, the thermal analysis curve should not be termed as thermogram [131].

Property	Technique and abbreviation		Notes
	Thermobarometry	–	Pressure exerted by dense sample on the walls of a constant volume cell is studied
Optical properties	Thermoptometry or thermos optical analysis	TOA	A family of techniques in which optical characteristics or property of the sample is studied
	Thermoluminescence	TL	Light emitted by the sample is measured
Acoustic properties	Thermosonimetry or thermoacoustimetry	TS	Emitted (sonimetry) or absorbed (acoustimetry) sound is measured.
Structure	Thermodiffractometry	–	Techniques where the compositional or chemical nature of the sample is studied.
	Thermospectrometry		

Table 1 Methods of thermal analysis according to measured property or physical quantity.

The measurement should be performed as follows:

1. **Combined**: the application of two or more techniques to different samples at the same time. This can include thermal and nonthermal analytical techniques.
2. **Simultaneous**: indicates the measurement of two or more properties of a single sample at the same time.

The sample-controlled method where the feedback used to control the heating is the rate of transformation is termed as controlled-rate thermal analysis (CRTA) [130].

Simultaneous thermogravimetry (thermogravimetric analysis) and differential thermal analysis (TG-DTA) are mostly used to investigate the course of synthesis and the characterization of prepared apatites or to investigate the process of thermal decomposition of apatites, i.e. the processes such as dehydroxylation, defluorination, decarbonation (thermal decomposition of carbonate-apatites), etc.

Optical properties

The analysis of optical properties is essential for each mineral examined, and through the use of microscopy, the optical properties of individual minerals may be interpreted in great detail. Optical mineralogy investigates the interaction of light (usually is limited to visible light) with minerals and rocks. Optical mineralogy concerns mainly the use of polarizing (petrographic) microscope which has two Nicol prisms, polarizer and analyzer (polarizer below the stage and the analyzer above the objective) [133]. Human eye is the most sensitive for viewing a solid in the wavelength symmetrically spread in intensity around 550 nm (**Fig. 11(a)**). Optical microscopy in visible light (from 700 nm (red) to 420 nm (violet), **Fig. 11(b,c)**) helps study the objects of smaller sizes up to lower limit of ~ 1 μm (c). **Fig. 11(d)** shows the scope of spectroscopic techniques associated with electromagnetic spectrum [134].

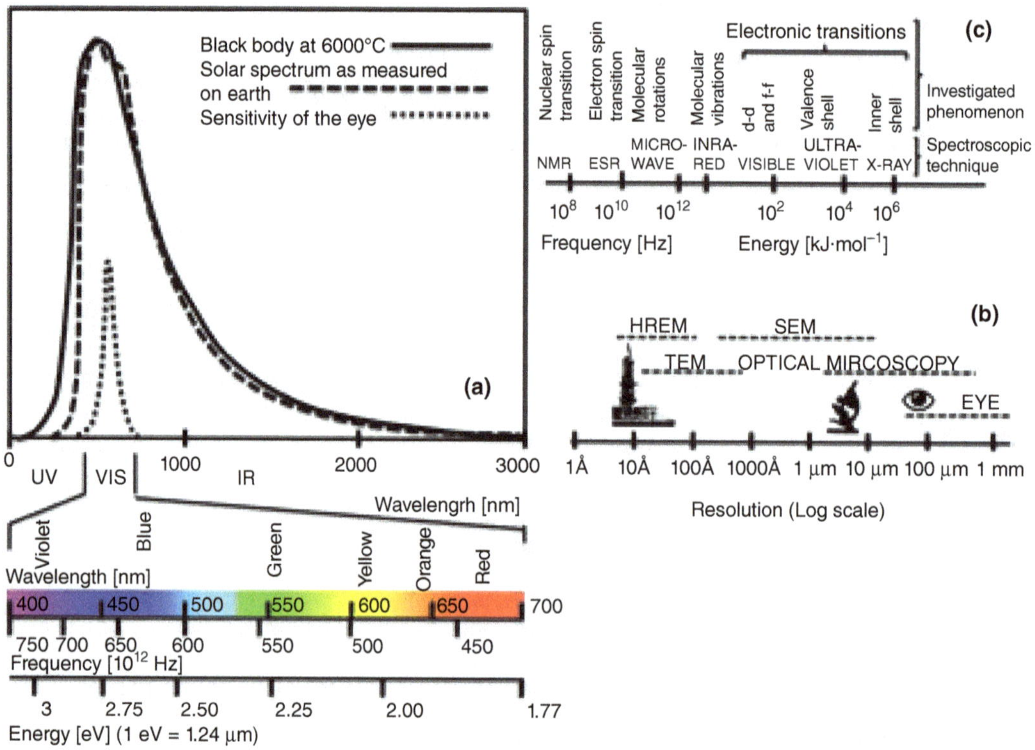

Fig. 11. Electromagnetic spectrum of solar radiation and black body (a), the resolution of various techniques (b) and the regions associated with spectroscopic techniques (c) [134].

Refractive index (n) is related to the angle of incidence (i) and the angle of refraction (r) according to the Snell's law:

$$n = \sin i / \sin r \tag{4}$$

The refractive index increases as the wavelength of light decreases [134]. The absorption coefficient is related to the imaginary part of the refractive index. It was found that since the refractive index of a medium depends on the density of electrons in that medium, the index increases with the density of matter [135].

The luster of mineral (R) depends of the way in which the light is reflected from the surface of a mineral. The reflection is again dependent on the refractive index. Normally, the greater is the index of refraction, the brighter is the luster. The luster is classified into two broad classes [134]:

a. Nonmetallic luster: results from the interaction of light with dielectric semiconducting and poorly conducting substances
b. Metallic luster: occurs in minerals having metallic bonding (e.g. native metals) or high degree of covalent bonding (sulfides, sulfosalts, etc.)

Optical properties of apatites can be determined using the complex dielectric function [136], [137],[138]

$$\varepsilon(\omega) = \varepsilon_1(\omega) + i\varepsilon_2(\omega) \tag{5}$$

in the range of linear response. By calculating the wave function matrix and using Kramers–Krönig relations, the imaginary and real part of the dielectric function $\varepsilon_1(\omega)$ and $\varepsilon_2(\omega)$ can be derived respectively as follows:

$$\varepsilon_2(\hbar\omega) = \frac{e^2}{\pi m \omega^2} \int_{BZ} dk^3 \sum_{n,l} \left| \langle \Psi_n(\bar{k},\bar{r}) | \rangle - i\hbar | \langle \Psi_1(\bar{k},\bar{r}) \rangle \right|^2 f_1(\bar{k}) \\ \left[1 - f_n(\bar{k})\right] \delta\left[E_n(\bar{k}) - E_1(\bar{k}) - \hbar\omega\right] \tag{6}$$

where $f(k)$ is the Fermi distribution function, ***l*** and ***n*** mark occupied state and unoccupied state, respectively. $\Psi_n(k,r)$ is the Bloch wave function for the *n*th band with the energy $E_n(k)$ at the Brillouin zone point ***k***. The matrix element momentum transition corresponds to the term $|\langle\Psi_n(\bar{k},\bar{r})\rangle| - i\hbar|\langle\Psi_1(\bar{k},\bar{r})\rangle|^2$.

For each apatite, the real part ε_1 and imaginary part ε_2 of dielectric function have similar features with some subtle differences. The real part ε_1 has two main peaks:

1. First peak lies near the energy of 7 eV. Obviously, the first peak is caused by the transitions from O–*2p* and P–*3p* levels.
2. The second peak lying at ~25 eV results from the transitions from Ca–*3p* levels.

The ε_2 curve in the energy range from 5 to 15 eV characterizes three main peaks:

1. The first peak lies in 7.10, 7.16 and 7.43 eV for FAP, ClAP and BrAP, respectively.
2. The second peak is located at 8.51, 8.62 and 8.76 eV for FAP, ClAP and BrAP, respectively.
3. The third peak is situated at 10.99, 11.08 and 11.45 eV for FAP, ClAP and BrAP, respectively.

The refractive index *n* can be obtained by $n = \sqrt{\varepsilon_1(0)}$. The $\varepsilon_1(0)$ for FAP, ClAP and BrAP are 1.38, 1.41 and 1.46, respectively. Therefore, the refractive indexes for FAP, ClAP and BrAP are 1.17, 1.19 and 1.21. Moreover, analogous dielectric function curves and similar refractive index values show that the optical property of each apatite has some independence from the *c*-axis ion [136].

Measuring of surface area, porosity and pore size distribution

There is a conventional mathematical idealization that asserts that a cube of edge length ***a*** possesses a surface area of $6a^2$ and that a sphere of radius ***r*** exhibits $4\pi r^2$. In reality, however, mathematical perfect or ideal geometric forms are unattainable (**Fig. 12**) since all real surfaces exhibit flaws under microscopic examination. Real surface irregularities (voids, pores, steps, etc.) make the real surface area greater than corresponding theoretical area. When the

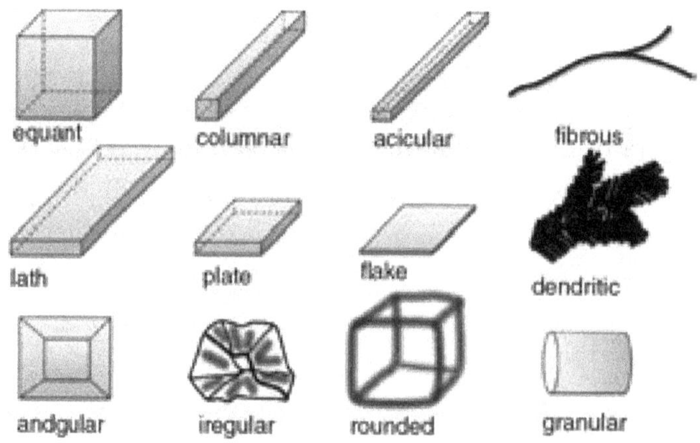

Fig. 12. Description of particle shapes [140].

cube, real or imaginary, of one meter edge length is subdivided into smaller cubes each one micrometer (10^{-6} m) in length, there will be 10^{18} particles formed, each exposing an area of 6×10^{-12} m^2. Thus, the total area of all particles is 6×10^6 m^2. This operation increases by millionfold the exposed area of fine powder compared to undivided material. Whenever the matter is divided into smaller particles, new surfaces must be produced with corresponding increase in surface area [139],[140].

The particle size distribution (PSD) was usually determined by sieve analysis, sedimentation methods (gravitational or centrifugal), microscopic techniques, light scattering, multiangle laser light scattering (MALLS), etc. [141].

The range of specific surface area, i.e. area per gram of matter, can vary widely depending on the particles' size, shape and porosity. The influence of pores can often overwhelm the size and external shape factors. The powder consisting of spherical particles exhibits total surface (S_t) and volume ($V = M/\rho$) [139]:

$$S_t = 4\pi\left(r_1^2N_1 + r_2^2N_2 + \ldots + r_n^2N_n\right) = 4\pi \sum_{i=1} r_i^2N_i \tag{7}$$

$$V = \frac{4}{3}\pi\left(r_1^3N_1 + r_2^3N_2 + \ldots + r_n^3N_n\right) = \frac{4}{3}\pi \sum_{i=1} r_i^3N_i \tag{8}$$

$$S = \frac{S_t}{M} = \frac{3\sum_{i=1} N_ir_i^2}{\rho\sum_{i=1} N_ir_i^3} \tag{9}$$

where r_i and N_i are the average radii and numbers of particles in the fraction i. For spheres of uniform size, **Eq. 9** becomes the law:

$$S = \frac{3}{\rho r} \quad (10)$$

Besides the calculation of specific surface from its geometry and PSD curve, the adsorption isotherm (predominantly Langmuir and BET[13]), air permeability methods,[14] and monolayer sorption methods were used to determine the specific surface area [142],[143].

The porosity is defined as the ration of pore volume to total volume. Porous material is defined as solids containing pores **Fig. 13(a)**, which are classified into two major types: open and closed pores. Penetrating open pores (interconnected pores) are permeable for fluid and therefore are important in applications such as filters. Pores accessible from only one end are referred to as dead-end pores. Noninterconnected (closed) pores are not accessible at all. The classification of pores according to their size is shown in **Fig. 13(b)**. Pores can be also classified as the pores among agglomerates and pores among primary particles (**Fig. 13(c)**) [144],[145].

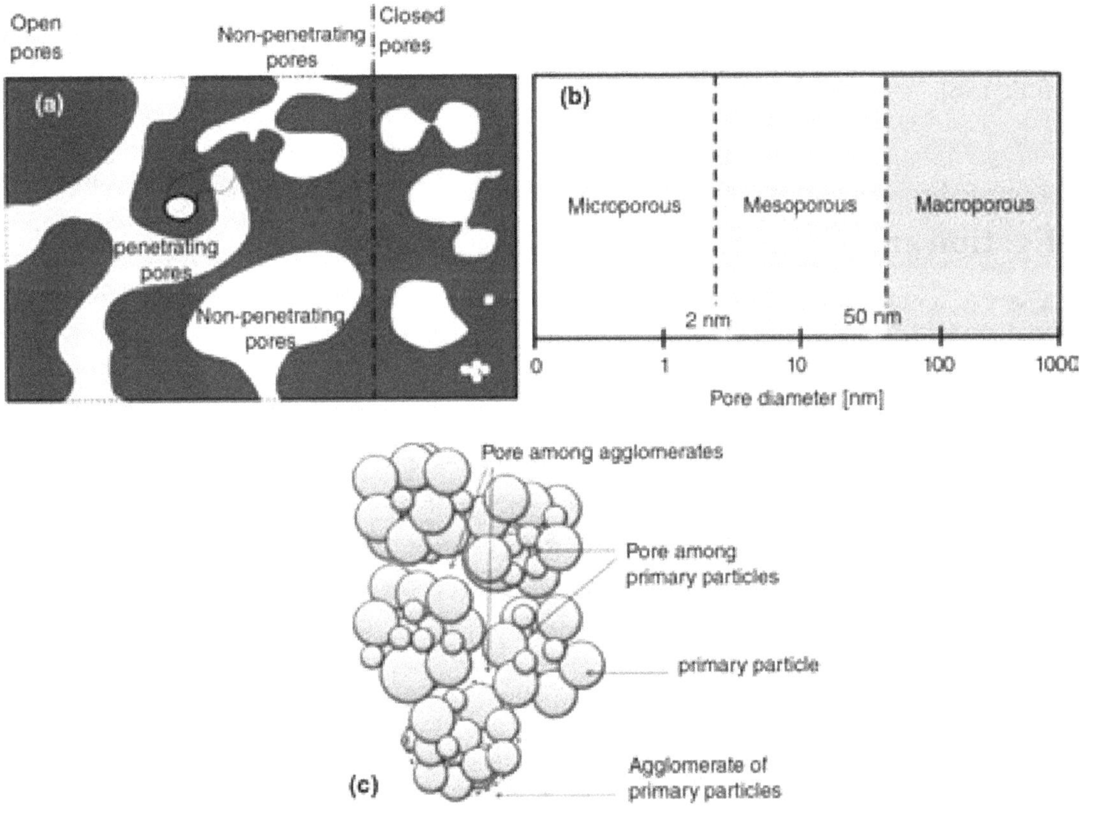

Fig. 13. Schematic illustration of different morphology of pores (a) and classification of porous materials based on pore size (b) and schematic illustration of pores among agglomerates and primary particles (c) [144].

[13] BRUNAUER–EMMETT–TELLER (BET) theory of multilayered physical adsorption of gas molecules on a solid surface [143].

[14] For example, the Blaine method, where fixed volume of air passes through the bed at steadily decreasing rate, which is controlled and measured by the movement of oil in a manometer, the time required being measured. The method is widely used for the determination of specific surface area of cements [143].

Pore size distribution and permeability[15] are a very desirable quantities for the characterization of structure of porous solids, which can be determined by the following [145],[146],[147]:

i. Stationary fluid (static) method

ii. Capillary pressure (HASSLER[16]) method

iii. Quasi-steady-state methods, including the gas drive method and the solution-gas drive method.

The surface area and the porosity of apatites have effect on the floatability (**Section**) of apatites of different type and origin [148]. The surface area, the porosity and the pore size distribution are properties of great importance for the preparation of biological apatites in tissue engineering (**Section**) and tailoring their mechanical properties, solubility and bioactivity [149],[150],[151],[152]. Ionic surfactants, such as decyltrimethylammonium bromide ($CH_3(CH_2)_9N(CH_3)_3(Br)$, C(10)TAB), hexadecyltrimethylammonium bromide ($(CH_3(CH_2)_{15}N\,(Br)(CH_3)_3$, C(16)TAB), as well as nonionic surfactant, can be used to control the pore (pore size and volume) and surface characteristics of mesoporous apatite materials under maintained pH [153]. The porosity also affects electrical properties of oxyapatites (**Chapter**) [154],[155],[156].

Investigation and characterization of apatite specimen

Identification of the specimen

Some techniques mentioned above will be demonstrated on the specimen of apatite sample (**Fig. 14(a)**) from Sljudjanka, Bajkal. The translucent specimen with glassy luster is greenish blue colored and brittle as can be seen from large amount of smoothly curving conchoidal fractures on the surface (please see also **Fig.** 20). When scratched by a single crystal of corundum (**Fig. 14(b)**), the sample shows white colored scratch. Since the sample surface can be also scratched by feldspar (**Fig. 14(c)**) but not by fluorite, the hardness in the Mohs scale[17] is equal to 5, i.e. corresponds to apatite.

---

[15] The flow of fluids through porous materials is of great importance in the fields of industrial chemistry, oil technology, and agriculture. In general, it may be stated that the principal interest is in the transport through reactive materials [147].

[16] Experimental techniques for the measurement of relative permeability can be divided to steady- (1) and unsteady-state (displacement) methods (2). The steady-state method was developed by HASSLER (1944). Semipermeable membranes are provided at each end which keeps the fluids separated, except inside the core where the fluids flow simultaneously. The pressure is measured in each phase through semipermeable barriers, and the pressure differences between the phases are maintained constant throughout the medium so as to eliminate the capillary end effect as well as to ensure a uniform saturation along the core. The saturation can be altered by applying capillary pressures across the nonwetting phase ports and wetting phase semipermeable membranes [146].

[17] The Mohs scale of mineral hardness is graduated as follows: talc (1), gypsum (2), calcite (3), fluorite (4), apatite (5), orthoclase (6), quartz (7), topaz (8), corundum (9), and diamond (10). Apatite should be also scratched by steel knife (up to 5.5) and glass (up to 6).

Fig. 14. Specimen of apatite from Sljudjanka, Bajkal (a) shows white color of scratch (b). The sample can be also scratched by minerals with hardness ≥6 (feldspar and higher) on the Mohs scale (c).

The microhardness of apatite sample was then determined by Vickers microhardness test.[18] **Fig. 15** shows the replica of diamond pyramid base[19] on the surface after the indentation of sample. The average hardness of sample was determined to be 552 (±30) HV 0.05/10. The formation of radial cracks on the corners indicates brittle material [157]. According to the mineral hardness conversion charts, the measured value is in good agreement with the tabular value of apatite (535 HV [158]).

Moreover, the sample does not show any luminescence when elucidated by long UV light (**Fig. 16(a)**). The specific gravity of the specimen was assessed by hydrostatic weighting (b) and pycnometric technique (c) to be 3.18 and 3.16 ± 0.20 g·cm^{-3}, respectively. These values are in a good agreement with average density of apatite (3.19 g·cm^{-3}, **Chapter**). All properties of investigated specimen mentioned above identify it as apatite, but the exact kind of apatite mineral and its chemical composition must still be determined yet.

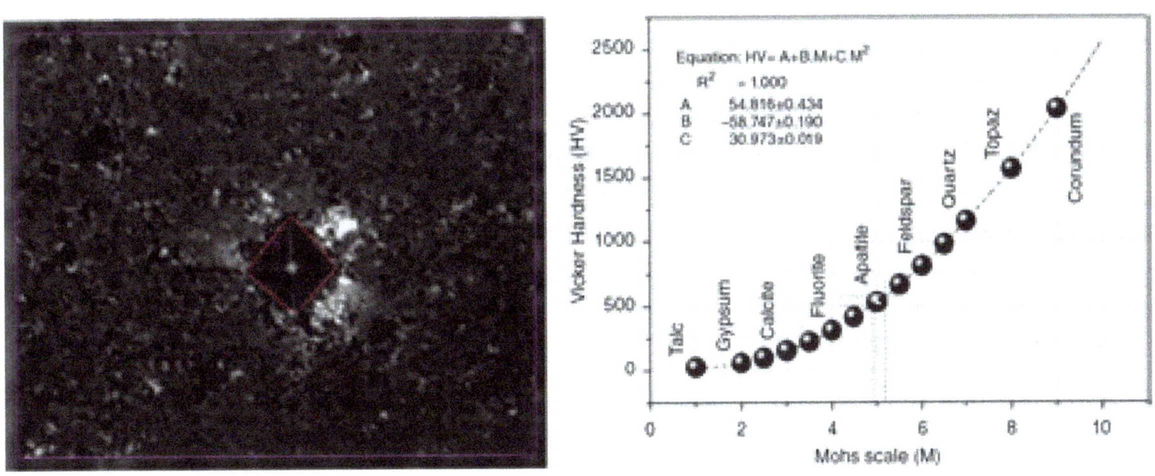

Fig. 15. The Vickers microhardness test with the load of 0.05 kgf for the time of 10 s and the correlation of results with the Mohs scale.

[18] Hardness tester LECO AMH 43. The method is also known as the Vickers pyramid number (HV) or the diamond pyramid hardness (DPH).

[19] Diamond pyramid with apical angle of 136°.

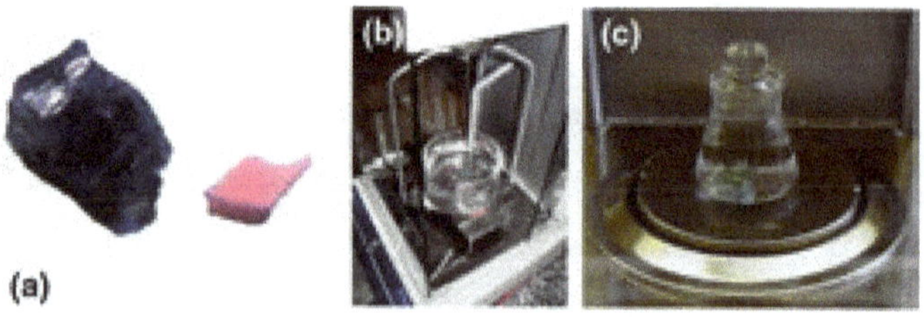

Fig. 16. The specimen of apatite under UV light (compared with the fragment of red luminescence of alumina).

The sample was next treated to fine powder via milling in stain-less steal laboratory vibration mill. The apatite mineral was then determined by X-ray diffraction analysis (**Fig. 17**), mid-infrared spectroscopy (**Fig. 18**) and EBDS (**Fig. 23**) as fluorapatite ($Ca_5(PO_4)_3F$, ref. [159]) with small amount (1%) of accessory mineral calcite[20] ($CaCO_3$). Since the crystallographic parameters of identified hexagonal apatite are $a = 9.3917$, $c = 6.8826$ Å and **Z** = 2, it is possible to calculate the axial ratio (**Eq. 11**), the volume of cell (**Eq. 12**) and the density (**Eq. 13**) as follows:

$$a{:}c = 1{:}0.7328 \tag{11}$$

$$V = a^2\ c\ \sin(60) = 9.3917^2 6.8826\ \sin(60) = 525.74\ \overset{\circ}{A}{}^3 \tag{12}$$

$$\rho_{\text{calculated}} = \frac{M_{\text{apatite}} Z}{V \dfrac{N_A}{1\times10^{24}}} = \frac{504.31\times 2}{525.74 \dfrac{6.025\times10^{23}}{1\cdot10^{24}}} = 3.19\ \text{g}\cdot\text{cm}^{3} \tag{13}$$

where M is the molecular weight of fluorapatite and N_A is the Avogadro constant. The reconstruction of the cell of investigated apatite specimen is shown in **Fig. 24**.

X-ray diffraction analysis

Powder X-ray diffraction analysis of the apatite specimen (**Fig. 17**) identified it as fluorapa-tite. According to quantitative Rietveld analysis, the sample contains 99% of fluorapatite. There is also small amount (1%) of calcite[21] that occurs on the surface of apatite specimen. Since the fluorapatite specimen (**Fig. 14**) is single crystal (**Section**), the crystal faces cannot be recognized, i.e. the crystal habit is anhedral (refer to **Footnote 2** in **Chapter**). Nevertheless, XRD pattern shows that the most intensive diffraction possesses the Miller index (211), which corresponds to the Miller-Bravais indices (21–31), i.e. dihexagonal dipyramid (**Chapter**).

[20] Since it is present as "free carbonate," the sample cannot be considered as carbonate-fluorapatite.

[21] As the results of thermal analysis revealed (**Section**), the Rietveld analysis slightly overestimates the content of free carbonate in the sample.

There is also basal pinacoid {1000}, first-order {10–10} and second-order {11–20} hexagonal prism and dihexagonal prism {21–30}.

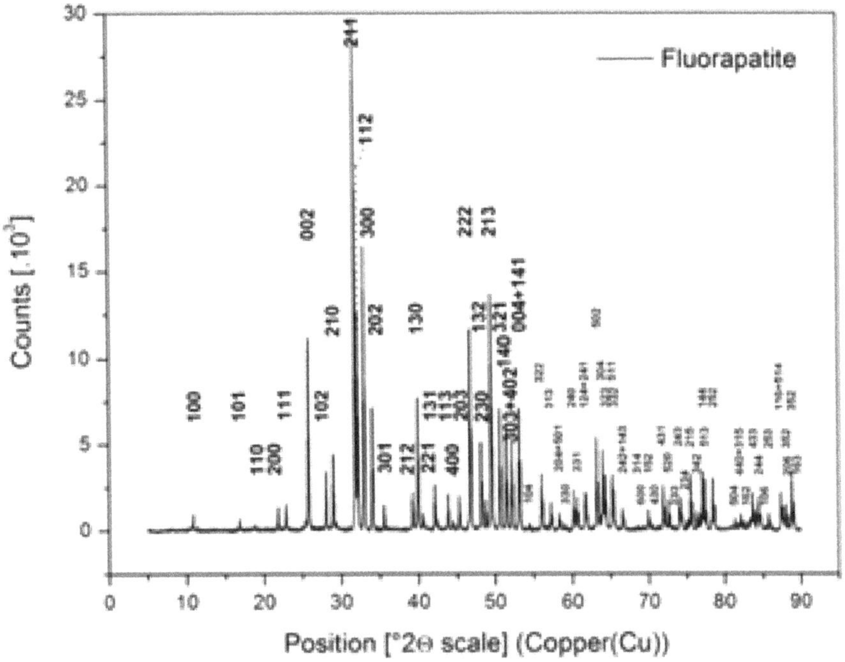

Fig. 17. X-ray diffraction analysis of investigated specimen of apatite.

It is obvious that fluorapatite belongs to the hexagonal-dipyramidal crystal system, but the estimation of crystal habit of corresponding euhedral crystals from these results is highly speculative due to possible combination of pinacoid (*c*), first-order (m) and second-order hexagonal prisms (a) and first-order (*p*) and second-order dipyramids (*s*) with dihexagonal dipyramid (*v*) faces in the single crystal.

Infrared and Raman spectroscopy

Infrared (mid-FT-IR[22]) and Raman spectrum of fluorapatite is shown in **Fig. 18**(**a**) and (**b**), respectively. The most expressive infrared bands are attributed to fundamental frequencies of tetrahedral phosphate ion $[PO_4]^{3-}$. The structure of apatite leads to the reduction of ion symmetry from T_d (four fundamental frequencies with IR inactive ν_1 mode) to C_s, where ν_1 mode becomes IR active [128],[160],[161],[162],[163]:

1. The $\nu_1(PO_4)$ mode is very weak (*vw*[23]) band that is related to symmetric stretching of phosphate ion.
2. Bending: ν_2 mode (*vw*).

[22] Baseline corrected spectrum measured by KBr technique.

[23] Abbreviation used for the expression of intensity and width of peak in the spectrum: very weak (*vw*), weak (*w*), middle (*m*), strong (*s*) and very strong (*vs*), shoulder (*sh*), broad (*b*), very broad (*vb*), and sharp (*sp*). Spectral bands related to impurities are abbreviated as *imp* [162].

3. Anti-symmetric stretching: ν_3 mode is the strongest (*vs, vb*) band in the infrared spectrum appearing in the spectral region from 1000 to 1150 cm^{-1}.

4. Bending: ν_4 mode (*m, vb*) is observed between 540 and 620 cm^{-1}.

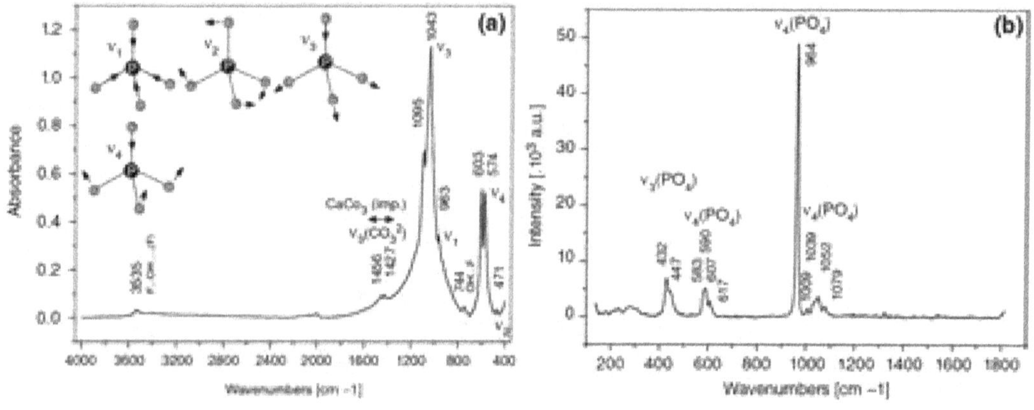

Fig. 18. Infrared (a) and Raman (b) spectrum of investigated specimen of fluorapatite.

The crystallinity of natural and synthetic apatite samples is often determined from the broadening of $\nu_4(PO_4)$ infrared absorption bands [160]. The assignment of bands in infrared and Raman spectrum of fluorapatite is listed in **Table** 2.

Vibration mode	Assignment C_{6h} factor group symmetry	Raman shift [cm^{-1}]	IR
ν_1	A_g, E_{2g}, E_{1u}	956	965
ν_2	E_{1g}	432	–
	E_{2g}	–	–
	A_g	449	–
	E_{1u}	–	460
	A_u	–	470
ν_3	A_u, E_{2g}	1034	1032
	E_{1g}, E_{1u}	1042	1040
	A_g	1053	–
	E_{2g}	1061	–
	A_g	1081	–
	E_{1u}	–	1090
ν_4	A_u	–	560
	E_{1u}	–	575
	E_{2g}	581	–

Vibration mode	Assignment C_{6h} factor group symmetry	Raman shift [cm^{-1}]	IR
	A_g, E_{1g}	592	–
	E_{1u}	–	601
	A_g	608	–
	E_{2g}	617	–

Table 2. The interpretation of infrared and Raman bands in the spectrum of stoichiometric fluorapatite [164].

Factor group analysis of the hexagonal P6$_3$/M space group fluorapatite structure (Z = 2) yields an irreducible representation for optically active vibration of [165]:

$$\Gamma = 12\ A_g(R) + 7A_u(\mathrm{IR}) + 8E_{1g}(R) + 11E_{1u}(\mathrm{IR}) + 13E_{2g}(R) \quad (14)$$

where IR and R denote infrared and Raman activity, respectively. The influence of pressure on the infrared and Raman spectra of fluorapatite was investigated by WILLAMS and KNITTLE [165]. Fluorapatite remains stable under pressures of at least 25 GPa at 300 K. Local environment of phosphate groups in fluorapatite becomes progressively less distorted from tetrahedral symmetry under the compression, as manifested by progressively smaller site-group.

The Davydov (factor group) splitting also decreases under the compression. This decrease is consistent with nondipole effects playing a primary role in the Davydov splitting of apatite; indeed, the magnitude of the Davydov splitting appears to be modulated by increases of the site symmetry of phosphate group under the compression [165].

The spectrum **Fig. 18(a)** shows weak peak located in the domain of OH stretching modes (from 3500 to 3600 cm^{-1}) at the wave number of 3535 cm^{-1}. This band belongs to the OH stretching mode in the hydrogen bond F...OH...(F) [166]. According to FREUND and KNOBEL [167], the band at ~744 cm^{-1} belongs to the vibration of OH...F bond. According to KNUBOVETS [168], the bands in the range from 745 to 720 cm^{-1} in apatite spectra could also be attributed to symmetric valence oscillations of the P-O-P bridge bonds, formed by the condensation of the PO_4^{3-} tetrahedron. The presence of calcite causes that antisymmetric stretching mode (ν_3) of planar CO_3^{2-} ion appears in the infrared spectrum of investigated sample [169].

Thermal analysis

The results of simultaneous TG-DTA of investigated fluorapatite specimen are shown in **Fig. 19**. The sample is heated with the rate of 10°C·min^{-1} up to the temperature of 1425°C. The mass of sample is reduced by 1.25% during TG-DTA when the final temperature is reached. The most important features are the thermal decomposition of $CaCO_3$ and the thermal decomposition of fluorapatite.

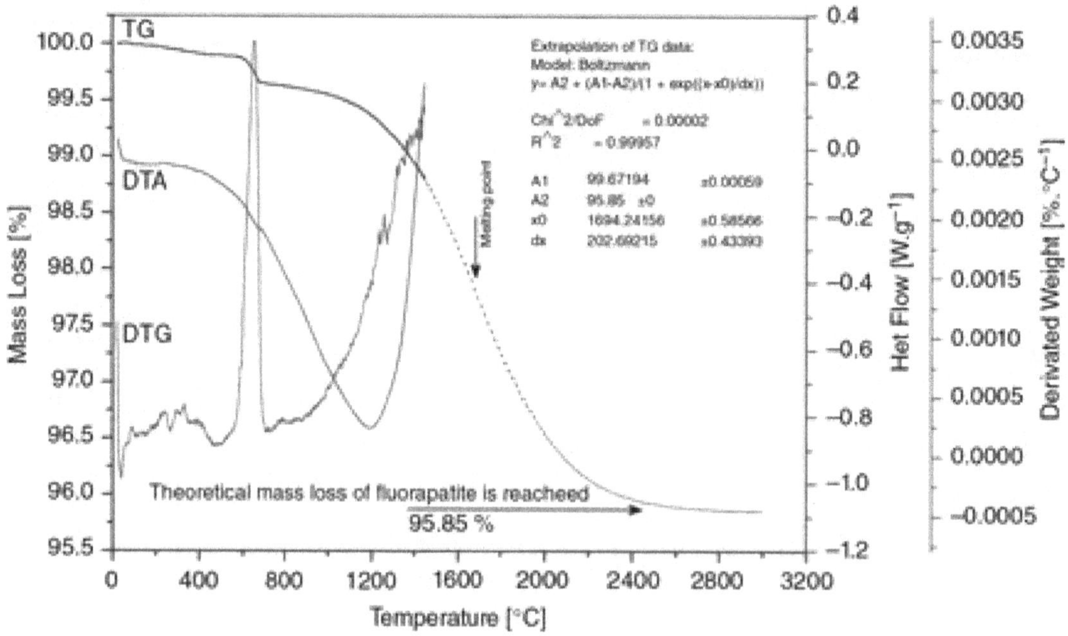

Fig. 19. Thermal analysis of investigated specimen of fluorapatite.

The content of calcite was verified by thermal analysis. The weight of sample was reduced by 0.26% during the thermal decomposition of calcite. Since the theoretical mass loss of calcite is (100 × 44.09) / 100.086 = 43.97%,[24] the content of calcite in the investigated sample of fluorapatite is (100 × 0.26) / 43.97 = 0.59%. Although this value is lower than the content of calcite determined by Rietveld analysis, there is still good agreement of both methods. The DTG curve shows that the process starts at the temperature of 565°C and wide of peak is of 135°C. The maximum rate decarbonation is reached at the temperature of 656°C.

At temperatures higher than 900°C, the weight of sample is reduced by the thermal decomposition of fluorapatite. The extrapolated beginning of the defluorination process (**Section**) was determined to be 1199°C. At the temperature of 1425°C, the defluorination process is still not complete. The extrapolation of experimental data[25] shows that the thermal decomposition is most probably not complete before the temperature of melting point is reached.

Scanning electron microscopy and WDX analysis

The microphotographs from SEM analysis of break plane of investigated fluorapatite specimen are shown in **Fig. 20**. The series of conchoidal fractures on the surface show brittle fracture as further characteristic properties of apatite.

---

[24] It was calculated as 100 × molar mass (molar weight) of CO_2/molar mass of $CaCO_3$.

[25] Theoretical mass loss (3.77%) of fluorapatite was used to set the fixed value of parameter A_2.

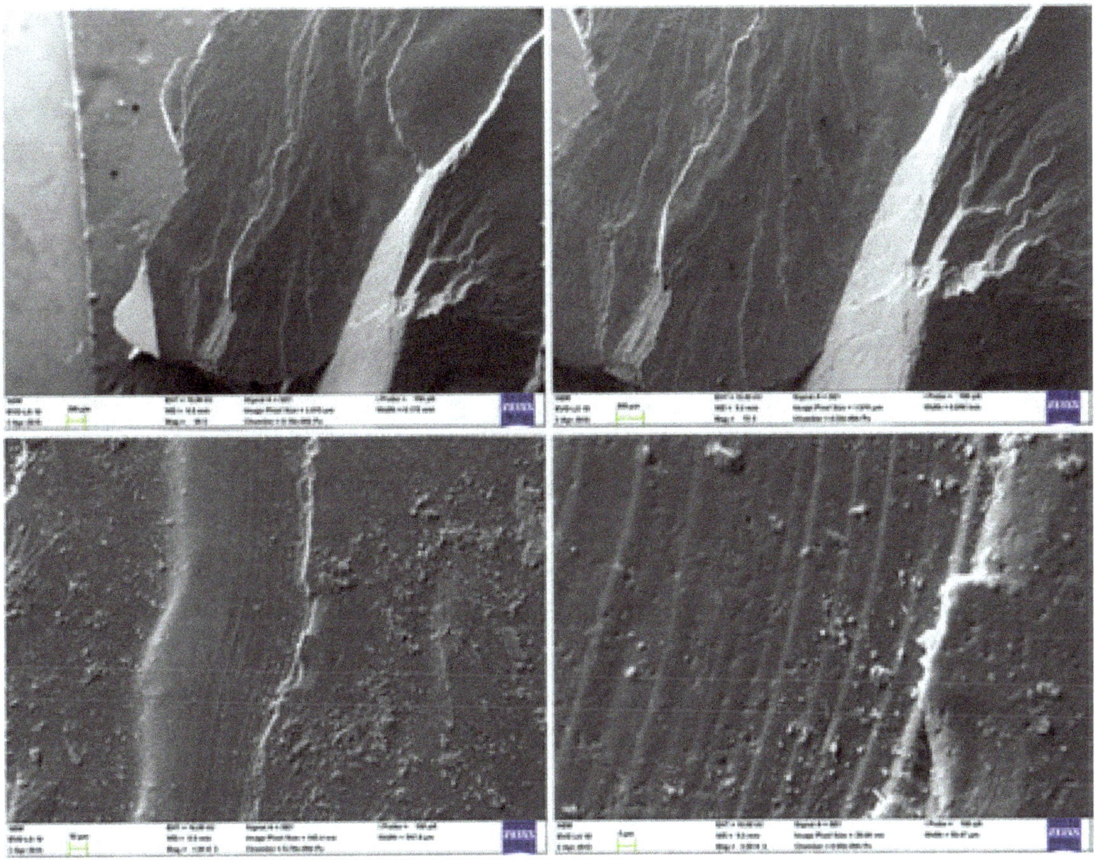

Fig. 20. SEM image of the break plane of investigated fluorapatite specimen: 50× (a), 75× (b), 1 000× (c) and 5 000× (d).

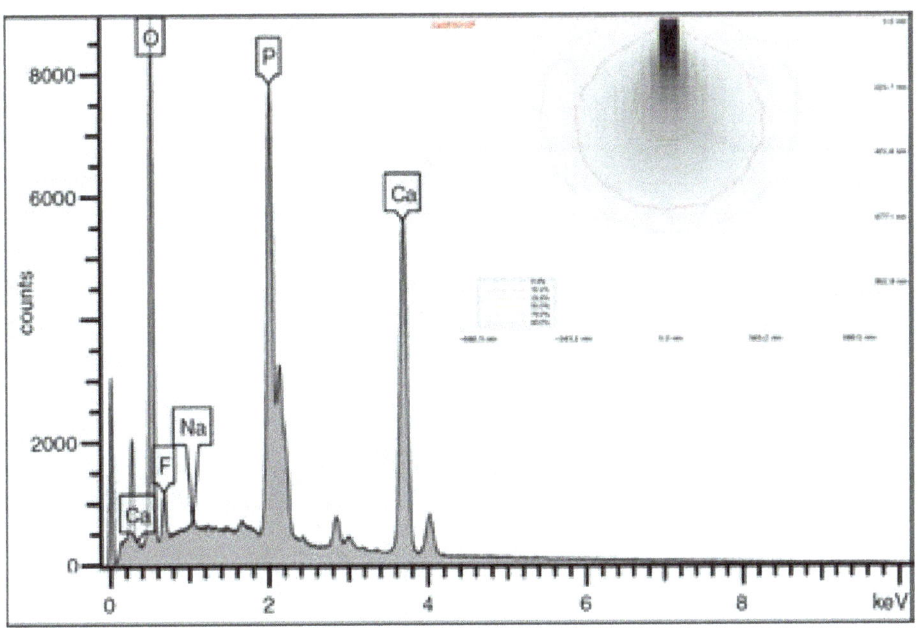

Fig. 21. Typical results of WDX analysis of the fluorapatite specimen and simulation of electron–specimen interaction.

The chemical composition of sample was determined by WDX analysis (**Fig. 21**). The average results are introduced in **Table** 3.

Element	Ca	P	O	F	Na
Weight [%]	38.64 ± 0.96	16.86 ± 0.47	40.55 ± 1.37	3.75 ± 0.08	0.23 ± 0.03
Ideal composition*	39.74	18.43	38.07	3.77	–
Atomic [%]	22.67 ± 0.85	12.81 ± 0.52	59.63 ± 1.31	4.65 ± 0.05	0.24 ± 0.03
Atomic ratio	4.91 ± 0.31	2.78 ± 0.18	12.77 ± 0.19	1	0.05 ± 0.01
Ideal composition[26]	5	3	12	1	–

Table 3. Average chemical composition of investigated specimen of fluorapatite.

The composition of fluorapatite ($Ca_5(PO_4)_3F \rightarrow Ca_5P_3O_{12}F$) corresponds to the element ratio: Ca:P:O:F = 5:3:12:1. The results of WDX analysis of investigated specimen (**Table** 3) are in good agreement with the composition of ideal fluorapatite and fulfill all the criteria mentioned.

EBDS analysis

The fragments of investigated fluorapatite specimens (**Fig. 22**) were further investigated by EBDS analysis.

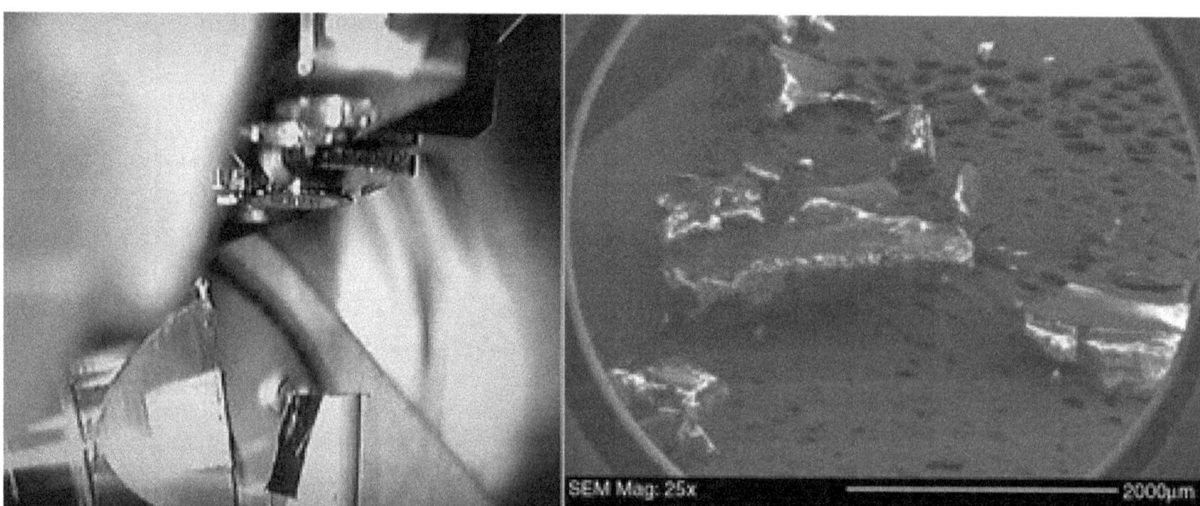

Fig. 22. Sample stage (a) and microphotograph of fluorapatite fragments used for EBDS (b).

The Kikuchi patterns confirm the sample as single crystal of fluorapatite **Fig. 23**.

[26]. Resulting from the apatite formula.

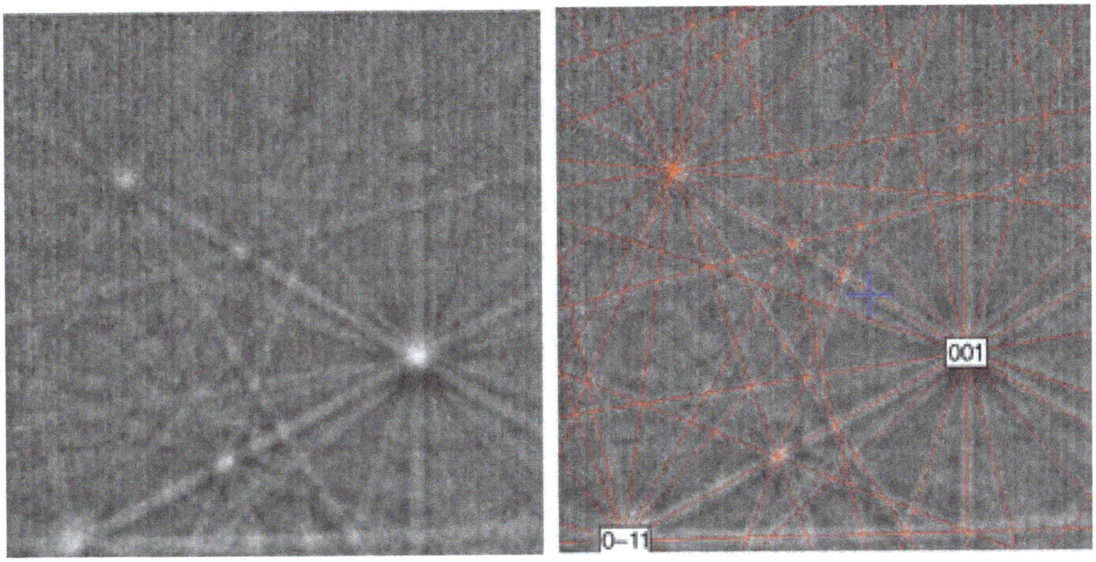

Fig. 23. EBDS analysis of investigated fluorapatite specimen.

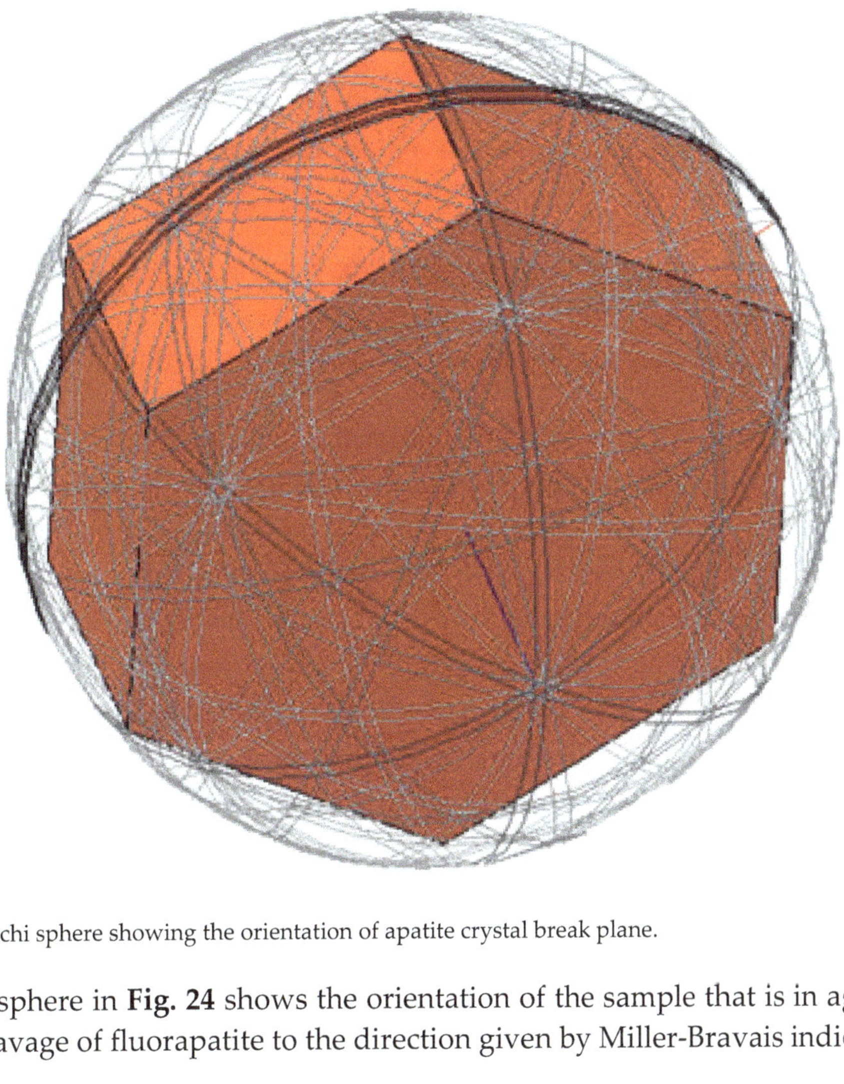

Fig. 24. The Kikuchi sphere showing the orientation of apatite crystal break plane.

The Kikuchi sphere in **Fig. 24** shows the orientation of the sample that is in agreement with indistinct cleavage of fluorapatite to the direction given by Miller-Bravais indices of {0001}.

Thermodynamic properties of apatite

DROUET [170] gives the comprehensive review on experimental and predicted thermodynamic properties of phosphate apatites and oxyapatites, where O^{2-} ion replaces $2X^-$ in general $M_{10}(PO_4)_6X_2$ formula of apatite phase and publishes the summary of available thermodynamic data including standard formation Gibbs energy ($\Delta G_f°$), $\Delta H_f°$ and $S°$ at the temperature of 298 K (25°C) and the pressure of 1 bar (10^5 Pa), which are listed in the periodic table of phosphate apatites in **Fig. 25.**

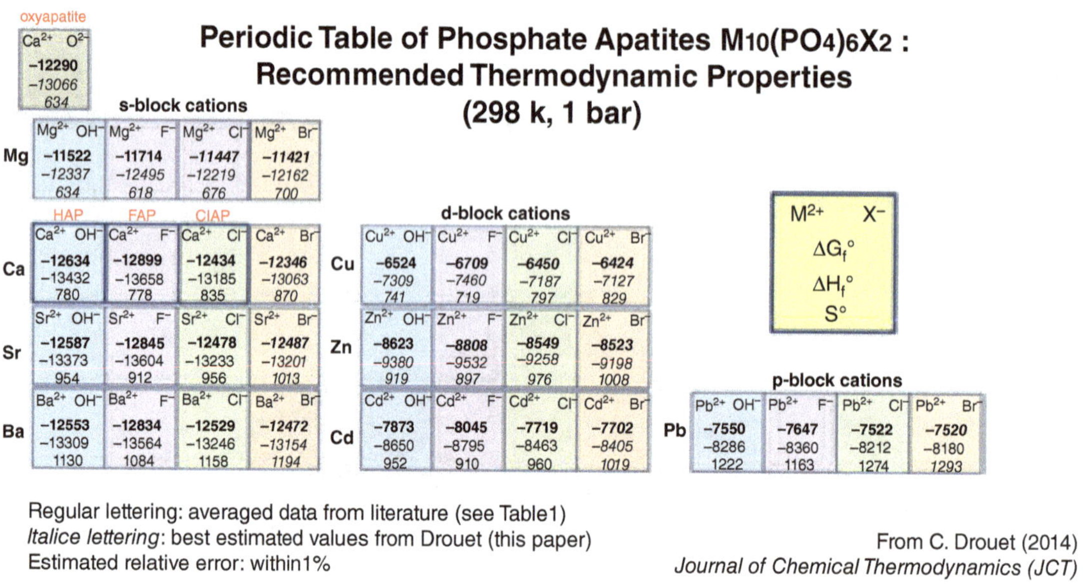

Fig. 25. Comprehensive periodic table of phosphate apatites provided by C. DROUET [170].

The discrepancies between published thermodynamic data probably arise from variable crystallinity states, polymorphs (either hexagonal or monoclinic, those not being systematically identified in literature reports), nonstoichiometry, hydration state and/or the presence of undetected impurities. When experimental-based data are not available (or are questionable), the so-called prediction of thermodynamic properties of solids becomes relevant. For example, it may allow an understanding of some unsuccessful experiment aiming at obtaining a desired hypothetical composition, or it may fill the gap between reported and needed thermodynamic values for the evaluation of equilibria constants or for the establishment of phase diagrams [170]. There are many methods developed for this purpose the summary of which can be found in works [171],[172].

For double oxides, $A_xB_yO_z$ in the system AO-BO was established the dependence [171]:

$$\Delta H^{\circ}_{f,cc} = \left(A_x B_y O_z\right) = f\left(\Delta \bar{H}^{\circ}_f\right) \tag{15}$$

where $\Delta H_{fcc}°(A_xB_yO_z)$ is the standard enthalpy of the formation of double oxide AxByOz from the component oxides AO and BO and $\Delta\bar{H}_f$ represents the sum of molar fraction enthalpies of component oxides AO and BO according to the following relationship:

$$\Delta\bar{H}_f^\circ = x_{AO}\Delta H_f^\circ(AO) + x_{BO}\Delta H_f^\circ(BO) \tag{16}$$

$\Delta H°(AO)$ and $\Delta H°(BO)$ are the standard enthalpies of the formation of component oxides from the elements, and X_{AO} and X_{BO} are the molar fractions of component oxides in the double oxide $A_xB_yO_z$ with a given composition [171].

The entropy of a solid compound is a function of masses of constituent atoms and the forces acting between these atoms: the greater the mass and the lower the force, the larger the entropy. The entropy of ionic solid will also depend upon the magnitude of the ionic charges. For compounds, the specific heat of which has reached the Dulong and Petit [173] value of 6 cal. per gram-atom [174],[175],[27] the mass is the principal factor, and in 1921, the authors gave an equation for the contribution of each element to the entropy of the compound [176].

$$S^\circ(298\,K) = \tfrac{3}{2}R\ \ln \text{at.wt.} - 0.94 \tag{17}$$

For simple salts, such as alkali halides, the entropy may be estimated with fair accuracy as the sum of the entropies of constituent elements as given by this equation. However, the forces in solid salts are largely the ionic attractions, and the effect of the ionic radii upon the force constants and the vibrational frequencies is appreciable; in general, the entropy of a large ion is increased and the entropy of a small ion is decreased compared to the values given by **Eq. 17.** [176].

Volume-based thermodynamic predictive method

The volume-based thermodynamic approach (VTB), the so-called first-order method, has especially received much attention because the method is rather easy to use and has been shown in some cases to lead to output data well related to experimental results [170].

[27] One calorie is 4.184 J (joules). Gram-atom [gm] (and gram-molecule) was used to specify the amount of chemical elements or compound. These units had a direct relation with "atomic weights" and "molecular weights," which are in fact relative masses. "Atomic weights" were originally referred to the atomic weight of oxygen, by general agreement taken as 16. Although physicists separated the isotopes in a mass spectrometer and attributed the value of 16 to one of the isotopes of oxygen, chemists attributed the same value to the (slightly variable) mixture of isotopes 16, 17, and 18, which was for them naturally occurring element oxygen. Finally, an agreement between the International Union of Pure and Applied Physics (IUPAP) and the International Union of Pure and Applied Chemistry (IUPAC) brought this duality to an end in 1959/1960. Physicists and chemists have ever since agreed to assign the value 12, exactly, to the so-called atomic weight of the isotope of carbon with the mass number 12 (carbon 12, ^{12}C), correctly called the relative atomic mass $Ar(^{12}C)$. The unified scale thus obtained gives the relative atomic and molecular masses, also known as the atomic and molecular weights, respectively [174]. The law is also known as Dulong and Petit principle, which can be expressed in modern unit as: atomic weigh × specific heat ≈ $\partial(3kTN_A)/\partial T\ 3kN_A \approx c_V \approx 25\ J{\cdot}K^{-1}mol^{-1}$, i.e. the atomic weight of solid element multiplied by its molar specific heat is a constant [175].

Additive estimation methods

Additive estimation or contributive methods are probably the simplest approach based on the following [170],[177]:

1. **Atomic and ionic contribution**: the technique based on the method proposed by KELLOGG [178]:

$$C^{\circ}_{\mathrm{pm}}\left(A_xB_y,298.15\ \mathrm{K}\right)=x\ C^{\circ}_{\mathrm{pm}}\left(A\right)+y\ C^{\circ}_{pm}\left(B\right)\quad \text{e.g.:}$$
$$C^{\circ}_{\mathrm{pm}}\left(\mathrm{BaCl_2},298.15\ \mathrm{K}\right)=C^{\circ}_{\mathrm{pm}}\left(\mathrm{Ba}\right)+2\ C^{\circ}_{\mathrm{pm}}\left(\mathrm{Cl}\right) \tag{18}$$

The approach was later revised by KUBASCHEWSKI [179],[180]. These authors also proposed the method for the estimation of parameters A, B and C in the temperature dependence of $C_{\mathrm{pm}}{}^{\circ}(\mathrm{T})$[28]:

$$C^{\circ}_{\mathrm{pm}}\left(T\right)=A+BT+\frac{C}{T^2} \tag{19}$$

$$A=\frac{10^{-3}T_m\left[C^{\circ}_{\mathrm{pm}}\left(298.15\ \mathrm{K}\right)+4.7n\right]-1.25n\cdot 10^5\left(T_m\right)^{-2}-9.05n}{10^{-3}T_m-0.298} \tag{20}$$

$$B=\frac{25.6n+4.2n\cdot 10^5\left(T_m\right)^{-2}-C^{\circ}_{\mathrm{pm}}\left(298.15\ \mathrm{K}\right)}{10^{-3}T_m-0.298} \tag{21}$$

$$C=-4.2n \tag{22}$$

where n is the number of ions (contributions) in the formula unit. The described approach is worthy for the substances with melting point temperatures (T_{m}) bellow 2300 K. The data on cationic and anionic contributions to heat capacity at 298 K are published in works [177], [179],[181],[182],[183].

For ionic compounds, the entropy can be calculated[29] from additive data given in **Table** 4, empirically found for cation and anion constituents of the compound (increments method of LATIMER [184]) [172].

	Element	$C_{\mathrm{pm}}{}^{\circ}$	$S_{\mathrm{m}}{}^{\circ}$	Element	$C_{\mathrm{pm}}{}^{\circ}$	$S_{\mathrm{m}}{}^{\circ}$	Element	$C_{\mathrm{pm}}{}^{\circ}$	$S_{\mathrm{m}}{}^{\circ}$
Contribution of cations	Ag	25.76	57.6	Hf	25.52	53,0	Pr	24.27	61.1

[28] The full equation for the temperature dependence is $C_{\mathrm{pm}}{}^{\circ}=\mathrm{A}+\mathrm{BT}+\mathrm{C/T^2}+\mathrm{DT^2}+\mathrm{F/T^{1/2}}$ [$\mathrm{J\cdot K^{-1}\cdot mol^{-1}}$].

	Element	C_{pm}°	S_m°	Element	C_{pm}°	S_m°	Element	C_{pm}°	S_m°
[J·K⁻¹·mol⁻¹]									
	Al	19.66	23.4	Hg	25.10	59.4	Rb	26.36	59.2
	As	25.10	45.2	Ho	23.01	56.0	Sb	23.85	58.9
	Au	–	58.5	In	24.27	55.0	Se	21.34	60.5
	B	–	23.5	Ir	(23.85)	50.0	Si	–	35.2
	Ba	26.36	62.7	K	25.94	46.4	Sm	25.10	60.2
	Be	(9.62)	12.6	La	(25.52)	62.3	Sn	23.43	58.2
	Bi	26.78	65.0	Li	19.66	14.6	Sr	25.52	48.7
	Ca	24.69	39.1	Lu	–	51.5	Ta	23.01	53.8
	Cd	23.01	50.7	Mg	19.66	23.4	Te	–	69.0
	Ce	23.43	61.9	Mn	23.43	43.8	Th	25.52	59.9
	Co	28.03	34.1	Mo	–	35.9	Ti	21.76	39.3
	Cr	23.01	32.9	Na	25.94	37.2	Tl	27.61	72.1

[29]**Table 4** [172] refers to the values of entropy contribution from work [176] where some data for PO_4^{3-} ion are given in brackets. In order to verify this value (for the charge of cation, i.e. Ca^{2+}, the contribution of PO_4^{3-} anion is 17 [calories] × 4.184 = 71.13 J·K⁻¹·mol⁻¹, it is possible to calculate it from recommended thermodynamic data from **Fig. 25**. If (for example) three apatite end members were used, it is possible to calculate the contribution to PO_4^{3-} anion in hydroxylapatite, fluorapatite, chlorapatite, and bromapatite as follows:

$$\text{HAP}: \quad (10\cdot 39.1)+(6x)+(2\cdot 18.83)=780 \quad \Rightarrow x=58.56 \text{ J}\cdot\text{K}^{-1}\cdot\text{mol}^{-1}$$

$$\text{FAP}: \quad (10\cdot 39.1)+(6x)+(2\cdot 17.00)=728 \quad \Rightarrow x=50.50 \quad \text{J}\cdot\text{K}^{-1}\cdot\text{mol}^{-1}$$

$$\text{ClAP}: \quad (10\cdot 39.1)+(6x)+(2\cdot 31.80)=835 \quad \Rightarrow x=63.40 \text{ J}\cdot\text{K}^{-1}\cdot\text{mol}^{-1}$$

$$\text{BrAP}: \quad (10\cdot 39.1)+(6x)+(2\cdot 45.70)=870 \quad \Rightarrow x=64.60 \text{ J}\cdot\text{K}^{-1}\cdot\text{mol}^{-1}$$

It is also possible to calculate it from the contribution data for $Ca_3(PO_4)_2$ or $Mg_3(PO_4)_2$, where S°(298.15K) = 235.998 and 189.2 J·K⁻¹mol⁻¹ (HSC software v.7.1), respectively:

$$(3\cdot 39.1)+(2x)=235.998 \quad \Rightarrow x=59.35 \quad \text{J}\cdot\text{K}^{-1}\cdot\text{mol}^{-1}$$

$$(3\cdot 23.40)+(2x)=189.2 \quad \Rightarrow x=59.5 \quad \text{J}\cdot\text{K}^{-1}\cdot\text{mol}^{-1}$$

Therefore, it is suggested to use average value from these calculations, i.e. PO_4^{3-} (for M^{2+} cation) ≈ 59 J·K⁻¹mol⁻¹. It is then possible that the application of contribution techniques to apatite leads to positive error in estimated thermodynamic data.

	Element	C_{pm}°	S_m°	Element	C_{pm}°	S_m°	Element	C_{pm}°	S_m°
	Cs	26.36	67.9	Nb	23.01	48.1	U	26.78	64.0
	Cu	25.10	44.0	Nd	24.27	60.7	V	22.18	36.8
	Fe	24.94	35.0	Ni	(27.61)	35.1	Y	(25.10)	50.4
	Ga	(20.92)	40.0	Os	–	50.0	Zn	21.76	42.8
	Gd	23.43	56.0	P	14.23	39.5	Zr	23.85	37.2
	Ge	20.08	49.8	Pb	26.78	72.2	–	–	–

	Anion	C_{pm}°	S_m° for oxidation charge of cation			
			I	II	III	IV
Contribution of anions J·K^{-1}·mol^{-1}]	H^-	8.79	–	–	–	–
	F^-	22.80	20.8	17.0	18.3	20.4
	Cl^-	24.69	36.3	31.8	30.3	34.4
	Br^-	25.94	50.3	45.7	44.7	50.8
	I^-	26.36	58.3	53.5	54.8	53.9
	O^{2-}	18.41	4.5	2.9	2.4	3.2
	S^{2-}	24.48	20.6	18.4	20.1	17.0
	Se^{2-}	26.78	35.5	32.8	34.1	30.9
	Te^{2-}	27.20	38.1	41.9	44.1	40.1
	OH^-	30.96	(20.9)	(18.8)	(12.6)	–
	SO_4^{2-}	76.57	80.0	69.5	64.2	(41.8)
	SO_3^{2-}	–	42.9	–	–	–
	NO_3^-	64.43	86.0	74.0	–	–
	NO_2^-	–	70.6	–	–	–
	CO_3^{2-}	58.58	62.4	46.6	–	–
	CrO_4^{2-}	90.76	–	–	–	–
	MoO_4^{2-}	90.37	–	–	–	–
	WO_4^{2-}	97.49	–	–	–	–
	SiO_3^{2-}	59.3	60.7	43.9	29.3	–
	SiO_4^{4-}	73.5	–	–	–	–
	PO_4^{3-}	73.90	(100.4)	(71.1)	(50.2)	–
	UO_4^{2-}	107.11	–	–	–	–

Table 4. The contribution of cations and anions to C_{pm}° and S_m° (298.15 K) [172],[180].

2. **Structural and simple oxide contribution**: the value of $C_{pm}°(298.15\ K)$ or the parameters of the temperature dependence in **Eq. 19** can be calculated from the contribution of constituent oxides (Neumann–Kopp rule, NKR):

$$x\ A(s) + yB(s) = A_xB_y(s), \quad \text{e.g.,} \quad \text{SrO}(s) + \text{CuO}(s) = \text{SrCuO}_2(s) \tag{23}$$

$$\begin{aligned} &C^{\circ}_{pm}(A_xB_y, 298.15\,\text{K}) = x\ C^{\circ}_{pm}(A, 298.15\,\text{K}) + y\ C^{\circ}_{pm}(B, 298.15\,\text{K}) \quad \text{e.g.,} \\ &C^{\circ}_{pm}(\text{SrCuO}_2, 298.15\,\text{K}) = C^{\circ}_{pm}(\text{SrO}, 298.15\,\text{K}) + C^{\circ}_{pm}(\text{CuO}, 298.15\,\text{K}) \end{aligned} \tag{24}$$

or from the structural contribution [179].

3. **Prediction method for homological series and groups of chemically related substances (oxides)**: based on the approach of ALDABERGENOV et al [185] and GOSPODINOV and MIHOV [186]. The molar heat capacity in homological series as $A_m(B_xO_y)_n$, is a linear function of n, i.e. the coefficient which specifies the number of complex anions $(B_xO_y)^{z-}$ in the formula unit. For example, for the series of alkaline aluminate, it can be written as

$$\text{AlO}_2 \overset{+\text{AlO}_2}{\rightarrow} \text{Al}_2\text{O}_4^{\ 2-} \overset{+\text{AlO}_2}{\rightarrow} \text{Al}_3\text{O}_6^{\ 3} \ \ldots \tag{25}$$

Since each higher anion is formed by the addition of primary ion $(AlO_2)^-$ unit, higher anion is considered to form n-multiples of primary ion, the value of which is determined from the available experimental data for $KAlO_2$, $LiAlO_2$ and $NaAlO_2$ and from ions contribution for cation K^+, Li^+ and Na^+ obtained from their standard entropies in an infinitely diluted solution [179].

Apatite phase may be treated in the first approximation as the sum of contributions arising from the constitution of binary oxides/compounds. For example, in the case of fluorapatite ($Ca_{10}(PO_4)_6F_2$), a decomposition into contribution of $9CaO + 3P_2O_5 + CaF_2$ could be considered. It can be generalized to any end-member in the form $9CaO + 3P_2O_5 + XF_2$ [170].

Dissolution of apatites

At the fundamental level, the reactions between solids and liquids involve a coupled sequence of mass transport, adsorption/desorption phenomena, heterogeneous reactions, chemical transformations of intermediates, etc., the identification, separation and kinetic quantification of which are all necessary if the mechanism of the process is to be fully understood and described [111],[187]. It was generally accepted that the process during the dissolution of lattice ions includes the following [187],[188]:

a. Detachment of species (ion) from a kink site

b. Surface diffusion from the crystal steps

c. Desorption from the surface

d. Diffusion into the bulk solution

The dissolution of apatites under steady-state conditions, in pure water or in aqueous acidic media, includes the following simultaneous steps [187],[189]:

i. Diffusion of chemical reagents (H^+) from bulk solution to the solid/liquid interface. In the case of acidic dissolution, the diffusion of acid anions (A^{n-}) must be taken into account as well.

ii. Adsorption of H^+ onto the surface of apatite.

iii. Chemical transformation of the surface.

iv. Desorption of ions of fluoride, calcium and phosphate from the crystal surface.

v. Adsorption of chemicals from solution back onto the surface of apatite.

These steps are likely to be more complicated, e.g. the processes I and V include chemical transformation of ionic species during diffusion because the pH of solution is known to depend on the distance from the solid/liquid interface. In other words, the value of pH is higher near to the surface of apatite and decreases with increasing distance from the surface.

When apatite gets in contact with undersaturated solution, the dissolution states from 1 to 5 mentioned above take place. In order to provide detailed description of the process, the following assumption must be introduced [187]:

a. Stoichiometric apatite is dissolved, and neither nonstoichiometric layer nor other ions except for calcium, phosphate, hydroxide and fluoride are initially present in the crystal lattice, whereas the volume and surface defects (dislocations and dislocations outlets, respectively) might be present and, when present, they are distributed randomly.

b. Except for the presence of dislocation outlets, the initial surface of apatite is perfect (molecularly sooth). Otherwise, each imperfection might be the dissolution nucleus.

c. Despite the limitations and drawbacks discussed above, all models are correct and complementary to each other.

d. In some cases, anions of acid might have an influence on apatite dissolution due to the specific affinity (e.g. citrate) by means of formation of insoluble compounds (e.g. sulfate). That is the reason why anions are not specified either here or below. It is just assumed that the dissolution of apatite happens in acid H_nA, where A^{n-} is an anion.

e. The hydration effect on all ions and molecules involved as well as that on crystal surface of apatite is omitted for simplicity.

f. All crystal faces of solid apatite are equal and have similar ionic arrangement. There is not any considered specific influence of different crystal faces on the dissolution mechanism.

Since the dissolution models have limitations and drawbacks, none of them was able to describe the dissolution of apatite in general. Furthermore, the most of models were elaborated for the apatite dissolution in slightly acidic or nearly neutral solution ($4 < pH < 8$), for relatively small values of solution undersaturation and for the temperatures in the range from 25°C to 37°C, nothing is known about their validity for the dissolution of apatite in strong inorganic acid such as HCl, HNO_3 and H_2SO_4 and at temperatures above 70°C [187].

The classification of congruent (stoichiometric)/incongruent (nonstoichiometric) dissolution is based on direct measurements of either ionic concentrations in the solution or the surface composition of apatite during the dissolution [187]:

1. **Congruent dissolution**: ions in solids are dissolved simultaneously with the dissolution rates proportional to their molar concentrations, e.g. for $Ca_5(PO_4)_3Z$ it should be written [190],[191]:

$$\frac{[\mathrm{Ca}]_t}{[\mathrm{P}]_t - [\mathrm{P}]_0} - R = 0 \tag{26}$$

   where $[Ca]_t$, $[P]_t$ and $[P]_0$ denotes actual (at time ***t***) concentration of calcium and phosphorus in the solution and initial concentration of phosphorus (at time $t = 0$), respectively. The value of R is given by ideal stoichiometric ration of Ca:P = 5/3 in the formula of apatite.

2. **Incongruent dissolution**: the dissolution rate is different for each ion. That leads to the formation of surface layer with chemical composition different from the bulk of solid apatite phase.

The behavior of surface of apatite during the dissolution according to Dorozhkin [112] is shown in **Fig. 26**. Fluorine from fluorapatite or hydroxyl from hydroxyapatite dissolves most probably as the first. This can be explained by their position in the channels of crystal lattice. The dissolution starts with replacements of fluorine for water. Proton(s), chemisorbed on the nearest phosphate group(s), most probably catalyze this process. Local positive charge on apatite is formed as the result (*b*). Obtained local positive charge is removed by the detachment of one of the nearest calcium cations: Ca(2) is more likely to be detached first (*c*) since Ca(1) is located rather far from the channel. Acidic anions present in the solution most probably participate in this process. Later, proton(s) from the bulk solution replace other calcium cation(s) around the nearest phosphate group. Very thin surface layer of acidic calcium phosphates is formed as the result [112].

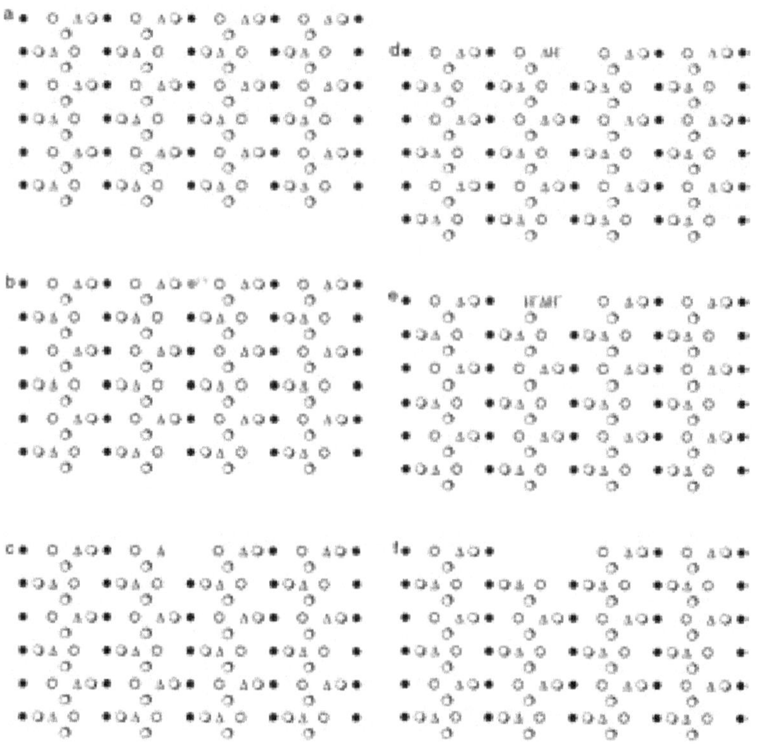

Fig. 26. Schematic illustration of the surface dissolution mechanism of apatite at the nanolevel: (a) part of the initial surface of apatite; (b) replacement of one fluorine (or hydroxyl) anion with water molecule resulting in local positive charge formation; (c) removal of one of the nearest calcium cations; (d) sorption of next proton; (e) removal of another calcium cation with simultaneous formation of acidic calcium phosphate; (f) detachment of one phosphate anion together (or simultaneously) with third calcium cation. A jump-wise shift of the dissolution step occurs simultaneously at stage f. (●) Fluorine for fluorapatite or hydroxyl for hydroxylapatite; (○) Ca(II) on the first plane, (○) Ca(II) on the back plane and Ca(I) on the back plane; (*(+)) molecule of water and local positive charge; (Δ) PO_4^{3-} tetrahedra; $H^+\Delta H^+$ and ΔH^+ represent the surface tetrahedral anions of $H_2PO_4^-$ and HPO_4^{2-}, respectively. Chemisorbed protons, water molecules and acidic anions are omitted for simplicity. Note that crystal structure of apatite is shown very schematically: it should be hexagonal, while here it looks more or less like cubic [112].

When all (or almost all) the nearest calcium cations have been replaced with protons according to the reactions [112]:

$$CaHPO_4 + H^+ \rightarrow Ca^{2+} + H_2PO_4^- \quad (27)$$

$$CaHPO_4 + 2H^+ \rightarrow Ca^{2+} + H_3PO_4 \quad (28)$$

phosphate anions ($H_2PO_4^-$, $CaH_2PO_4^+$, or H_3PO_4) also detach (*f*). As the result, the dissolution step moves forward jump-wise over a distance equal to the dimension of phosphate anion, of approximately 3 Å. The detachment of phosphate anions and calcium cations results in the formation of hole. The dimension of this hole should be close to the lattice parameters of apatite. Most probably, it is a dissolution nucleus on which the polynuclear dissolution mechanism is based [112].

Dissolution of fluorapatite

The adsorption of H^+ onto the surface of apatite (**Fig. 26(b)–(d)**) resulted in the aqueous pH increasing from 5.60 to 8.45 within the first hour of dissolution. Ions of H^+ were adsorbed onto oxygen ions of phosphate groups as well as onto ions of fluoride [187],[191].

The reaction of stoichiometric (congruent) dissolution of pure stoichiometric apatite can be expressed by the reaction[30] [112],[187],[192]:

$$Ca_5(PO_4)_3Z \rightarrow 5\ Ca^{2+} + 3\ PO_4^{3-} + Z^- \tag{29}$$

where **Z** = OH^- and F^-. Assuming the unit activity of the solid phase ($a_{Ca_5(PO_4)_3Z}$ = 1), the equilibrium constant of dissolution (*K*) can be expressed via the solubility product or ion activity product (IAP) of apatite[31] [187],[193],[194],[195],[196]:

$$K = \mathrm{IAP} = a^5_{Ca^{2+}}\, a^3_{PO_4^3}\, a_Z \qquad [\mathrm{eq.}] \tag{30}$$

where a_i denotes the thermodynamic activity of aqueous species. The standard Gibbs (free) energy of the reaction related to the standard temperature (298.15 K) and pressure (0.101 MPa) is given by the formula[32]:

---

[30] Reaction **Eq. 7** and **Eq. 12** are widely used for the description of dissolution process of apatite [112] using stoichiometry pertinent single or double apatite formula and Z = F or OH.

[31] Out of equilibrium state, *IAP* is not equal to *K* (see the discussion to **Eq. 20**). Double formula of apatite is assumed then with respect to the apatite stoichiometry; the law for ionic activity product has the following form:

$K_{sp} = a_{(Ca^{2+})}{}^{10}\, a_{(PO_4^{3-})}{}^{6}\, a_{(Z-)}{}^{2}$.

For example, in hydroxylapatite, where $Z^- = OH^-$, the activity of OH^- anion can be expressed by using ionic product of water (25°C): $K_w = a_{(H^+)}\, a_{(OH^-)} = 1\cdot10^{-14}\ (\mathrm{mol\cdot dm^{-3}})^2$ and then $a_{(OH^-)} = K_w/\, a_{(H^+)}$. Since $pH = -log\ [H^+]$ and $pOH = -log\ [OH^-]$ the $[H^+] = 10^{-pH}$ and $[OH^-] = 10^{-pOH}$ and $pK_w = pH + pOH = 14$:

$K_{sp} = a_{(Ca^{2+})}{}^{10}\, a_{(PO_4^{3-})}{}^{6}\, 10^{-2pOH} = a_{(Ca^{2+})}{}^{10}\, a_{(PO_4^{3-})}{}^{6}\, 10^{-2(14-pH)} = a_{(Ca^{2+})}{}^{10}\, a_{(PO_4^{3-})}{}^{6}\, K_w{}^2 10^{2pH}$.

The activity of ionic species is the product of ion molar concentration ([196]) and ion activity coefficient, e.g. ($a_{(Ca^{2+})}$ = ($[Ca^{2+}]$ / $[Ca^{2+}]^\circ$) $\gamma_{(Ca^{2+})}$, where the standard state $[Ca^{2+}]^\circ = 1\ \mathrm{mol\cdot dm^{-3}}$ can be chosen), which can be calculated, e.g. the example via **Debye–Hückel**, **extended Debye–Hückel**, or **modified Davies equation** (in dependence on ionic strength):

$\log \gamma_i = -AZ_i^2 \sqrt{I}\ (I < 10^{-3})$,

$\log \gamma_i = [-AZ_i^2 \sqrt{I} / (1+B{\cdot}\alpha_i \sqrt{I})]\ (10^{-3} < I < 0.1)$,

or $\log \gamma_i = [-AZ_i^2 \sqrt{I} / (1+B{\cdot}\alpha_i \sqrt{I})] + 0.3I\ (I > 0.1)$, respectively.

A and *B* are the temperature-dependent constants, Z_i is the charge number of *i*th ion, α_i is the radius of hydrated *i*th ion, and *I* is the number of ions: $I = ½ \sum ([C_i]Z_i^2)$.

The solubility product defined as the product of concentration of compound constituents ions, which are released during the dissociation is often used: $K_{sp} = [Ca^{2+}]^{10}\, [PO_4^{3-}]^6\, [Z^-]^2$.

[32] Δ_rG° = -RT ln *K* (in the equilibrium state) and ln *K* = ln 10 log *K*. From that, it should be derived that Δ_rG° = -8.314·298.15 ln 10 log *K* = -5708 log *K* [J·mol⁻¹], where ln 10 ≈ 2.303. Since $\log_{10}K = \log K = \lg K$, it can also be written as $\Delta_rG^\circ = -5708 \lg K$.

$$\Delta_r G^\circ = -5.707 \ \log \ \mathrm{K} \qquad \left[\mathrm{kJ \cdot mol^{-1}}\right] \tag{31}$$

For reaction 29, the following equation[33] can be derived [187]:

$$\begin{aligned}\Delta_r G^\circ &= 5 \ \Delta G_f^\circ\left(\mathrm{Ca^{2+}}\right) + 3 \ \Delta G_f^\circ\left(\mathrm{PO_4^{3-}}\right) + \Delta G_f^\circ\left(\mathrm{Z^-}\right) \\ &-\Delta G_f^\circ\left(\mathrm{Ca_5\left(PO_4\right)_3 Z}\right)\end{aligned} \tag{32}$$

Eq. 32 can be further treated as follows:

$$\begin{aligned}\Delta G_f^\circ\left(\mathrm{Ca_5\left(PO_4\right)_3 Z}\right) &= 5 \ \Delta G_f^\circ\left(\mathrm{Ca^{2+}}\right) + 3 \ \Delta G_f^\circ\left(\mathrm{PO_4^{3-}}\right) \\ &+ \Delta G_f^\circ\left(\mathrm{Z^-}\right) - \Delta_r G^\circ\end{aligned} \tag{33}$$

HAROUIYA et al [197] assumes that the dissolution of apatite in the temperature range from 5°C to 50°C, and the pH from 1 to 6 can be expressed by the following formula:

$$\mathrm{Ca_5\left(PO_4\right)_3} F + 3 \ \mathrm{H^+} \rightarrow 5 \ \mathrm{Ca^{2+}} + 3 \ \mathrm{HPO_4^{2-}} + F^- \tag{34}$$

With regard to assumed standard state, the equilibrium constant of reaction (**Eq. 34**) can be written as[34]

$$K' = a_{\mathrm{Ca^{2+}}}^5 \, a_{\mathrm{HPO_4^{2-}}}^3 \, a_{\mathrm{F}} \, a_{\mathrm{H^+}}^{-3} \tag{35}$$

The chemical affinity[35] (A) of **Eq. 34** is given by the law:

$$A = -RT \ \ln\left(\frac{K' a_{\mathrm{H^+}}^3}{a_{\mathrm{Ca^{2+}}}^5 \, a_{\mathrm{HPO_4^{2-}}}^3 \, a_{\mathrm{F}}}\right) \tag{36}$$

In the closed-system experiment, the dissolution rates are generally obtained from the slope of concentration of reactive solution versus the time:

---

33 $\Delta_r G^\circ = \sum \nu_i \, \Delta G^\circ_{f,i}$, where ν_i denotes the stoichiometric coefficient of given species and $\Delta G^\circ_{f,i}$ its standard enthalpy of formation.

34 Since the saturation of solution with respect to $Ca_5(PO_4)_3Z$ means that $K' = 0$ (equilibrium state), it can be derived that $a_{(Ca^{2+})}^5 \, a_{(PO_4^{3-})}^3 \, a_{(Z-)} - K' a_{(H+)}^3 = 0$ and then $\Delta_r G^\circ = -RT \ln (a_{(Ca^{2+})}^5 \, a_{(PO_4^{3-})}^3 \, a_{(Z-)} / K' a_{(H^+)}^3)$. Since $A = -\Delta_r G^\circ$ (please see note 35), $A = -RT \ln (K' a_{(H^+)}^3 / a_{(Ca^{2+})}^5 \, a_{(PO_4^{3-})}^3 \, a_{(Z-)})$.

35 The relationship between the reaction Gibbs energy and chemical affinity: $A = -\Delta_r G^\circ$ was introduced by T. DE DONDER.

$$r = \frac{\partial c_i}{\partial t} \frac{M_r}{v_i s} \tag{37}$$

where r refers to the dissolution rate of apatite, c_i denotes the concentration of ith element, t is the time, M_r designates the mass of fluid in the reactor, v_I is the stoichiometric coefficient and s designates the total surface area of sample in the reactor. The slope of the plot may not be constant and may increase or decrease with time from the following reasons [195],[197]:

1. Changes in the reactive fluid volume, which may occur due to the evaporation of solvent or regular sampling of reactive fluid
2. Nonzero order reaction kinetics
3. Approach to equilibrium, where the dissolution rate decreases and reaches zero at equilibrium. This approach is described by the transition state theory as follows:

$$r = r_+ \left(1 - \exp\left[\frac{-A}{\sigma RT}\right]\right) \tag{38}$$

The symbol r_+ symbolizes the far from equilibrium dissolution rate, which may depend on the composition of solution, A is the affinity of reaction of dissolution, σ stands for the Temkin's average stoichiometric number equal to the ration of rate of destruction of the activated or precursor complex relative to overall rate, R designates universal gas constant and T is the temperature on the absolute scale. Overall rate (r) is equal to forward rate (r_+) when $A >> \sigma RT$. As one of the approaches of equilibrium, overall rates gradually decrease and reach zero at equilibrium where $A = 0$. The value of r is within 10% of r_+ when $A/\sigma RT > 2.3$ which is equivalent to $A > 1.36\ \sigma$ kcal·mol^{-1}. It indicates that the parameter σ plays a crucial role in the variation of dissolution rates at near to equilibrium conditions [195],[197].

The value of r_+ depends on the pH according to the following equation [197],[198]:

$$r_+ = k\ \mathrm{pH}^{-n} \tag{39}$$

where k refers to the tare constant and n stands for the reaction order determined as the slope of linear dependence of $ln\ r_+$ on pH. The dependence of k on the temperature is given by the Arrhenius law:

$$k = A_{\mathrm{A}}\ \exp\left(\frac{-E_{\mathrm{A}}}{\mathrm{RT}}\right) \tag{40}$$

where A_A is the preexponential (frequency) factor and E_A is the activation energy of the process. The combination of **Eqs. 38, 39** and **40** leads to the equation for the dissolution rate as follows [197]:

$$r = A_A a^n_{H^+} \exp\left(E_A / RT\right)\left(1 - \exp\left[\frac{-A}{\sigma RT}\right]\right) \tag{41}$$

For acidic dissolution of calcium fluorapatite, ions of *F* were found to dissolve faster (or prior to) when compared to calcium and phosphate. A similar phenomenon of prior (or faster) dissolution of calcium when compared to that of phosphate was also found [187],[199]. The release of calcium and phosphate ions from the surface of apatite seems to be affected by the presence of salts, such as Na_2SO_4, CH_3COONa, or NaCl, in the solution. The concentration of phosphate in the solution increases in the following order [187]:

$$Na_2SO_4 > CH_3COON_a > NaCl.$$

On the contrary, the concentration of calcium ions decreases in the same order.

The undersaturation (US) and relative undersaturation (US_r) of apatite solvent dissolved upon is defined as follows[36] [193]:

$$US_r = 1 - US = 1 - \left(\frac{IAP}{K}\right)^{1/18} \tag{42}$$

where *K* is the equilibrium constant of reaction 29 and IAP is the ion activity product. The law is written with regard to the stoichiometry of double formula of apatite, where $\sum \nu_i$ in **Eq. 30** is 2 × (5 + 3 + 1) = 18. The value of *IAP*/*K* ratio is as follows:

a. IAP/*K* > 1, the reaction proceeds to the left (precipitation, supersaturated solution)

b. IAP/*K* = 1, the reaction is in the equilibrium state (saturated solution)

c. IAP/*K* < 1, the reaction proceeds to the right (dissolution, undersaturated solution)

This ration is also used to calculate the saturation index (SI) for the reaction of dissolution [200]:

$$SI = \log\left(\frac{IAP}{K}\right) \tag{43}$$

Depending on the saturation index, the following states of solutions are recognized:

---

[36] Since the system is not in the equilibrium state IAP ≠ K.

i. SI < 0 undersaturation

ii. SI = 0 saturation, i.e. mineral[37] or salt is in equilibrium with solution

iii. SI > 0 supersaturation

Classification of the dissolution models

The models, which are usually applied for the description of dissolution of apatites, include the following [187]:

1. **Diffusion and kinetically controlled models**: the dissolution of apatite was found to be the diffusion (transport) controlled in some cases [201],[202], kinetically (surface) controlled in other ones [203],[204], and even intermediate [205], i.e. both diffusion and kinetically controlled. Both models usually operate with the so-called driving force which means either the concentration gradient within the Nernst diffusion layer (the diffusion controlled model) or the gradient of ionic chemical potentials between the apatite crystal surface and bulk solution (the kinetically controlled model). Moreover, the results obtained on these models are valid only within the experimental conditions studied; no extrapolation can be made beyond the tested ranges. For example, after, let's say, a slight agitation decrease or temperature increase, an initially kinetically controlled dissolution might be controlled by the diffusion. Thus, high sensitivity to applied experimental conditions appears to be the main drawback of these models [187].

2. **Polynuclear model**: is based on the study of dissolution and kinetics of growth of apatite under constant composition conditions [191],[193],[206],[207],[208]. Polydispersed samples of apatite were put into a stirred undersaturated (for dissolution experiments) or supersaturated (for those on crystal growth) solutions, and the pH of solution and the amount of added chemicals (an acid for the dissolution experiments and a base for those on crystal growth) were permanently recorded as the functions of time. The results obtained were plotted versus either undersaturation or supersaturation values: straight lines were obtained in the specific logarithmic coordinates typical for this model. According to the model, the dissolution nuclei, i.e., the collections of vacant sites for Ca^{2+}, PO_4^{3-}, and OH^- ions, are formed on the crystal surface of apatite and spread over the surface with a definite lateral rate.

   The nucleation rate is assumed as a function of mean ion activity. The lateral growth rate of nuclei is assumed proportional to the difference between total concentration of calcium ions in the saturated solution and in a solution, while the rate constant is related to the frequency for calcium ions to make a diffusion jump into a kink and, simultaneously, partly dehydrate. Recent investigation reveals that the rate-determining step was not the diffusion but two-dimensional surface nucleation [187].

3. **Self-inhibition model**: assumes the formation of self-inhibition calcium-rich layer on the surface of apatite during the dissolution. According to this model, apatite is dissolved by

[37] The ranges of Si near zero are generally considered to be within the equilibrium zone for the mineral. The ranges of SI = 0 ± 0.5 and 0 ± (5%) (lg $K_{mineral}$) were used in various studies [200].

ionic detachment of calcium and phosphate ions from the surface to a solution. When an initial portion of apatite has been dissolved, some amount of calcium cations (probably, in connection with anionic counter ions) is returned from the solution and adsorbed back onto the surface of apatite. The latter process results in the formation of a semipermeable ionic membrane, which is formed from positively charged layer containing strongly adsorbed calcium ions, i.e. calcium-rich layer [187], [201],[209],[210],[211].

4. **Congruent and incongruent dissolution model**: was already described above.

5. **Chemical model**: This model was developed from the self-evident supposition that it would be highly unlikely if apatite were dissolved by the detachment of "single molecules" equal to the unit cells and consisting of 18 ions. Moreover, in the crystal lattice, practically all ions are shared with neighboring unit cells and often cannot be attributed to given "single molecule." Based on the experimental results obtained on one hand, and on analysis of the data found in references on the other hand, a sequence of four successive chemical reactions was proposed to describe the process of apatite dissolution [112], [187],[212],[213]:

$$Ca_5(PO_4)_3(F,OH) + H_2O + H^+ \rightarrow Ca_5(PO_4)_3(H_2O^+) + HF,\ H_2O \tag{44}$$

$$2\ Ca_5(PO_4)_3(H_2O^+) \rightarrow 3\ Ca_3(PO_4)_2 + Ca^{2+} + 2\ H_2O \tag{45}$$

$$Ca_3(PO_4)_2 + 2H^+ \rightarrow Ca^{2+} + 2\ CaHPO_4 \tag{46}$$

$$CaHPO_4 + H^+ \rightarrow Ca^{2+} + H_2PO_4 \tag{47}$$

Eqs. 44–47 can be used instead well-known net reactions [187]:

$$\begin{aligned} &Ca_5(PO_4)_3(F,OH) + H_2O + 7H^+ \rightarrow 5\ Ca^{2+} + 3\ HPO_4 \\ &+HF,\ H_2O \end{aligned} \tag{48}$$

$$Ca_5(PO_4)_3(F,OH) \rightarrow 5\ Ca^{2+} + 3\ PO_4^{3-} + F,\ OH^- \tag{49}$$

In principle, the dissolution process could also happen according to reaction 49 followed by chemical interaction in the solution among ions of apatite and acid near the crystal surface [214]:

$$5\ Ca^{2+} + 3\ PO_4^{3-} + F^-, OH^- + 7\ H^+ \rightarrow 5\ Ca^{2+} + 3\ H_2PO_4^- + HF,\ H_2O \tag{50}$$

1. **Etch pit formation**: the process of etch pit formation describes the dissolution of apatite crystals containing structural defects (dislocations and inclusions). The presence of dislocations accelerates the dissolution because they give rise to continuous steps on the surface and the strain energy, which they cause in crystals, favors the etch pit formation. The pits appear at the dislocation outlets; they are 0.1–10 pm in size (the dimensions depend on the dissolution kinetics and dissolution time: they increase when the dissolution progresses) and usually have a hexagonal shape according to the crystal symmetry **$P6_3/M$** of pure apatite [187],[215],[216],[217],[218],[219].
2. **Ion exchange model**: based on a supposition about the adsorption of protons and anions of acid (e.g. citrate anions) from a solution onto the surface of apatite and removing of ions of calcium and phosphate into the solution instead [187].
3. **Hydrogen catalytic model**: based on a reasonable suggestion about the adsorption of protons onto negatively charged oxygen ions of phosphate groups of apatite. The sorption of protons results in the transformation of surface PO_4^{3-} groups into HPO_4^{2-} and catalyzes the dissolution process [187].

Methods for the evaluation of reactivity of phosphate rocks

Chemical methods are used for the evaluation of reactivity of different phosphate rocks from which the fertilizers are manufactured for their possible direct application as fertilizers via empirical solubility test. Citric acid, formic acid, neutral ammonium citrate,[38] and alkaline ammonium citrate are used as solvents for the extraction of P_2O_5. The latter is used mainly for the evaluation of calcined aluminum phosphates. Most of these reagents were not originally intended to evaluate the reactivity of phosphate rocks. For instance, neutral and alkaline ammonium citrate solutions were originally intended to separate chemical reaction products in superphosphate and other fertilizers from unreacted rock on the assumption that unreacted rock was insoluble in these reagents. The citric acid extraction was developed to evaluate basic slag, a popular fertilizer material in European countries. The formic acid extraction was developed specifically for phosphate rocks [220].

Nearly all extraction methods use the ratio of sample weight to extraction volume 1 g:100 ml. [39] The extraction time usually ranges from 30 min to 1 hour. The temperature and the agitation during extraction test may be specified. For example, the AOAC method uses neutral ammonium citrate[40] of specified concentration (1 g of sample and 100 ml of solution) with the extraction time of 30 min at 65°C. The Wagner method uses 2% solution of citric acid, the

[38] Neutral ammonium citrate is prepared by dissolving required amount of citric acid and neutralizing it with ammonium hydroxide. The pH of the reagent is adjusted to neutral [221].

[39] The amount of used solution is also expressed in the name of the method, e.g. 100 ml method or 150 ml method [222].

[40] The neutral ammonium citrate test was used as the official method in the United States, and the test by acidic acid was developed for the comparison [222].

extraction time of 30 min and the temperature of 17.5°C [220],[221],[222],[223]. Unavailable phosphoric acid is usually expressed as the portion of fertilizer, which is *insoluble* in neutral ammonium citrate [769]. Phosphate removed during the neutral ammonium test is termed as *citrate-soluble*. The sum of *water-soluble* and *citrate-soluble* phosphate is termed as *available* [224].

Fused magnesium phosphate (FMP) is highly soluble in 2% citric acid[41] but is less soluble in neutral citrate, while calcined defluorinated phosphate (CDP) and Thomas slag (Thomas phosphate) are fairly soluble in both citric acid and citrate. Actually, FMP dissolves fairly rapidly in neutral citrate at the beginning, but the dissolution is hindered by gelatinous silica, which forms on the surface of the FMP particles. This layer can be broken by vigorous stirring [222].

One disadvantage of all these methods is that the percentage of leached P_2O_5 depends on the grade of the rock, especially when the rock contains inert gangue minerals such as silica. In order to eliminate the adventitious effect of grade, the concept of absolute citrate solubility index (ASC) was developed [225],[226]:

$$\mathrm{ASC} = \frac{\text{AOAC citrate solubility } \mathrm{P_2O_5}\,[\%]}{\text{Theoretical } \mathrm{P_2O_5} \text{ in apatite } [\%]} \tag{51}$$

The percentage of dissolved P_2O_5 is expressed as the gangue-free apatite [220]. If rocks contain free calcium of magnesium carbonate, these carbonates should be removed by the extraction with a suitable reagent before carrying out the test in order to obtain correct indication of reactivity [227].

It was also found that the length of the *a*-axis in the apatite unit cell (a_0) is statistically related to the ASC according to the relationship [225]:

$$\mathrm{ASC} = 421.4\,(9.369 - a_0) \tag{52}$$

Author details

Petr Ptáček

Brno University of Technology, Czech Republic

[41] It is believed that weak citric acid solution imitates the condition near the plant roots [222].

References

[1] Straaten PV. Rocks for Crops, Agro Minerals of Sub-Sahara Africa. (CD), ICRAF, Nairobi, Kenya, 2002. ISBN: 0-88955-512-5.

[2] Abouzeid Abdel-Z.M. Physical and thermal treatment of phosphate ores—an overview. International Journal of Mineral Processing 2008;85(4) 59–84.

[3] Comodi P, Liu Y, Zanazzi PF, Montagnoli M. Structural and vibrational behaviour of fluorapatite with pressure. Part 1: in situ single-crystal X-ray diffraction investigation. Physics and Chemistry of Minerals 2001;28(4) 219–224.

[4] Wilson RM, Elliott JC, Dowker SEP, Smith RI. Rietveld structure refinement of precipitated carbonate apatite using neutron diffraction data. Biomaterials 2004;25(11) 2205–2213.

[5] Smith F. Industrial Applications of X-Ray Diffraction. CRC Press, 1999. ISBN: 978-0824719920

[6] Elliott JC. Structure and Chemistry of the Apatites and Other Calcium Orthophosphates, Elsevier Press, Amsterdam, 1994. ISBN: 978-0-444-81582-8

[7] Svergun DI, Koch MHJ, Timmins PA, May RP. Small Angle X-Ray and Neutron Scattering from Solutions of Biological Macromolecules. IUCr (International Union of Crystallography) texts on crystallography—Volume 19. Oxford University Press, 2013. ISBN: 978-0199639533

[8] Nitske WR. The Life of Wilhelm Conrad Röntgen, Discoverer of the X ray. University of Arizona Press, 1971. ISBN: 978-0816502592

[9] Daubrée A. Experiences sur la production artficielle de l'apatite, de la topaze, at de quelques autres métax fruorifés. Comptes rendus de l'Académie des Sciences 1851;32 625.

[10] Röntgen WC. On a new kind of rays. Nature 1896;53 274–276. doi: 10.1038/053274b0

[11] Weber AS. Nineteenth-Century Science: An Anthology. Broadview Press, 2000. ISBN: 978-1551111650

[12] Rosenthal-Schneider I. Begegnungen mit Einstein, von Laue und Planck: Realität und wissenschaftliche Wahrheit, Facetten der Physik—Volume 12. Springer-Verlag, 2013. ISBN: 978-3663018841

[13] Chatterji T. Neutron Scattering from Magnetic Materials. Gulf Professional Publishing, 2005. ISBN: 978-0080457055

[14] Liang L, Rinaldi R, Schober H. Neutron Applications in Earth, Energy and Environmental Sciences. Neutron Scattering Applications and Techniques. Springer Science & Business Media, 2008. ISBN: 978-0387094168

[15] Kolb U, Shankland K, Meshi L, Avilov A, David WIF. Uniting Electron Crystallography and Powder Diffraction, NATO Science for Peace and Security Series B: Physics and Biophysics. Springer, 2012. ISBN: 978-9400755802

[16] Fitter J, Gutberlet T, Katsaras J. Neutron Scattering in Biology: Techniques and Applications. Biological and Medical Physics, Biomedical Engineering. Springer Science & Business Media, 2006. ISBN: 978-3540291114

[17] Guillot S, Beaudet-Savignat S, Lambert S, Vannier Rose-N, Roussel P, Porcher F. Evidence of local defects in the oxygen excess apatite $La_{9.67}(SiO_4)_6O_{2.5}$ from high resolution neutron powder diffraction. Journal of Solid State Chemistry 2009;182(12) 3358–3364.

[18] Lambert S, Vincent A, Bruneton E, Beaudet-Savignat S, Guillet F, Minot B, Bouree F. Structural investigation of $La_{9.33}Si_6O_{26}$- and $La_9AESi_6O_{26+\delta}$-doped apatites-type lanthanum silicate (AE = Ba, Sr and Ca) by neutron powder diffraction. Journal of Solid State Chemistry 2006;179(8) 2602–2608.

[19] Sansom JHJ, Richings D, Slater PR. A powder neutron diffraction study of the oxide-ion-conducting apatite-type phases, $La_{9.33}Si_6O_{26}$ and $La_8Sr_2Si_6O_{26}$. Solid State Ionics 2001;139(3–4) 205–210

[20] Matsushita Y, Izumi F, Kobayashi K, Igawa N, Kitazawa H, Oyama Y, Miyoshi S, Yamaguchi S. Powder neutron diffraction of La-apatite under low temperature. Nuclear Instruments and Methods in Physics Research Section A: Accelerators, Spectrometers, Detectors and Associated Equipment 2009;600(1) 319–321.

[21] Tolchard JR, Slater PR. A high temperature powder neutron diffraction structural study of the apatite-type oxide ion conductor, $La_{9.67}Si_6O_{26.5}$. Journal of Physics and Chemistry of Solids 2008;69(10) 2433–2439

[22] Mazza D, Tribaudino M, Delmastro A, Lebech B. Synthesis and Neutron Diffraction Study of $La_5Si_2BO_{13}$, an Analog of the Apatite Mineral. Journal of Solid State Chemistry 2000;155(2) 389–393.

[23] O'Donnell MD, Hill RG, Fong SK. Neutron diffraction of chlorine substituted fluorapatite. Materials Letters 2009;63(15) 1347–1349.

[24] Masubuchi Y, Higuchi M, Takeda T, Kikkawa S. Oxide ion conduction mechanism in $RE_{9.33}(SiO_4)_6O_2$ and $Sr_2RE_8(SiO_4)_6O_2$ (RE = La, Nd) from neutron powder diffraction. Solid State Ionics 2006;177(3–4) 263–268.

[25] Leventouri Th, Bunaciu CE, Perdikatsis V. Neutron powder diffraction studies of silicon-substituted hydroxyapatite. Biomaterials 2003;24(23) 4205–4211.

[26] Alonso JA, Martínez-Lope MJ, Aguadero A, Daza L. Neutron powder diffraction as a characterization tool of solid oxide fuel cell materials. Progress in Solid State Chemistry 2008;36(1–2) 134–150.

[27] Loong C-K, Rey C, Kuhn LT, Combes C, Wu Y, Chen S-H, Glimcher MJ. Evidence of hydroxyl-ion deficiency in bone apatites: an inelastic neutron-scattering study. Bone 2000;26(6) 599–602.

[28] Young RA, Spooner S. Neutron diffraction studies of human tooth enamel. Archives of Oral Biology 1970;15(1) 47–63.

[29] Ibrahim DM, Mostafa AA, Korowash SI. Chemical characterization of some substituted hydroxyapatites. Chemistry Central Journal 2011;5(1) 74.

[30] Lachance GR, Claisse F. Quantitative X-ray fluorescence analysis: theory and application. Wiley, 1995. ISBN: 978-0471951674

[31] Muller R. Spectrochemical Analysis by X-Ray Fluorescence. Springer Science & Business Media, 2012. ISBN: 978-1468417975

[32] Crompton TR. Polymer Reference Book. Smithers Rapra Publishing, 2006. ISBN: 978-1859574928

[33] Nadkarni RA. Modern Instrumental Methods of Elemental Analysis of Petroleum Products and Lubricants, ASTM special technical publication—Volume 1109. ASTM (American Society for Testing and Materials) International, 1991. ISBN: 978-0803114166

[34] Tang YH, Han CM, Bao ZK, Huang YY, He W, Hua W. Analysis of apatite crystals and their fluid inclusions by synchrotron radiation X-ray flourescence microprobe. Spectrochimica Acta Part B: Atomic Spectroscopy 2005;60(4) 439–446.

[35] Uda M, Demortier G, Nakai I. X-rays for Archaeology. NATO Science Series: Mathematics, Physics, and Chemistry—Volume 224. Springer Science & Business Media, 2005. ISBN: 978-1402035814

[36] Nutman AP, Friend CRL. Petrography and geochemistry of apatites in banded iron formation, Akilia, W. Greenland: consequences for oldest life evidence. Precambrian Research 2006;147(1–2) 100–106.

[37] Chen Ch, Chai Z, Gao Y. Nuclear Analytical Techniques for Metallomics and Metalloproteomics. Royal Society of Chemistry, 2010. ISBN: 978-1847559012

[38] Bunker G. Introduction to XAFS: A Practical Guide to X-ray Absorption Fine Structure Spectroscopy. Cambridge University Press, 2010. ISBN: 978-0521767750

[39] Parat F, Dungan MA, Streck MJ. Anhydrite, pyrrhotite, and sulfur-rich apatite: tracing the sulfur evolution of an Oligocene andesite (Eagle Mountain, CO, USA). Lithos 2002;64(3–4) 63–75.

[40] Klemm W. Anomale Wertigkeiten. Angewandte Chemie 1951;63(17–18) 396–402.

[41] Chauhan P, Chauhan RP, Gupta M. Estimation of naturally occurring radionuclides in fertilizers using gamma spectrometry and elemental analysis by XRF and XRD techniques. Microchemical Journal 2013;106; 73–78.

[42] Hasikova J, Sokolov A, Titov V, Dirba A. On-line XRF analysis of phosphate materials at various stages of processing. Procedia Engineering 2014;83 455–461.

[43] Zhou F, Wang L, Xu Z, Liu Q, Chi R. Reactive oily bubble technology for flotation of apatite, dolomite and quartz. International Journal of Mineral Processing 2015;134 74–81.

[44] Oliveira MS, Santana RC, Ataíde CH, Barrozo MAS. Recovery of apatite from flotation tailings. Separation and Purification Technology 2011;79(1) 79–84.

[45] Taylor HE. Inductively Coupled Plasma–Mass Spectrometry: Practices and Techniques. Academic Press, 2001. ISBN: 978-0126838657

[46] Montaser A. Inductively Coupled Plasma Mass Spectrometry. John Wiley & Sons, 1998. ISBN: 978-0471186205

[47] Levinson R. More Modern Chemical Techniques. Royal Society of Chemistry, 2001. ISBN: 978-0854049295

[48] Lajunen LHJ, Perämäki P. Spectrochemical Analysis by Atomic Absorption and Emission. Royal Society of Chemistry, 2004. ISBN: 978-0854046249

[49] Hasebe N, Barbarand J, Jarvis K, Carter A, Hurford AJ. Apatite fission-track chronometry using laser ablation ICP-MS. Chemical Geology 2004;207(3–4) 135–145.

[50] Johnstone S, Hourigan J, Gallagher Ch. LA-ICP-MS depth profile analysis of apatite: Protocol and implications for (U–Th)/He thermochronometry. Geochimica et Cosmochimica Acta 2013;109 143–161.

[51] Guzmics T, Zajacz Z, Kodolányi J, Halter W, Szabó C. LA-ICP-MS study of apatite- and K feldspar-hosted primary carbonatite melt inclusions in clinopyroxenite xenoliths from lamprophyres, Hungary: Implications for significance of carbonatite melts in the Earth's mantle. Geochimica et Cosmochimica Acta 2008;72(7) 1864–1886.

[52] Chew DM, Sylvester PJ, Tubrett MN. U–Pb and Th–Pb dating of apatite by LA-ICPMS. Chemical Geology 2011;280(1–2) 200–216.

[53] Donelick RA, O'Sullivan PB, Ketcham RA, Hendriks BWH, Redfield TF. Relative U and Th concentrations from LA-ICP-MS for apatite fission-track grain-age dating. Geochimica et Cosmochimica Acta 2006;70(18) A143.

[54] Fisher ChM, McFarlane ChRM, Hanchar JM, Schmitz MD, Sylvester PJ, Lam R, Longerich HP. Sm-Nd isotope systematics by laser ablation-multicollector-inductively coupled plasma mass spectrometry: methods and potential natural and synthetic reference materials. Chemical Geology 2001;284(1–2) 1–20.

[55] Yang Yue-H Wu Fu-Y, Xie Lie-W, Chu Zhu-Y, Yang Jin-H. Re-evaluation of interferences of doubly charged ions of heavy rare earth elements on Sr isotopic analysis using multi-collector inductively coupled plasma mass spectrometry. Spectrochimica Acta Part B: Atomic Spectroscopy 2014;97 118–123.

[56] Yang Yue-H Wu Fu-Y, Chu Zhu-Y, Xie Lie-W, Yang Jin-H. High-precision simultaneous determination of 147Sm/144Nd and 143Nd/144Nd ratios in Sm-Nd mixtures using multi-collector inductively coupled plasma mass spectrometry and its comparison to isotope dilution analysis. Spectrochimica Acta Part B: Atomic Spectroscopy 2013;79–80 82–87.

[57] Pourmand A, Dauphas N, Ireland TJ. A novel extraction chromatography and MC-ICP-MS technique for rapid analysis of REE, Sc and Y: revising CI-chondrite and Post-Archean Australian Shale (PAAS) abundances. Chemical Geology 2012;291 38–54.

[58] Boutton T. Mass Spectrometry of Soils (Books in Soils, Plants, and the Environment). CRC Press, 1996. ISBN: 978-0824796990

[59] Casas JS, Sordo J. Lead: Chemistry, analytical aspects, environmental impact and health effects. Elsevier, 2011. ISBN: 978-0080463889

[60] Becker S. Inorganic Mass Spectrometry: Principles and Applications. John Wiley & Sons, 2008. ISBN: 978-0470517208

[61] Hellebrand SJE. Textural, geochronological and chemical constraints from polygenetic titanite and monogenetic apatite from a mid-crustal shear zone: an integrated EPMA, SIMS, and TIMS study. Chemical Geology 2007;241(1–2) 88–107.

[62] Schoene B, Latkoczy C, Schaltegger U, Günther D. A new method integrating high-precision U–Pb geochronology with zircon trace element analysis (U–Pb TIMS-TEA). Geochimica et Cosmochimica Acta 2010;74(24) 7144–7159.

[63] Teixeira W, D'Agrella-Filho MS, Hamilton MA, Ernst RE, Girardi VAV, Mazzucchelli M, Bettencourt JS. U–Pb (ID-TIMS) baddeleyite ages and paleomagnetism of 1.79 and 1.59 Ga tholeiitic dyke swarms, and position of the Rio de la Plata Craton within the Columbia supercontinent. Lithos 2013;174 157–174.

[64] Metcalfe I, Crowley JL, Nicoll RS, Schmitz M. High-precision U-Pb CA-TIMS calibration of middle Permian to lower Triassic sequences, mass extinction and extreme climate-change in eastern Australian Gondwana. Gondwana Research, In Press, Corrected Proof, Available online 16 September 2014. doi:10.1016/j.gr.2014.09.002

[65] Yang Yue-H Wu Fu-Y, Yang Jin-H, Chew DM, Xie Lie-W, Chu Zhu-Y, Zhang Yan-B, Huang Ch. Sr and Nd isotopic compositions of apatite reference materials used in U–Th–Pb geochronology. Chemical Geology 2014;385 35–55.

[66] Cherepin VT. Secondary Ion Mass Spectroscopy of Solid Surfaces. VSP, 1987. ISBN: 978-9067640787

[67] O'Connor J, Sexton BA, Smart RSC. Surface Analysis Methods in Materials Science Physics and Astronomy. Springer Series in Surface Sciences—Volume 23. Springer Science & Business Media, 2003. ISBN: 978-3540413301

[68] van der Heide P. Secondary Ion Mass Spectrometry: An Introduction to Principles and Practices. John Wiley & Sons, 2014. ISBN: 978-1118916773

[69] Kuo J. Electron Microscopy: Methods and Protocols—Volume 369. Springer Science & Business Media, 2007. ISBN: 978-1597452946

[70] Microelectronics Failure Analysis: Desk Reference. ASM International, 2004. ISBN: 978-0871708045

[71] Watson LL, Hutcheon ID, Epstein S, Stolper EM, Water on Mars: clues from deuterium/hydrogen andwater contents of hydrous phases in SNCmeteorites. Science 1994;265 86–90.

[72] Barnes JJ, Franchi IA, Anand M, Tartèse R, Starkey NA, Koike M, Sano Y, Russell SS. Accurate and precise measurements of the D/H ratio and hydroxyl content in lunar apatites using NanoSIMS. Chemical Geology 2013;337–338 48–55.

[73] Gotliv Bat-A, Robach JS, Veis A. The composition and structure of bovine peritubular dentin: mapping by time of flight secondary ion mass spectroscopy. Journal of Structural Biology 2006;156(2) 320–333.

[74] Stephan T. TOF-SIMS in cosmochemistry. Planetary and Space Science 200149(1) 859–906.

[75] Li Qiu-L, Li Xian-H, Wu Fu-Y, Yin Qing-Z, Ye Hai-M, Liu Y, Tang Guo-Q, Zhang Chuan-L. In-situ SIMS U-Pb dating of phanerozoic apatite with low U and high common Pb. Gondwana Research 2012;21(4) 745–756.

[76] Schaltegger U, Schmitt AK, Horstwood MSA. U-Th-Pb zircon geochronology by ID-TIMS, SIMS, and laser ablation ICP-MS: recipes, interpretations, and opportunities. Chemical Geology 2015;402 89–110.

[77] Ling Xiao-X, Schmädicke E, Li Qiu-L, Gose J, Wu Rui-H, Wang Shi-Q, Liu Y, Tang Guo-Q, Li Xian-H. Age determination of nephrite by in-situ SIMS U–Pb dating syngenetic titanite: a case study of the nephrite deposit from Luanchuan, Henan, China. Lithos 2015;220–223 289–299.

[78] Friedbacher G, Bubert H. Surface and Thin Film Analysis: A Compendium of Principles, Instrumentation, and Applications. John Wiley & Sons, 2011. ISBN: 978-3527636945

[79] Lee MS. Mass Spectrometry Handbook. Wiley Series on Pharmaceutical Science and Biotechnology: Practices, Applications and Methods—Volume 7. John Wiley & Sons, 2012. ISBN: 978-1118180723

[80] Dambach S, Fartmann M, Kriegeskotte C, Brüning C, Wiesmann HP, Lipinsky D, Arlinghaus HF. Laser-SNMS analysis of apatite formation in vitro. Applied Surface Science 2004;231–232 506–509.

[81] Rabi II, Zacharias S, Millman P. Kusch. A new method of measuring nuclear magnetic moment. Physical Review 1938;53(4) 318.

[82] Weil JA, Bolton JR. Electron Paramagnetic Resonance: Elementary Theory and Practical Applications. 2nd ed., John Wiley & Sons, 2007. ISBN: 978-0470084984

[83] Brustolon M. Electron Paramagnetic Resonance: A Practitioner's Toolkit. John Wiley & Sons, 2009. ISBN: 978-0470432228

[84] Mabbs FE, Collison D. Electron Paramagnetic Resonance of d Transition Metal Compounds Studies in Inorganic Chemistry. Elsevier, 2013. ISBN: 978-1483291499

[85] Misra SK. Handbook of Multifrequency Electron Paramagnetic Resonance: Data and Techniques. John Wiley & Sons, 2014. ISBN: 978-3527672455

[86] Moens PDW, Callens FJ, Verbeeck RMH, Naessens DE. An EPR spectrum decomposition study of precipitated carbonated apatites (NCAP) dried at 25°C: adsorption of molecules from the atmosphere on the apatite powders. Applied Radiation and Isotopes 1993;44(1–2) 279–285.

[87] de Oliveira LM, Rossi AM, Lopes RT. Dose response of A-type carbonated apatites prepared under different conditions. Radiation Physics and Chemistry 2001;61(3–6) 485–487.

[88] Schramm DU, Rossi AM. Electron spin resonance (ESR) studies of CO_2^- radicals in irradiated A and B-type carbonate-containing apatites. Applied Radiation and Isotopes 2000;52(5) 1085–1091.

[89] Vugman NV, Rossi AM, Rigby SEJ. EPR dating CO_2^- sites in tooth enamel apatites by ENDOR and triple resonance. Applied Radiation and Isotopes 1995;46(5) 311–315.

[90] Fattibene P, Callens F. EPR dosimetry with tooth enamel: a review. Applied Radiation and Isotopes 2010;68(11) 2033–2116.

[91] Rudko VV, Ishchenko SS, Vorona IP, Baran NP. Isotropic radical image in biological apatites. Radiation Measurements 2007;42(9) 1580–1582.

[92] Ishchenko SS, Vorona IP, Okulov SM, Baran NP. ^{13}C hyperfine interactions of CO_2^- in irradiated tooth enamel as studied by EPR. Applied Radiation and Isotopes 2002;56(6) 815–819.

[93] Nosenko VV, Vorona IP, Baran NP, Ishchenko SS, Vysotskyi BV, Krakhmalnaya TV, Semenov YA. Comparative EPR study CO_2^- radicals in modern and fossil tooth enamel. Radiation Measurements 2015;78 53–57.

[94] Rudko VV, Vorona IP, Baran NP, Ishchenko SS. Thermally stimulated transformation of the EPR spectra in γ-irradiated bone tissue. Radiation Measurements 2009;44(3) 239–242.

[95] Ishchenko SS, Vorona IP, Okulov SM, Baran NP, Rudko VV. ENDOR study of CO_2^- radicals in hydroxyapatite of γ-irradiated bone. Radiation Measurements 2011;46(1) 37–39

[96] Köseogˇlu R, Köseogˇlu E, Köksal F, Başaran E, Demirci D. EPR of some irradiated renal stones. Radiation Measurements 2005;40(1) 65–68.

[97] Fleet ME. Infrared spectra of carbonate apatites: ν2-Region bands. Biomaterials 2009;30(8) 1473–1481.

[98] Yu H, Zhang H, Wang X, Gu Z, Li X, Deng F. Local structure of hydroxy–peroxy apatite: a combined XRD, FT-IR, Raman, SEM, and solid-state NMR study. Journal of Physics and Chemistry of Solids 2007;68(10) 1863–1871.

[99] Klinowski J. New Techniques in Solid-State NMR, Volume 246. Springer Science & Business Media, 2004. ISBN: 978-3540221685

[100] Duer MJ. Solid State NMR Spectroscopy: Principles and Applications. John Wiley & Sons, 2008. ISBN: 978-0470999387

[101] Ohannesian L, Streeter A. Handbook of Pharmaceutical Analysis Drugs and the Pharmaceutical Sciences. CRC Press, 2001. ISBN: 978-0824741945

[102] Bakhmutov VI. Solid-State NMR in Materials Science: Principles and Applications. CRC Press, 2011. ISBN: 978-1439869635

[103] Burnell E, de Lange CA. NMR of Ordered Liquids. Springer Science & Business Media, 2003. ISBN: 978-1402013430

[104] Braun M, Jana C. 19F NMR spectroscopy of fluoridated apatites. Chemical Physics Letters 1995;245(1) 19–22.

[105] McElderry John-DP, Zhu P, Mroue KH, Xu J, Pavan B, Fang M, Zhao G, McNerny E, Kohn DH, Franceschi RT, Holl MMB, Tecklenburg MMJ, Ramamoorthy A, Morris MD. Crystallinity and compositional changes in carbonated apatites: Evidence from 31P solid-state NMR, Raman, and AFM analysis. Journal of Solid State Chemistry 2013;206 192–198.

[106] Kaflak A, Kolodziejski W. Phosphorus-31 spin-lattice NMR relaxation in bone apatite and its mineral standards. Solid State Nuclear Magnetic Resonance 2007;31(4) 174–183.

[107] Yesinowski JP, Eckert H. Hydrogen environments in calcium phosphates: proton MAS NMR at high spinning speeds. Journal of the American Chemical Society 1987;109(21) 6274–6282.

[108] Goldstein J, Newbury DE, Joy DC, Lyman ChE, Echlin P, Lifshin E, Sawyer L, Michael JR. Scanning Electron Microscopy and X-ray Microanalysis: Third Edition. Springer Science & Business Media, 2012. ISBN: 978-1461502159

[109] Reimer L. Scanning Electron Microscopy: Physics of Image Formation and Microanalysis. Springer Science & Business Media, 1998. ISBN: 978-3540639763

[110] Krinsley DH, Pye K, Boggs S, Tovey NK Jr. Backscattered Scanning Electron Microscopy and Image Analysis of Sediments and Sedimentary Rocks. Cambridge University Press, 2005. ISBN: 978-0521019743

[111] Compton RG, Harding MS, Pluck MR, Atherton JH, Brennan CM. Journal of Physical Chemistry 1993;97(40) 10416–10420.

[112] Dorozhkin SV. Surface Reactions of Apatite Dissolution. Journal of Colloid and Interface Science 1997;191(2) 489–497.

[113] Xu Wen-G, Fan Hong-R, Hu Fang-F, Santosh M, Yang Kui-F, Lan Ting-G. In situ chemical and Sr–Nd–O isotopic compositions of apatite from the Tongshi intrusive complex in the southern part of the North China Craton: implications for petrogenesis and metallogeny. Journal of Asian Earth Sciences 2015;105 208–222.

[114] Xing Chang-M, Wang ChY, Li C. Trace element compositions of apatite from the middle zone of the Panzhihua layered intrusion, SW China: insights into the differentiation of a P- and Si-rich melt. Lithos 2014;204 188–202.

[115] Roman-Lopez J, Correcher V, Garcia-Guinea J, Prado-Herrero P, Rivera T, Lozano IB. Effect of the chemical impurities on the luminescence emission of natural apatites. Spectrochimica Acta Part A: Molecular and Biomolecular Spectroscopy 2014;126 142–147.

[116] Kamiishi E, Utsunomiya S. Nano-scale reaction processes at the interface between apatite and aqueous lead. Chemical Geology 2013;340 121–130.

[117] Stuart BH. Infrared Spectroscopy: Fundamentals and Applications. Vydavatel John Wiley & Sons, 2004. ISBN: 978-0470011133

[118] Chalmers JM, Edwards HGM, Hargreaves MD. Infrared and Raman Spectroscopy in Forensic Science. John Wiley & Sons, 2012. ISBN: 978-0470749067

[119] McCreery RL. Raman Spectroscopy for Chemical Analysis. Chemical Analysis: A Series of Monographs on Analytical Chemistry and Its Applications—Volume 225. John Wiley & Sons, 2005. ISBN: 978-0471231875

[120] Ferraro JR, Nakamoto K. Introductory Raman Spectroscopy. Academic Press, 2012. ISBN: 978-0323137904

[121] Madou MJ. Solid-State Physics, Fluidics, and Analytical Techniques in Micro- and Nanotechnology. Fundamentals of Microfabrication and Nanotechnology—Volume 1. CRC Press, 2011. ISBN: 978-1420055115

[122] Kumar ChSSR. Raman Spectroscopy for Nanomaterials Characterization. Springer Science & Business Media, 2012. ISBN: 978-3642206207

[123] Fleet ME, Liu X. Local structure of channel ions in carbonate apatite. Biomaterials 2005;26(36) 7548–7554.

[124] Errassifi F, Sarda S, Barroug A, Legrouri A, Sfihi H, Rey C. Infrared, Raman and NMR investigations of risedronate adsorption on nanocrystalline apatites. Journal of Colloid and Interface Science 2014;420 101–111.

[125] Antonakos A, Liarokapis E, Leventouri T. Micro-Raman and FTIR studies of synthetic and natural apatites. Biomaterials 2007;28(19) 3043–3054.

[126] Gómez-Morales J, Iafisco Mi, Delgado-López JM, Sarda S, Drouet C. Progress on the preparation of nanocrystalline apatites and surface characterization: overview of

fundamental and applied aspects. Progress in Crystal Growth and Characterization of Materials 2013;59(1) 1–46.

[127] Grunenwald A, Keyser C, Sautereau AM, Crubézy E, Ludes B, Drouet C. Revisiting carbonate quantification in apatite (bio)minerals: a validated FTIR methodology. Journal of Archaeological Science 2014;49 134–141.

[128] Adler HH. Infrared spectra of phosphate minerals: symmetry and substitutional effects in the pyromorphite series. American Mineralogist 1964;49 1002–1015.

[129] Bencnn RM. A relation between internuclear distances and bond force constants. Journal of Chemical Physiscs 1934;2 128–131.

[130] Lever T, Haines P, Rouquerol J, Charsley EL, Van Eckeren P, Burlett DJ. ICTAC nomenclature of thermal analysis, 2014 (IUPAC Recommendations 2014). Pure and Applied Chemistry 2014;86(4) 545–553. DOI 10.1515/pac-2012-0609

[131] Brown ME. Introduction to Thermal Analysis: Techniques and Applications. Hot Topics in Thermal Analysis and Calorimetry—Volume 1. Springer Science & Business Media, 2001. ISBN: 978-1402002113

[132] Brown ME, Gallagher PK. Handbook of Thermal Analysis and Calorimetry: Recent Advances, Techniques and Applications. Handbook of Thermal Analysis and Calorimetry—Volume 5. Elsevier, 2011. ISBN: 978-0080556314

[133] Kerr PF. Optical Mineralogy. 4nd ed., McGraw-Hill, 1977. ISBN: 978-0070342187

[134] Mitra S. Fundamentals Of Optical, Spectroscopic And X-Ray Mineralogy. New Age International, 1996. ISBN: 978-8122409826

[135] Jonasz M, Fournier GR. Chapter —Refractive indices and morphologies of aquatic particles. Light Scattering by Particles in Water, 2007; 447–558.

[136] Li Cheng-X, Duan Yong-H, Hu Wen-Ch. Electronic structure, elastic anisotropy, thermal conductivity and optical properties of calcium apatite $Ca_5(PO_4)_3X$ (X=F, Cl or Br). Journal of Alloys and Compounds 2015;619 66–77.

[137] Duan YH, Sun Y. First-principles calculations of optical properties of Mg_2Pb. Science China Physics, Mechanics and Astronomy 2014;57(2) 233–238.

[138] Rulis P, Ouyang L, Ching WY. Electronic structure and bonding in calcium apatite crystals: Hydroxyapatite, fluorapatite, chlorapatite, and cromapatite. Physical Review B 70, 155104 (2004). DOI: 10.1103/PhysRevB.70.155104.

[139] Lowell S, Shields JE, Thomas MA, Thommes M. Characterization of Porous Solids and Powders: Surface Area, Pore Size and Density. Particle Technology Series—Volume 16. Springer Science & Business Media, 2012. ISBN: 978-1402023033

[140] Merkus HG. Particle Size Measurements: Fundamentals, Practice, Quality Particle Technology Series—Volume 17. Springer Science & Business Media, 2009. ISBN: 978-1402090165

[141] Allen T. Powder Sampling and Particle Size Determination. Elsevier, 2003. ISBN: 978-0080539324

[142] LeRoy Jackson M, Barak P. Soil Chemical Analysis: Advanced Course. UW-Madison Libraries Parallel Press, 2005. ISBN: 978-1893311473

[143] Taylor HFW. Cement Chemistry. Revised edition, Thomas Telford, 1997. ISBN: 978-0727725929

[144] Ishizaki K, Komarneni S, Nanko M. Porous Materials: Process Technology and Applications. Materials Technology Series—Volume 4. ISBN: 978-1461558118

[145] Borch J, Lyne MB, Mark RE, Habeger Ch. Handbook of Physical Testing of Paper, Volume 2. 2nd ed., CRC Press, 2001. ISBN: 978-0203910498

[146] Donaldson EC, Chilingarian GV, Yen TF. Enhanced Oil Recovery, I: Fundamentals and Analyses. Developments in Petroleum Science. Elsevier, 1985. ISBN: 978-0080868721

[147] Millington RJ, Quirk JP. Permeability of porous solids. Transactions of the Faraday Society 1961;57 1200–1207.

[148] Zhong K, Vasudevan TV, Somasundaran P. Floatability of apatites of different type and origin: role of surface area and porosity. International Journal of Mineral Processing 1993;38(3–4) 177–188.

[149] Cardoso L, Fritton SP, Gailani G, Benalla M, Cowin SC. Advances in assessment of bone porosity, permeability and interstitial fluid flow. Journal of Biomechanics 2013;46(2) 253–265.

[150] Tang H, Guo Y, Jia D, Zhou Y. High bone-like apatite-forming ability of mesoporous titania films. Microporous and Mesoporous Materials 2010;131(1–3) 366–372.

[151] Habibovic P, Juhl MV, Clyens S, Martinetti R, Dolcini L, Theilgaard N, van Blitterswijk CA. Comparison of two carbonated apatite ceramics in vivo. Acta Biomaterialia 2010;6(6) 2219–2226.

[152] Liu Dean-M. Influence of porosity and pore size on the compressive strength of porous hydroxyapatite ceramic. Ceramics International 1997;23(2) 135–139.

[153] Mohammad NF, Othman R, Yeoh FY. Controlling the pore characteristics of mesoporous apatite materials: hydroxyapatite and carbonate apatite. Ceramics International, In Press, Accepted Manuscript, Available online 9 May 2015. doi:10.1016/j.ceramint.2015.04.162

[154] Panteix PJ, Julien I, Abélard P, Bernache-Assollant D. Influence of porosity on the electrical properties of $La_{9.33}(SiO_4)_6O_2$ oxyapatite. Ceramics International 2008;34(7) 1579–1586.

[155] Pons A, Jouin J, Béchade E, Julien I, Masson O, Geffroy PM, Mayet R, Thomas P, Fukuda K, Kagomiya I. Study of the formation of the apatite-type phases $La_{9.33+x}(SiO_4)_6O_{2+3x/2}$

synthesized from a lanthanum oxycarbonate $La_2O_2CO_3$. Solid State Sciences 2014;38 150–155.

[156] Li B, Liu W, Pan W. Synthesis and electrical properties of apatite-type $La_{10}Si_6O_{27}$. Journal of Power Sources 2010;195(8) 2196–2201.

[157] Medeiros EE, Dias AMS. Experimental and numerical analysis of Vickers hardness testing. IJRRAS 2013;17(1) 9–18.

[158] CiDRA Precision Services, LLC. Mineral Hardness Conversion Chart. http://www.cidraprecisionservices.com/mohs-conversion.html

[159] Saenger AT, Kuhs WF. Structural disorder in hydroxyapatite. Zeitschrift für Kristallographie 1992;199 123–148.

[160] Bhatnagar VM. IR-spectra of fluorapatite and fluorchlorapatite. Experientia 1967;24(1) 10–12

[161] Balan E, Delattre S, Roche D, Segalen L, Morin G, Guillaumet M, Blanchard M, Lazzeri M, Brouder Ch, Salje EKH. Line-broadening effects in the powder infrared spectrum of apatite. Physics and Chemistry of Minerals 2011;38(2) 111–112.

[162] Foil AM, Wilkins ChH. infrared spectra and characteristic frequencies of inorganic ions. Their use in qualitative analysis. Analytical Chemistry 1952;24(8) 1253–1294.

[163] Bonner CE Jr, Chess ChC, Meegoda Ch, Stefanos S, Loutts GB. Volume effects on the Raman frequencies of phosphate ions in fluorapatite crystals. Optical Materials 2004;26 17–22.

[164] Tressaud A, Haufe G. Fluorine and Health: Molecular Imaging, Biomedical Materials and Pharmaceuticals. 1st ed., Elsevier, 2008. ISBN: 978–0444530868

[165] Williams Q, Knittle E. Infrared and raman spectra of $Ca_5(PO_4)_3F_2$-fluorapatite at high pressures: compression-induced changes in phosphate site and Davydov splittings. Journal of Physics and Chemistry of Solids 199657(4) 417–422.

[166] Veiderma M, Knubovets R, Tõnsuaadu K. Structural properties of apatites from Finland studied by FTIR spectroscopy. Bulletin of the Geological Society of Finland 1998;1–2 69–75.

[167] Freund F, Knobel RM. Distribution of fluorine in hydroxyapatite studied by infrared spectroscopy. Journal of the Chemical Society, Dalton Transactions 1977;11 1136–1140.

[168] Knubovets R. Structural mineralogy and properties of natural phosphates. Reviews in Chemical Engineering 1993;9(3–4) 161–216.

[169] Ptáček P, Bartoníčková E, Švec J, Opravil T, Šoukal F, Frajkorová F. The kinetics and mechanism of thermal decomposition of $SrCO_3$ polymorphs. Ceramics International 2015;41(1) 115–126.

[170] Drouet Ch. A comprehensive guide to experimental and predicted thermodynamic properties of phosphate apatite minerals in view of applicative purposes. Journal of Chemical Thermodynamics 2015;81 143–159.

[171] Moiseev G, Leitner J, Sestak J, Zhukovsky V. Empirical dependences of the standard enthalpy of formation for related inorganic compounds enhancing glass formers. Thermochimica Acta 1996;280–281 511–521.

[172] Moiseev GK, Sestak J. Some calculations methods for estimation of thermodynamical and thermochemical properties of inorganic compounds. Progress in Crystal Growth and Characterization of Materials 1995;30(1) 23–81.

[173] Petit A.-T, Dulong P-L. Recherches sur quelques points importants de la Théorie de la Chaleur. Annales de Chimie et de Physique (in French) 1819;10 395–413.

[174] Bureau International des Poids et Mesures the International System of Units (SI). 8th ed., Organisation Intergouvernementale de la Convention du Mètre, STEDI Media, Paris, 2006. ISBN: 92-822-2213-6

[175] Rogers D. Einstein's Other Theory: The Planck-Bose-Einstein Theory of Heat Capacity. Princeton University Press, 2005. ISBN: 978-0691118260

[176] Latimer WM. Methods of estimating the entropies of solid compounds. Journal of the American Chemical Society 1951;73(4) 1480–1482.

[177] Leitner J, Chuchvalec P, Sedmidubský D, Strejc A, Abrman P. Estimation of heat capacities of solid mixed oxides. Thermochemica Acta 2003;395 47–46.

[178] Kellogg HH, in Fitterer GR (Ed.). Problems in the non-ferrous industries and a note on estimation of heat capacity, in: Applications of Fundamental Thermodynamics to Metallurgical Processes. Gorden and Breach, London, 1967, p.357

[179] Kubaschewski O, Ünal H. An empirical estimation of the heat capacities of inorganic compounds. High Temperatures—High Pressures 1977;9(3) 361–365.

[180] Kubaschewski O, Alock CB, Spencer PJ. Materials thermochemistry. Materials Science Monographs. 6th ed., Pergamon Press, 1993. ISBN: 9780080418896

[181] Spencer PJ. Estimation of thermodynamic data for metallurgical applications. Thermochimica Acta 1988;314(1–2) 1–21.

[182] Moiseev GK, Šesták J. Some calculations methods for estimation of thermodynamical and thermochemical properties of inorganic compounds. Progress in Crystal Growth and Characterization of Materials 1995;30 23–81.

[183] Kumok VN. Problem of Correlation Methods for Evaluation of Thermodynamic and Characteristics, Direct and Reverse Task of Chemical Thermodynamics. Nauka, Novosibirsk, 1987, p.108.

[184] Latimer WM. Methods of estimating the entropies of solid compounds. Journal of the American Chemical Society 1951;73 1480–1482.

[185] Aldabergenov MK, Balakaeva GT, Kokibasova GT. Calculation of thermodynamic functions of aluminates of alkaline and alkaline earth elements. Journal of Physical chemistry 1998;72(5) 808–811.

[186] Gospodinov GG, Mihov DI. Molar heat capacities of the tellurites $MnTeO_3$, $MnTe_2O_5$, and $Mn_2Te_3O_8$. Journal of Chemical Thermodynamics 1993;25(10) 1249–1252.

[187] Dorozhkin SV. A review on the dissolution models of calcium apatites. Progress in Crystal Growth and Characterization of Materials 2001;44(1) 45–61.

[188] Zhang J, Nancolias GH. Unexpected pH dependence of dissolution kinetics of dicalcium phosphate dihydrate. Journal of Physical Chemistry 1994;98(6) 1689–1694.

[189] Zhu Y, Zhang X, Chen Y, Xie Q, Lan J, Qian M, He N. A comparative study on the dissolution and solubility of hydroxylapatite and fluorapatite at 25°C and 45°C. Chemical Geology 2009;268(1–2) 89–96.

[190] Patel PR, Gregory TM, Brown WE. Solubility of $CaHPO_4{\cdot}2H_2O$ in the quaternary system $Ca(OH)_2$-H_3PO_4-NaCl-H_2O at 25°C. Journal of Research of the National Bureau of Standards—A. Physics and Chemistry 1974;78(6) 675–681.

[191] Christoffersen J, Christoffersen MR. Kinetics of dissolution of calcium hydroxyapatite V. The acidity constant for the hydrogen phosphate surface complex. Journal of Crystal Growth 1982;57 21–26.

[192] McDowell H, Gregory TM, Brown WE. Solubility of $Ca_5(PO_4)_3OH$ in the System $Ca(OH)_2$-H_3PO_4-H_2O at 5, 15, 25 and 37°C. Journal of Research of the National Bureau of Standards —A. Physics and Chemistry 1977;81(2–3) 273–281.

[193] Tang R, Henneman ZJ, Nancollas GH. Constant composition kinetics study of carbonated apatite dissolution. Journal of Crystal Growth 2003;249(3–4) 614–624.

[194] Shellis RP, Lee AR, Wilson RM. Observations on the apparent solubility of carbonate-apatites. Journal of Colloid and Interface Science 1999;218(2) 351–358.

[195] Chaïrat C, Schott J, Oelkers EH, Lartigue Jean-E, Harouiya N. Kinetics and mechanism of natural fluorapatite dissolution at 25°C and pH from 3 to 12. Geochimica et Cosmochimica Acta 2007;71(24) 5901–5912.

[196] Wang LK, Hung Yung-T, Shammas NK. Physicochemical Treatment Processes. Handbook of Environmental Engineering—Volume 3. Springer Science & Business Media, 2007. ISBN: 978-1592598205

[197] Harouiya N, Chaïrat C, Köhler SJ, Gout R, Oelkers EH. The dissolution kinetics and apparent solubility of natural apatite in closed reactors at temperatures from 5 to 50°C and pH from 1 to 6. Chemical Geology 2007;244(3–4) 554–568.

[198] Chaïrat C, Oelkers EH, Schott J. Lartigue J-E Harouiya N. Fluorapatite surface composition in aqueous solution deduced from potentiometric, electrokinetic, and solubility measurements, and spectroscopic observations. Geochimica Cosmochimica Acta 2007;71(24) 5888–5900.

[199] Smith AN, Posner AM, Quirk JP. Incongruent dissolution and surface complexes of hydroxyapatite. Journal of Colloid and Interface Science 1974;48(3) 442–449.

[200] Deutsch WJ, Siegel R. Groundwater Geochemistry: Fundamentals and Applications to Contamination. CRC Press, 1997. ISBN: 978-0873713085

[201] Thomann JM, Voegel JC, Gramain P. Kinetics of dissolution of calcium hydroxyapatite powder. III: pH and sample conditioning effects. Calcified Tissue International 1990;46(2) 121–129.

[202] Chin KOA, Nancollas GH. Dissolution of fluorapatite. A constant-composition kinetics study. Langmuir 1991;7 2175–2179.

[203] Margolis HC, Moreno EC. Kinetics of hydroxyapatite dissolution in acetic, lactic, and phosphoric acid solutions. Calcified Tissue International 1992;50(2) 137–143.

[204] Paschalis EP, Wikiel K, Nancollas GH. Dual constant composition kinetics characterization of apatitic surfaces. Journal of biomedical materials research 1994;28(12) 1411–1418.

[205] Nelson DGA, Featherstone JDB, Duncan JF, Cutress TW. Effect of carbonate and fluoride on the dissolution behaviour of synthetic apatites. Caries Research 1983;17(3) 200–211.

[206] Christoffersen MR, Christoffersen J. Possible mechanisms for the growth of the biomaterial, calcium hydroxyapatite microcrystals. Journal of Crystal Growth 1992;121(4) 617–630.

[207] Christoffersen J, Christoffersen MR, Johansen T. Kinetics of growth and dissolution of fluorapatite. Journal of Crystal Growth 1996;163(3) 295–303.

[208] Christoffersen MR, Dohrup J, Christoffersen J. Kinetics of growth and dissolution of calcium hydroxyapatite in suspensions with variable calcium to phosphate ratio. Journal of Crystal Growth 1998;186 283–290.

[209] Mafe S, Manzanares JA, Reis H, Thomann JM, Gramain P. Model for the dissolution of calcium hydroxyapatite powder. Journal of Physical Chemistry 1992;96(2) 861–866.

[210] Gasser P, Voegel JC, Gramain P. Surface reactions on hydroxyapatite in the presence of fluoride ions 1. Saturated and congruent conditions. Colloids and Surfaces A: Physicochemical and Engineering Aspects 1993;74(2–3) 275–286.

[211] Schaad P, Poumier F, Voegel JC, Gramain P. Analysis of calcium hydroxyapatite dissolution in non-stoichiometric solutions. Colloids and Surfaces A: Physicochemical and Engineering Aspects 1997;121(2–3) 217–228.

[212] Dorozhkin SV. Acidic dissolution mechanism of natural fluorapatite. II. Nanolevel of investigations. Journal of Crystal Growth 1997;182(1–2) 133–140.

[213] Dorozhkin SV. Inorganic chemistry of the dissolution phenomenon: the dissolution mechanism of calcium apatites at the atomic (ionic) level. Comments on Inorganic Chemistry 1999;20(4–6) 285–299.

[214] Pearce EIF. On the dissolution of hydroxyapatite in acid solutions. Journal of Dental Research 1988;67(7) 1056–1058.

[215] Lowell LC. Dislocation etch pits in apatite. Acta Metallurgica 1958;6(12) 775–778.

[216] Jongebloed WL, Molenaar I, Arends J. Orientation-dependent etchpit penetration and dissolution of fluoroapatite. Caries Research 1973;7(2) 154–165.

[217] Jongebloed WL, van den Berg PJ, Arends J. The dissolution of single crystals of hydroxyapatite in citric and lactic acids. Calcified Tissue Research1974;15 1–9.

[218] Arends J, Jongebloed WL. Dislocations and dissolution in apatites: theoretical considerations. Caries Research 1977;11(3) 186–188.

[219] Arends J, Jongebloed WL. Ultrastructural studies of synthetic apatite crystals Journal of Dental Research 1979;58 (Spec. Issue B) 837–843.

[220] Hignett TP. Developments in Plant and Soil Sciences—Volume 15. Springer Science & Business Media, 2013. ISBN: 978-9401715386

[221] Somani L.L. Dictionary of agricultural and allied sciences—Volume 4. Mittal Publications, 1989. ISBN: 978-8170991410

[222] Nielsson FT. Manual of Fertilizer Processing. Fertilizer Science and Technology—Volume 5. CRC Press, 1986. ISBN: 978-0824775223

[223] Manual Fertilizer, UN Industrial Development Organization, International Fertilizer Development Center. Springer Science & Business Media, 1998. ISBN: 978-0792350323

[224] Naidu S. International Encyclopaedia of Agricultural Science and Technology: Soils and composts. International Encyclopaedia of Agricultural Science and Technology—Volume 4. Mittal Publications, 2008. ISBN: 978-8183242233

[225] Espinosa J. Influence of Rock Phosphate on Available Phosphorus as Measured by Plant Uptake and Soil Extractants. Thesis, INIAP Archivo Historico, 1979

[226] Espinosa J. Abstract Influence of Rock Phosphate on Available Phosphorus as Measured by Plant Uptake and Soil Extractants. Thesis, INIAP Archivo Historico.

[227] Shemilt LW. Chemistry and World Food Supplies: The New Frontiers, Chemrawn II: Invited papers presented at the International Conference on Chemistry and World Food Supplies, Manila, Philippines, 1982. ISBN: 0-08-029243-7.

Permissions

All chapters in this book were first published in ATSASSPA, by InTech Open; hereby published with permission under the Creative Commons Attribution License or equivalent. Every chapter published in this book has been scrutinized by our experts. Their significance has been extensively debated. The topics covered herein carry significant findings which will fuel the growth of the discipline. They may even be implemented as practical applications or may be referred to as a beginning point for another development.

The contributors of this book come from diverse backgrounds, making this book a truly international effort. This book will bring forth new frontiers with its revolutionizing research information and detailed analysis of the nascent developments around the world.

We would like to thank all the contributing authors for lending their expertise to make the book truly unique. They have played a crucial role in the development of this book. Without their invaluable contributions this book wouldn't have been possible. They have made vital efforts to compile up to date information on the varied aspects of this subject to make this book a valuable addition to the collection of many professionals and students.

This book was conceptualized with the vision of imparting up-to-date information and advanced data in this field. To ensure the same, a matchless editorial board was set up. Every individual on the board went through rigorous rounds of assessment to prove their worth. After which they invested a large part of their time researching and compiling the most relevant data for our readers.

The editorial board has been involved in producing this book since its inception. They have spent rigorous hours researching and exploring the diverse topics which have resulted in the successful publishing of this book. They have passed on their knowledge of decades through this book. To expedite this challenging task, the publisher supported the team at every step. A small team of assistant editors was also appointed to further simplify the editing procedure and attain best results for the readers.

Apart from the editorial board, the designing team has also invested a significant amount of their time in understanding the subject and creating the most relevant covers. They scrutinized every image to scout for the most suitable representation of the subject and create an appropriate cover for the book.

The publishing team has been an ardent support to the editorial, designing and production team. Their endless efforts to recruit the best for this project, has resulted in the accomplishment of this book. They are a veteran in the field of academics and their pool of knowledge is as vast as their experience in printing. Their expertise and guidance has proved useful at every step. Their uncompromising quality standards have made this book an exceptional effort. Their encouragement from time to time has been an inspiration for everyone.

The publisher and the editorial board hope that this book will prove to be a valuable piece of knowledge for researchers, students, practitioners and scholars across the globe.

Index

A
Alkaline Aluminum Phosphate Glasses 11, 42
Anisotropic Displacement, 55, 71
Anti-scuff Film, 22
Apatite Crystal, 59, 173, 187
Atomistic Simulation, 17
Authigenic Phosphorite Genesis, 106
Auxiliary Chelating Molecules, 21-22

B
Bioactive Glass-ceramics, 11, 42-43
Bioactivity, 11, 13, 27, 35, 38, 44, 50, 164
Biocoatings, 1, 38, 43
Bone Tissue Engineering, 1, 26, 32, 46-48
Bone-bonding Ability, 11
Burner Quarls, 6

C
Cation Vacancies, 55, 57-58, 60, 65, 71
Chemically Bonded Ceramics, 1, 3, 39
Chitosan, 32, 37, 50
Citrate Solubility, 142-143, 190
Collagen Apatite Composites, 1, 32, 150
Conglomeratic Phosphorites, 102-103
Corrosion Resistance, 2, 6, 22, 34
Coupled Plasma Spectrometry, 146
Crystalline Matrix, 5
Cytotoxicity, 37-38

E
Electron Paramagnetic Resonance 151-152, 196-197
Extracellular Matrix, 27, 33, 37, 151
Extraterrestrial Apatite, 122

F
Fluorosilicate Apatite, 20, 45
Freeze-drying, 35-36

G
Gelation, 36, 75
Glaser-hay Coupling, 21
Granitoid Rocks, 40

H
Heterogeneous Catalysis, 21, 45
Hydrothermal Synthesis, 65, 91
Hydroxyapatite Ceramics, 14, 45, 94

I
Inductively Coupled Plasma, 138, 146-148, 194-195
Interatomic Distances, 92, 144
Isotope Dilution, 148, 195

K
Kaolinite, 5, 109

L
Lanthanium Silicates, 52-53, 56, 59
Laser Ablation, 148, 194, 196
Lithium Vanado-phosphate, 12, 43

M
Metamorphic Fluids, 117
Monazite, 95, 128-129, 132, 141
Mullite, 6-8, 14, 38, 42, 44

N
Nanoscale Crystallization, 14
Neutron Powder Diffraction, 86, 144, 192
Noncentrosymmetric Borate, 68, 90
Nuclear Magnetic Resonance, 86 151-152, 198

O
Organometallic Molecules, 21
Orthophosporic Acid, 3
Osteoinductivity, 27
Oxide Ion Conductivity, 55, 65, 68 85-86, 89
Oxyapatite, 14-15, 18, 44, 51-52, 58-60 65, 70, 73, 76-77, 88, 90, 92-93, 174, 201

P
Phosphate Cement, 1, 8, 40-42, 101
Phosphate Ceramics, 2, 34, 39
Phosphate Conversion Coatings, 22-23, 46
Phosphate Impregnation, 8, 41
Post-milling Thermal Treatment, 52-53
Praseodymium, 57, 69-71, 84

R
Refractive Index, 142
Refractories, 1, 4, 6-7, 40-41

S
Scattering Amplitude, 144
Scattering Vector, 144
Silicon Carbide, 6, 40
Solid Oxide Fuel Cell, 15-16, 44, 52 145, 192
Structural Ceramics, 4
Surface Nucleation, 15, 187
Synthetic Apatite, 24, 168, 206

T
Terminal Alkynes, 21, 46
Tetraethyl Orthosilicate, 56
Thermal Ionization Mass Spectroscopy 148
Thermal Shock Resistances, 6
Triclinic Symmetry, 65, 73

V
Vanadocalcic Apatite, 51, 84
Vascularization, 32

W
Water Fugacities, 123, 140
Waveguide Fabrication, 11
Wollastonite, 13-14, 25, 38, 42-44, 81

X
X-ray Diffraction, 91, 94, 142-143, 152, 166-167, 191
X-ray Fluorescence, 142, 145-146, 193

Y
Yttrium Phosphates, 51, 84

www.ingramcontent.com/pod-product-compliance
Lightning Source LLC
LaVergne TN
LVHW080649210926
848485LV00003B/353